Feature Papers in Agriculture Technology—Using Computer Simulation for Agricultural Machinery Design and Development

Feature Papers in Agriculture Technology—Using Computer Simulation for Agricultural Machinery Design and Development

Guest Editor

Mustafa Ucgul

MDPI

Basel • Beijing • Wuhan • Barcelona • Belgrade • Novi Sad • Cluj • Manchester

Guest Editor
Mustafa Ucgul
Faculty of Science and Engineering
Southern Cross University
Lismore
Australia

Editorial Office
MDPI AG
Grosspeteranlage 5
4052 Basel, Switzerland

This is a reprint of the Special Issue, published open access by the journal *Agriculture* (ISSN 2077-0472), freely accessible at: https://www.mdpi.com/journal/agriculture/special_issues/850C9KQGW6.

For citation purposes, cite each article independently as indicated on the article page online and as indicated below:

Lastname, A.A.; Lastname, B.B. Article Title. *Journal Name* **Year**, *Volume Number*, Page Range.

ISBN 978-3-7258-3925-4 (Hbk)
ISBN 978-3-7258-3926-1 (PDF)
https://doi.org/10.3390/books978-3-7258-3926-1

Contents

About the Editor

Mustafa Ucgul

Dr. Mustafa Ucgul is a mechanical engineer with extensive expertise in computational mechanics and its applications in agricultural machinery design. With 18 years of research experience across Turkey and Australia, Dr. Ucgul has specialised in discrete element modelling of soil-engaging tools, the design of agricultural machinery, and the thermal analysis of agricultural greenhouses and desalination systems. Since completing his PhD, he has contributed to multiple industry-linked projects.

Preface

This Special Issue showcases research on the transformative role of computer simulation in advancing agricultural machinery. Given the growing demands for productivity, sustainability, and resource efficiency, simulation technologies have become essential for enhancing modern machinery.

The articles feature methods like discrete element modelling (DEM), computational fluid dynamics (CFD), fuzzy control, and artificial intelligence, tackling issues like harvester design, robotic manipulation, force optimisation, and intelligent control integration. Highlights include deep learning models for detecting corn kernel breakage, the genetic algorithm-based optimisation of cultivation frames, and the implementation of innovative controls for sprayers and transplanters. Other studies improve harvesting robotics, soil-tool modelling, and machinery evaluations, with reviews on potato harvesting and tractor transmissions.

Together, these works demonstrate how simulation drives efficiency, adaptability, and innovation in agricultural machinery, benefiting researchers and industry professionals alike. We thank all contributors and reviewers for their efforts in making this collection a vital resource for simulation-driven agricultural engineering.

Mustafa Ucgul
Guest Editor

agriculture

MDPI

Article

Simulation and Evaluation of the Performance of Pneumatic Residual Film Recycler Comb Teeth

Sen Jiang [1,2,3], Baiyu Chen [1,2,3], Haojie Jiang [1,2,3], Pengfei Guo [1,2,3], Xufeng Wang [1,2,3], Can Hu [1,2,3] and Wensong Guo [1,2,3,*]

1 College of Mechanical and Electrical Engineering, Tarim University, Alar 843300, China; senjiang78@gmail.com (S.J.); by987557@gmail.com (B.C.); jianghaojie0204@gmail.com (H.J.); pf8765445@gmail.com (P.G.); wxf@taru.edu.cn (X.W.); 120140004@taru.edu.cn (C.H.)

2 Modern Agricultural Engineering Key Laboratory, Universities of Education Department of Xinjiang Uygur Autonomous Region, Alar 843300, China

3 Xinjiang Production and Construction Corps (XPCC) Key Laboratory of Utilization and Equipment of Special Agricultural and Forestry Products in Southern Xinjiang, Alar 843300, China

* Correspondence: 120120004@taru.edu.cn

Abstract: The interaction law between soil and tillage components is the basis for designing and selecting soil tillage components. This paper uses the discrete element method to explore the soil penetration performance of the comb teeth of a pneumatic film-stripping tillage residual film recycler under different structural and working state parameters. The soil particle contact model is set up, the virtual prototype of the comb roller is established, and EDEM (Version 2018, DEM Solutions Company, Edinburgh, UK) discrete element software is applied to simulate the interaction between the comb roller and the soil particles during the residual film recycler's operation. Simulation and test results show that using a spiral arrangement of tooth comb knives (Alar, 843300, China, Zhongyuan Stainless Steel Bending Manufacturing Co.) can reduce the impact load on the machine, improving the soil disturbance and facilitating the penetration of soil mulch. The composite force on the combing roller increases with comb depth in the soil for a combing roller depth of 6–18 cm. Moreover, the rotational speed varies within the range of 60–120 r/min. The forward speed of the recycling machine significantly affects the soil penetration performance of the comb roller; the power it consumes increases with forward speed. This study can provide a reference for the structural design and optimization of working parameters of future deep tillage machines.

Keywords: residual film recycler; discrete element method; simulation test; comb roller

Academic Editor: Mustafa Ucgul

Received: 27 February 2025
Revised: 2 April 2025
Accepted: 7 April 2025
Published: 8 April 2025

Citation: Jiang, S.; Chen, B.; Jiang, H.; Guo, P.; Wang, X.; Hu, C.; Guo, W. Simulation and Evaluation of the Performance of Pneumatic Residual Film Recycler Comb Teeth. *Agriculture* **2025**, *15*, 811. https://doi.org/10.3390/agriculture15080811

1. Introduction

Soil is a non-renewable resource for all humanity, as important as water and air. In China, especially in Xinjiang, mulching technology has brought great economic benefits to agricultural production. However, it has also caused serious "white pollution" to the agroecological environment [1]. Farmers are most worried about the high cost of recycling residual film and the low performance of recycling equipment, resulting in a large part of the film not being recycled; instead, it is buried in the soil with the cultivation of farmland [2]. Residual film materials in agricultural fields, primarily made of polyethylene, take almost a century to decompose under natural conditions. Hence, the movement of water, gases, and minerals in the soil is disturbed. Moreover, micro-organism activity, the germination and growth of seeds, and the efficacy of soil working machinery are also disturbed [3]. According to a survey conducted by the Department of Agriculture and Rural Development of the Xinjiang Uygur Autonomous Region, the amount of film residue on arable soil in Xinjiang, where cotton is grown all year round, has reached 17.91 kg/ha or an area of 2460 m^2 [4]. Pollution from plastic sheeting has seriously threatened the safety of Xinjiang's agricultural soil resources [5] (Figure 1 shows the contamination of residual film in arable soil).

Figure 1. Residual film pollution in plow layer soil.

Many domestic research institutes, universities, and enterprises have conducted significant research on agricultural residual film-recycling machinery and mechanisms in response to the issue of agricultural film pollution [6]. They have developed a wide range of residual film-recycling machines. The author's team employed the "comb through hair" mimicry principle. Here, rotating comb teeth are employed to loosen soil and recycle film, including brush-type film separators, film separators with belts, and pneumatic film separators, i.e., three types of machines for tillage film recycling (Figure 2).

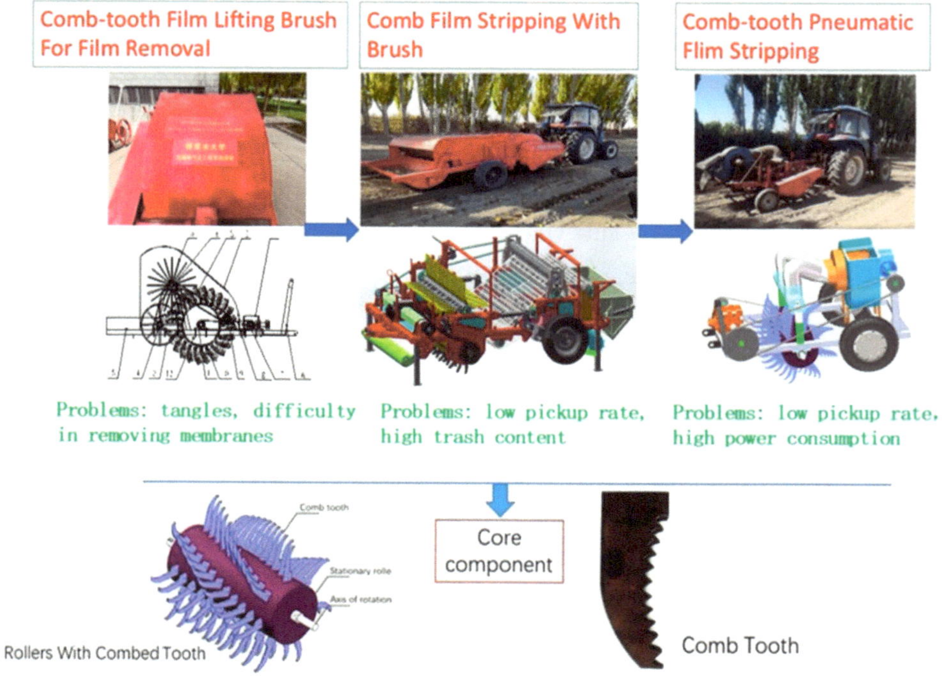

Figure 2. Different types of comb tillage film recyclers.

Researchers in agricultural engineering have prioritized drag reduction technologies for soil-contacting machinery parts, as soil–implement interactions critically influence resistance dynamics and energy utilization during field operations. Hence, the power consumption and efficiency of the machine are unstable. Investigating the interactions between the implements and the soil is conducive to the optimization of the structure of the implements and the work of parameters [7–12]. Aikins et al. described the application of the discrete element method to tillage and furrowing simulation, providing an overview of its application to tool optimization design [13]. Kesner et al. used a numerical model to simulate the stresses acting on the chisel shank during tillage [14]. Ucgul et al. utilized EDEM software to analyze the nonlinear contact model, hysteretic spring contact model (HSCM), and linear adhesion model. Their results demonstrate that the discrete element method can simulate tillage implements operating in soil [15].

Numerical simulation of agriculture machinery–soil interactions via a discrete element method is gaining popularity among researchers from different disciplines [16,17]. Wu numerically simulated the interaction process between the excavator shovel and soil based on DEM. The author investigated the dynamic performance of the cable shovel during the use of excavating robots [18]. Han used the discrete element simulation software EDEM to establish a contact mechanics model between manure particles and implements. Moreover, the author carried out an optimization of the structure and parameters of orchard fertilizer applicator devices [19]. Chong investigated the disturbance influence law of chisel-type plow shovels on soil based on EDEM simulation [20]. This software has become commercially used for discrete element analysis in soil modeling and machine–soil interaction simulation [21].

This paper investigates a pneumatic stripping-type deep plowing residual film recycler. A discrete element simulation test is carried out to investigate the performance of the key working part of the machine (the comb roller) in soil penetration. Lastly, a comparative analysis of the soil penetration performance of comb rollers under different structures and operating parameters is conducted to verify the power consumption of the recovery machine.

2. Materials and Methods

2.1. Structure and Working Principle

Many domestic research institutes, universities, and enterprises have conducted significant research on agricultural residual film-recycling machinery and mechanisms in response to the issue of agricultural film pollution. They have developed a wide range of residual film-recycling machines. The author's team employed the "comb through hair" mimicry principle. Here, rotating comb teeth are employed to loosen soil and recycle film, including brush-type film separators, film separators with belts, and pneumatic film separators, i.e., three types of machines for tillage film recycling (Figure 2).

Figure 3 shows the research object in this paper, i.e., a comb film-lifting, pneumatic film-stripping tillage residual film-recycling machine. The working principle is as follows. The comb teeth are first used to pick out the residual film in the soil of the cultivated layer. Then, the negative-pressure air flow of the film suction mouth is used to suck out the residual film picked up by the comb teeth. The wind speed decreases due to the considerable cross-sectional area of the piping in the gravity deposition chamber. Impurities with high suspended velocities, such as stones and cotton rods, cannot be drawn into the suction duct. Hence, residual film sucked by the negative wind will be blown by the centrifugal blower outlet and the screen installation at the outlet. Lastly, the recycled residual film can be collected [22].

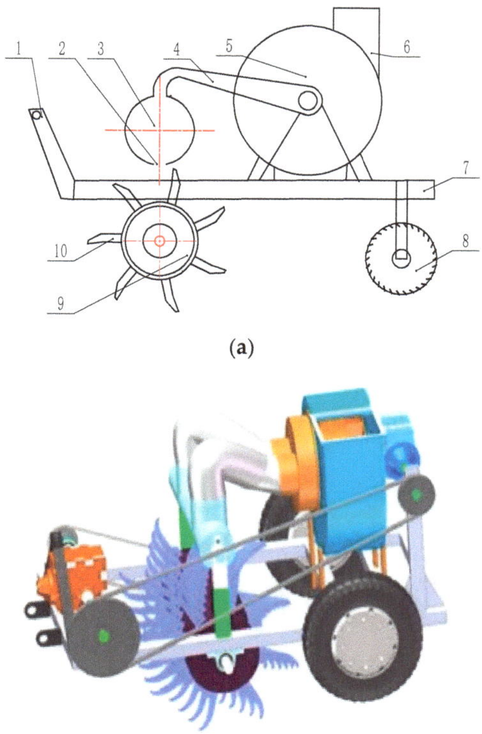

(**a**)

(**b**)

Figure 3. Working principle diagram of pneumatic tillage residual film recycler. (**a**) Working principle diagram; (**b**) 3D structural diagram. (1) Traction frame; (2) suction port; (3) gravitational settling chamber; (4) suction duct; (5) centrifugal fan; (6) residual film outlet; (7) rackmount; (8) travel wheel; (9) tollers with comb teeth; (10) comb tooth.

2.2. Simulation Test Method of Comb Roller Performance in Soil

2.2.1. Simulation Test Program

The main factors affecting the performance of comb teeth in the soil include the following [22]:

1. The arrangement of comb teeth. Comb teeth in the roller have various arrangements, such as parallel or spiral. The spiral arrangement can be mainly divided into a one-way single-row spiral arrangement, a one-way multi-row spiral arrangement, and a two-way multi-row spiral arrangement.
2. The structure of the comb teeth. The combing mechanism can be divided into flat and toothed knives.
3. The depth of comb tooth penetration.
4. The rotation speed of comb rollers.
5. The forward speed of comb rollers.

The main purpose of the simulation test is to explore the influence of the above five factors on the working performance of the comb roller in the soil and determine the comb roller's working parameters via simulation. The working performance of the comb roller in the soil mainly refers to the forward resistance, vertical force, resistance torque, and lateral force that the comb roller is subjected to when it enters the soil, as shown in Figure 4. Variable parameters of the simulation are shown in Table 1.

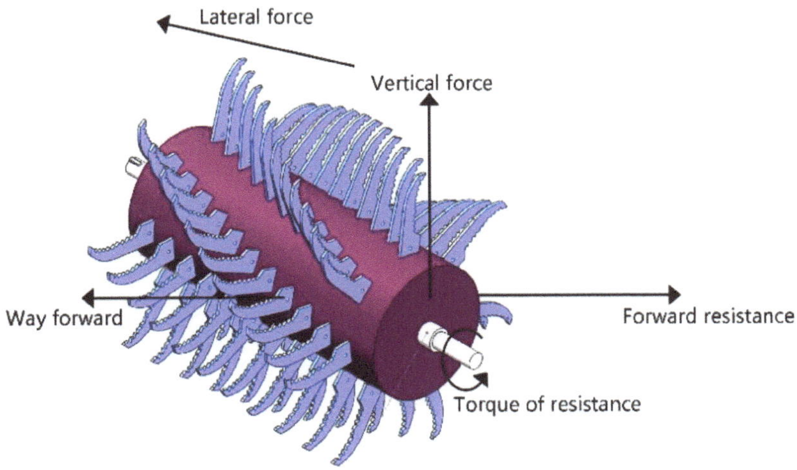

Figure 4. Force analysis of combing roller in the soil.

Table 1. Main variables affecting the performance of combs in soil.

Item	Variable			
Comb arrangement	Spiral	Parallel		
Comb structure	Toothed	Flat		
Depth of penetration of the comb in the soil (mm)	6	12	14	18
Speed of the roller (km/h)	60	80	100	120
Forward speed of the roller (km/h)	2	3	4	5

2.2.2. Contact Modeling of Soil Particles

This simulation of the soil penetration performance of the comb roller uses the discrete element method, which equates the soil as discrete spherical particles instead of a continuum. The execution of this simulation algorithm is achieved using the EDEM2018 software, which neglects the bonding between the soil and sets the contact model between the soil particles as the Hertz–Mindlin (no-slip) model. The model is efficient and accurate in calculating forces. The normal force component of the model is based on the Hertzian contact model [23]. In contrast, the tangential force component is based on the Middlin–Deresiewicz research theory [24].

The normal contact force F_n is calculated via Equation (1) [24]:

$$F_n = \frac{4}{3}E^{*}\sqrt{R^{*}}\delta_n^{\frac{3}{2}}, \quad (1)$$

where E^{*} is the equivalent of Young's modulus, R^{*} is the equivalent radius, and δ_n is the particle normal phase overlap. Parameters E^{*} and R^{*} are defined in Equations (2) and (3):

$$\frac{1}{E^*} = \frac{\left(1-v_i^2\right)}{E_i} + \frac{\left(1-v_j^2\right)}{E_j}, \quad (2)$$

$$\frac{1}{R^*} = \frac{1}{R_i} + \frac{1}{R_j}, \quad (3)$$

where Young's modulus, Poisson's ratio, and sphere radius of the spherical particles in contact are E_i, v_i, R_i and E_j, v_j, R_j, respectively. The normal damping force F_n^d is

$$F_n^d = -2\sqrt{\frac{5}{6}}\beta\sqrt{s_n m^*}\vec{v_n^{rel}}, \quad (4)$$

where m^* is the equivalent mass, $\vec{v_n^{rel}}$ is the normal component of the relative velocity between particles, and s_n is the normal stiffness. The expressions for the above parameters are shown in Equations (5)–(7).

$$m^* = \left(\frac{1}{m_i} + \frac{1}{m_j}\right)^{-1}, \tag{5}$$

$$\beta = \frac{lne}{\sqrt{ln^2 e + \pi^2}}, \tag{6}$$

$$s_n = 2E^* \sqrt{R * \delta_n}, \tag{7}$$

where e is the recovery coefficient. The tangential force F_t is calculated as [25]

$$F_t = -S_t \delta_t, \tag{8}$$

where S_t is the tangential stiffness coefficient, and δ_t is the tangential overlap of contacting particles. The tangential stiffness coefficient can be calculated as

$$S_t = 8G^* \sqrt{R^* \delta_n}, \tag{9}$$

where G^* is the equivalent shear modulus. The tangential damping F_t^d can be expressed as

$$F_t^d = -2\sqrt{\frac{5}{6}}\beta\sqrt{s_t m^*}\vec{v_t^{rel}}. \tag{10}$$

2.2.3. Simulation Parameterization

This simulation test mainly includes the following parameters:

1. Structural parameters of the soil film-recycling machine;
2. Working parameters of the recycling machine;
3. Soil parameters;
4. Comb parameters;
5. Soil and comb contact parameters.

The main technical parameters are also presented in Table 2.

Table 2. The main technical parameters of the recycling machine.

Item	Description/Value
Length of the virtual comb roller (m)	1
Length of the combs	20
Thickness of the combs	10
Distance between the combs	10
Number of comb rows in the circumferential direction of the carding roller	7
Number of each column of combs	11
Materials of the roller	Steel 45
Helix angle of the roller (rad)	0.13–0.21
Diameter of the roller (mm)	120
Length of the roller (mm)	1000
Forward speed of the recycling machine (km/h)	1–7
Rotating speed of the comb roller (r/min)	40–140
Working range of the comb entry depth (cm)	0–20
Suction pressure (Pa)	5740

Since most of the soils in the southern Xinjiang region are sandy soils, the main operation object of the comb tooth film-lifting pneumatic film-stripping-type tillage residual film recycler investigated as the focus of this research is the recycling of the residual film in sandy soils. Therefore, the calibration of soil parameters, comb tooth parameters, and soil–comb tooth contact parameters is conducted according to [25]. The standard ball calibration was selected, the collision recovery factor was set to 0.15, the static friction factor was set to 0.8, and the rolling friction factor was set to 0.2. The specific simulation parameters were designed as shown in Table 3.

Table 3. Virtual simulation parameters.

Parameters	Numerical Value
Dimensions of virtual soil tank (L × W × H), (mm × mm × mm)	1020 × 1020 × 832
Soil particle density (kg/m^3)	2630
Poisson's ratio of soil particles (v)	0.4
Soil particle shear modulus, G/Pa	1×10^6
65 M steel density (kg/m^3)	7830
65 M steel Poisson's ratio	0.35
65 M steel shear modulus	7.27×10^{10}
Soil–soil recovery factor e1	0.3
Soil–soil dynamic friction factor e2	0.2
Soil–soil static friction factor e3	0.05
Coefficient of recovery between soil and 65 M steel f1	0.3
Dynamic friction factor between soil and 65 M steel f2	0.5
Static friction factor between soil and 65 M steel f3	0.05
Filling unit radius, r/mm	6
Number of soil particles, n	212,786
Gravitational acceleration, g/(m·s^{-2})	9.81
Simulation time, s	4

2.2.4. Comb Roller Virtual Prototype

The final comb roller virtual prototype is shown in Figure 5.

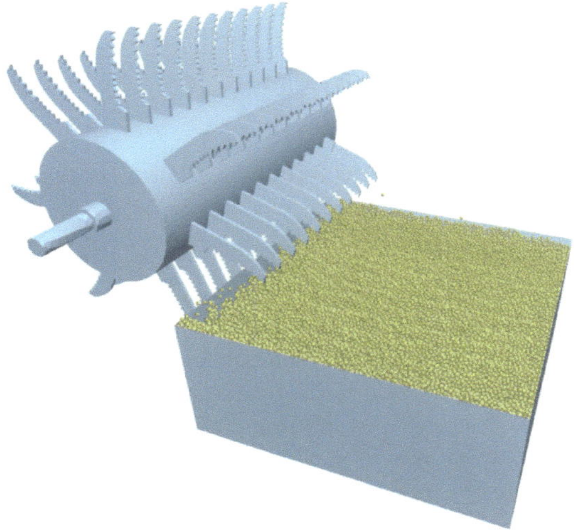

Figure 5. Virtual prototype operation diagram of comb roller.

This virtual simulation sets the length of the soil groove in the forward direction of the comb to 1 m to simplify the numerical calculation, as shown in Figure 6. The comb entry process of this virtual simulation experiment is divided into three main processes [26]:

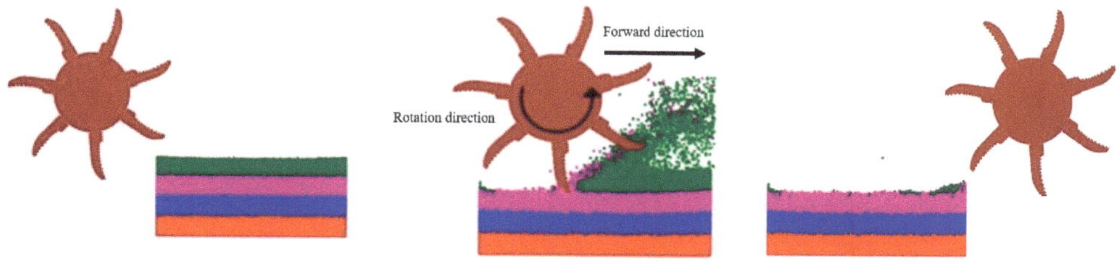

(a) (b) (c)

Figure 6. Working progress of comb teeth. (**a**) Before entering the working area; (**b**) in the working area; (**c**) leaving the working area.

3. Results

3.1. Influence of Comb Arrangement on Soil Penetration Performance

3.1.1. Test Conditions

The arrangement of teeth on the roller is important in determining how it performs on the ground. Figure 7 shows parallel and spiral arrangements of the comb rollers. This experiment was conducted to simulate the comb teeth of these two structural modes and compare the results to analyze their soil penetration performance. Simulation conditions are as follows: forward speed is 3 km/h, rotation speed is 150 r/min, and soil penetration depth is 12 cm [27].

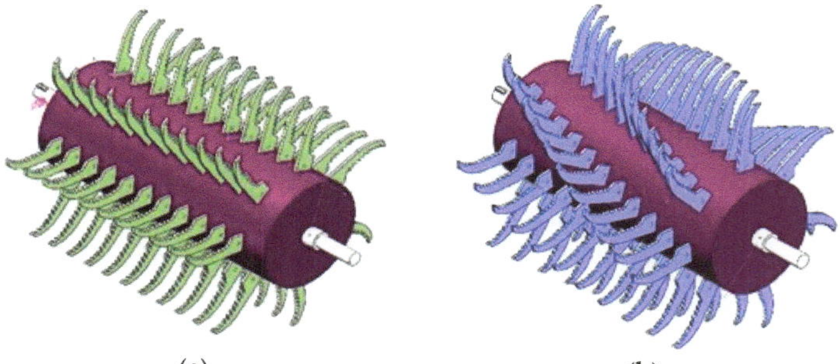

(a) (b)

Figure 7. Arrangement of comb teeth. (**a**) Parallel arrangement of the comb rollers; (**b**) spiral arrangement of the comb rollers.

3.1.2. Simulation Results

The effect of comb arrangement on soil penetration performance is shown in Figure 8.

The following can be observed according to Figure 8. The lateral force suffered by the spiral arrangement of the comb roller is high when entering the ground. In contrast, the lateral force suffered by the parallel arrangement of the comb roller is relatively low and stable. The comb teeth are arranged in a spiral manner, resulting in a structure similar to that of a spiral conveyor. This structure generates a lateral force on the comb rollers. The comb teeth arranged in a spiral configuration underwent minor changes in vertical force, forward resistance, and resisting moment during soil penetration. The maximum difference value for force was 1700 N, and the corresponding moment was 1000 Nm.

In contrast, the comb teeth arranged in a parallel configuration underwent significant changes at these same forces and torques. The maximum force difference value was 6000 N, and the maximum torque difference value was 2500 Nm. One possible explanation is as follows. The comb roller's parallel arrangement enters the soil at fixed intervals with a row of teeth. In contrast, the spiral arrangement continuously and uninterruptedly inserts numerous teeth into the soil, resulting in a smaller impact force during the insertion.

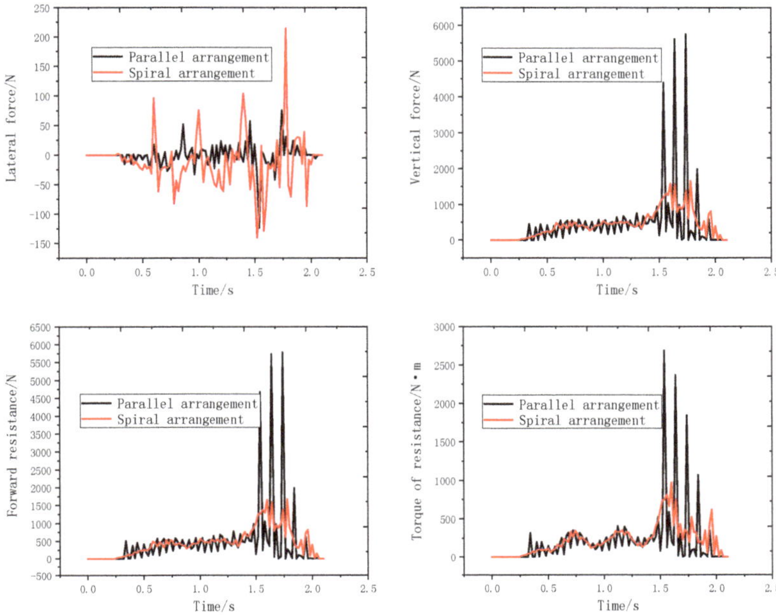

Figure 8. Influence of comb arrangement on soil entry performance.

In the parallel arrangement of the comb roller structure, a large impact force and impact torque are not conducive to the structural stability and operational stability of the recycling machine. Consequently, the entire drive system is somewhat harmed. Therefore, this paper chooses the spiral arrangement to install the comb teeth on the roller.

3.2. Influence of Comb Structure in Soil on Soil Penetration Performance

3.2.1. Test Conditions

Combs that create a small, serrated structure on the working surface are known as toothed combs. On the other hand, combs with smooth working surfaces with no additional structural modifications are called flat combs. As indicated by Figure 9, the comb roller displays a flat knife spiral arrangement and a toothed knife spiral arrangement. In this section, a comparative simulation study is conducted to investigate the law that influences these two structures in the performance of soil penetration. The arrangement is a spiral arrangement, the forward speed is 3 km/h, the rotational speed is 150 r/min, and the depth of soil entry is 12 cm [28].

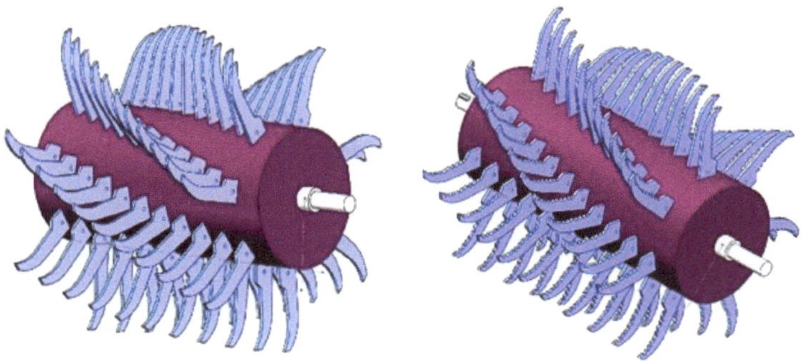

(**a**) (**b**)

Figure 9. Structure comparison of flat knife and toothed knife. (**a**) Flat knife spiral arrangement; (**b**) toothed knife spiral arrangement.

3.2.2. Simulation Results

A comparison and analysis of the lateral force, vertical force, forward resistance, and resistance moment curves of the flat knife and the toothed knife during soil penetration reveals that both knives similarly influence soil penetration performance. This observation is based on Figure 10.

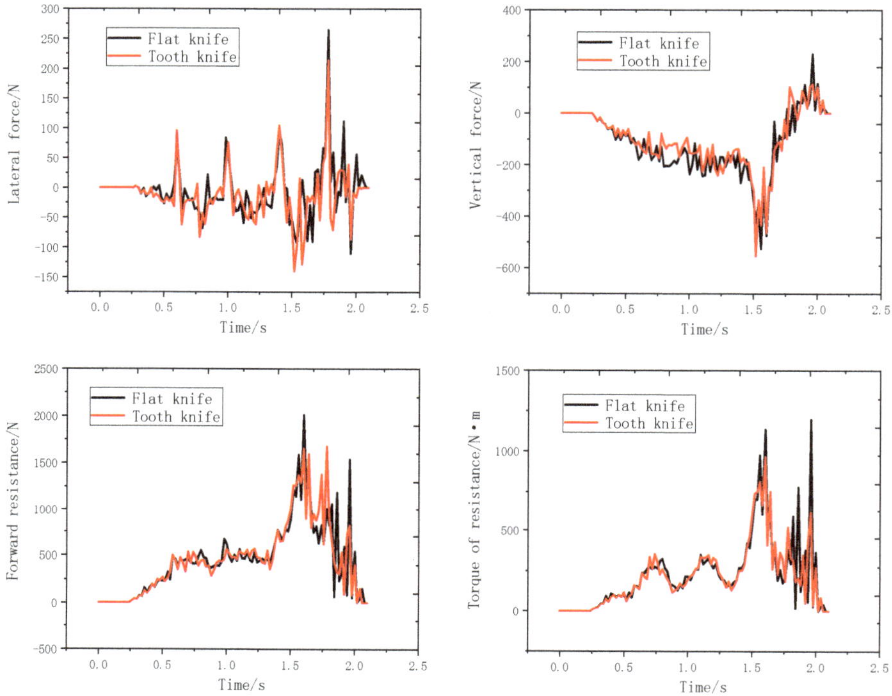

Figure 10. Influence of comb tooth structure on soil penetration performance.

A simulation test is conducted at standard depths of 6 cm, 10 cm, 12 cm, 14 cm, and 18 cm to examine the impact of the depth of the comb teeth on soil penetration efficiency. The simulation test includes a forward speed of 3 km/h, a rotational speed of 150 r/min, and toothed knife comb teeth. The comb teeth in the rollers follow a spiral arrangement. Figure 11 displays the simulation outcomes.

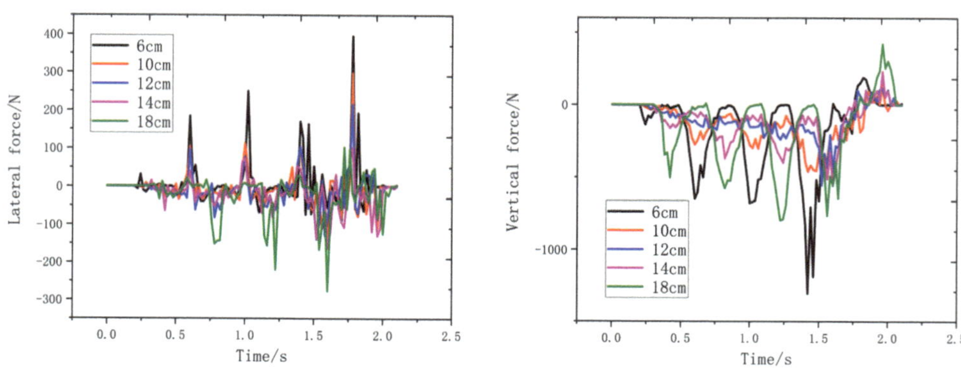

Figure 11. *Cont.*

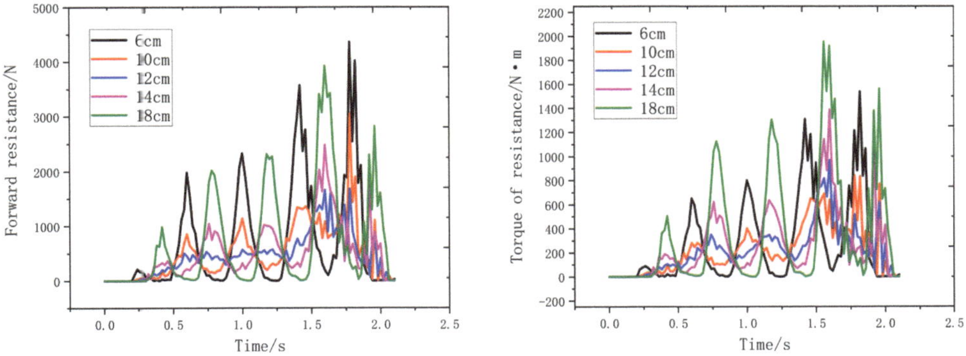

Figure 11. Influence of the depth of combing teeth in the soil.

According to Figure 11, when the depth of the comb in the soil varies within the range of 6–18 cm, the lateral force, vertical force, forward resistance, and resistance moment values suffered by the comb roller increase with soil depth. Moreover, the power consumption of the recycling machine increases with soil depth. The deeper the comb teeth are sunk into the soil, the greater the contact area between the comb teeth and the soil. Consequently, the soil has a greater resistance to the comb teeth. When checking and designing the structural strength of the tillage residual film-recycling machine, the values of force and moment received at the maximum forward speed should be used to design and check the parameters of the frame, shaft, and power output.

3.3. Influence of Comb Speed in Soil on Soil Penetration Performance

The simulation test will be set to the rotational speed standard to investigate the effect of the rotational speed of the combing roller on the performance of soil penetration, i.e., 60 r/min, 80 r/min, 100 r/min, and 120 r/min. Simulation test conditions are as follows: the forward speed is 3 km/h, the depth of soil penetration is 12 cm, the shape of the comb teeth is a toothed knife, and the arrangement of the comb teeth in the roller is a spiral arrangement. The simulation results are shown in Figure 12.

According to Figure 12, when the speed of the comb roller varies within the range of 60–120 r/min, the lateral force, vertical force, forward resistance, and resisting torque acting on the comb roller barrel increase with the depth of the comb in the soil. However, the increase is relatively small. Moreover, the effect of the speed of the comb roller on its performance in the soil is relatively low. The optimal value of comb roller speed should be obtained by further tests and combined with the tillage layer's residual film pick-up rate index.

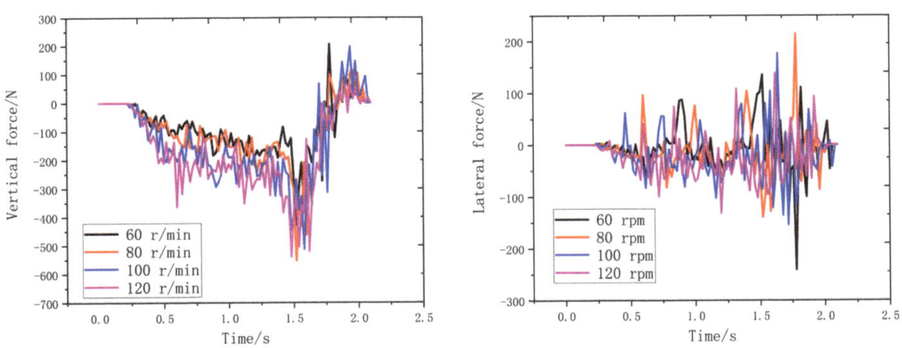

Figure 12. *Cont.*

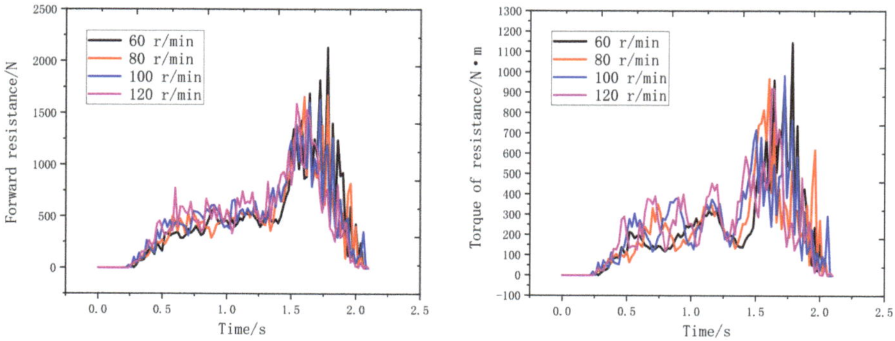

Figure 12. Influence of the speed of the combing roller on soil penetration performance.

3.4. Influence of Forward Speed on Soil Penetration Performance

The simulation test sets the forward speed standard as 2 km/h, 3 km/h, 4 km/h, and 5 km/h to explore the influence of the forward speed of the recycling machine on the performance of soil penetration [29]. The simulation test conditions are as follows: the rotational speed is 150 r/min, the depth of soil penetration is 12 cm, the shape of the comb teeth is a toothed knife, and the arrangement of the comb teeth in the roller is a spiral arrangement. The simulation results are shown in Figure 13.

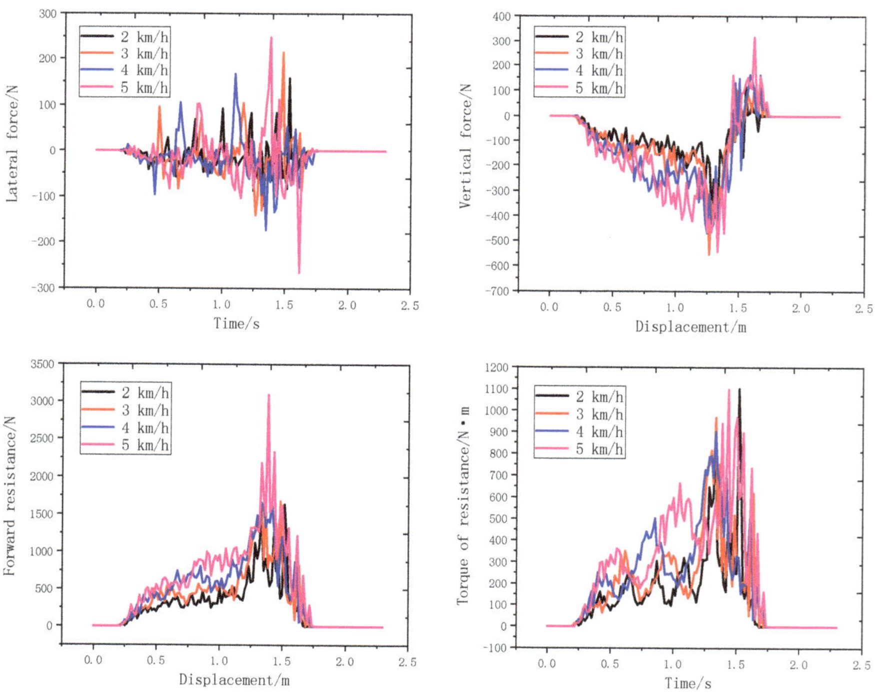

Figure 13. Influence of forward speed on soil penetration performance.

According to Figure 13, the forward speed of the recycling machine significantly impacts the performance of the comb roller in the soil. The comb roller is subjected to lateral, vertical, and forward resistance. In contrast, the resistance torque is increased with forward speed. The faster the forward speed of the recycling machine, the higher the power consumption. The recycling machine should be combined with power consumption and residual film pick-up rate to choose the optimal value of forward speed.

3.5. Experimental Validation

A comb performance test rig was constructed to validate the simulation results in 2024, as illustrated in Figure 14. The validation test was conducted in the soil tank, with the entire machine's traction force and the comb roller's torque serving as the response indexes. The test conditions were as follows: forward speed of 3 km/h, comb roller speed of 150 r/min, soil humidity (RH) in the range of 19–23%, comb arrangement in a spiral arrangement, and comb depth in the soil of 100 mm.

(a)

(b)

Figure 14. Test on soil trough bench of film-lifting device with rotary tillage cutter teeth; (**a**) 3D structural diagram; (**b**) comb performance test rig. (1) Tension sensor; (2) signal transmitter; (3) electric motor; (4) torque sensor; (5) comb tooth; (6) adjustable traveling wheels; (7) prop shaft.

The resistance torque and forward resistance were measured despite the technical limitations of the equipment. As demonstrated in Table 4, the difference between the simulated and actual comb torque is 74 Nm, and the discrepancy between the simulated and actual traction force is 0.47 kN. The results of the field tests show that the discrete element method can be used to investigate the in-soil performance of the structure of the agricultural tillage machinery.

Table 4. Comparison of simulation test data with actual test data for torque.

	Simulation Results	Actual Result	Numeric Difference
Resistance torque (Nm)	836	910	74
Forward resistance (kN)	2.75	3.22	0.47

4. Conclusions

(1) The results of the comb arrangement on the performance of soil penetration are as follows. The lateral force suffered by the combing roller in the spiral arrangement increases The vertical force, forward resistance, and resisting moment of the spirally arranged combs in soil penetration are relatively small. The comb teeth arranged in parallel are subjected to a large variation of the above forces and moments, with a maximum difference in the force of 6000 N and a maximum difference in the moment of 2500 Nm. The spiral arrangement of comb teeth is the most favorable for the combing roller to penetrate the soil.

(2) The results of the influence of the comb structure on the performance of soil penetration are as follows. The lateral force, vertical force, forward resistance, and resistance moment acting on the comb for flat and toothed knives during soil penetration are the same. However, the toothed knife is more conducive to separating the residual film from the soil. Therefore, this paper chooses the toothed knife structure.

(3) The results of the effect of comb entry depth on the entry performance are as follows. The values of lateral force, vertical force, forward resistance, and resisting moment suffered by the comb roller increase with the entry depth.

(4) The results of the effect of comb roller speed on the performance of soil penetration are as follows. The lateral force, vertical force, forward resistance, and resisting moment suffered by the comb roller increased with the soil penetration depth. However, the increase and the overall degree of influence were small.

(5) The effect of forward speed on the soil entry performance was as follows. The lateral force, vertical force, forward resistance, and the value of the resisting moment experienced by the comb rolls increased with forward speed.

(6) A comparison of the results of the field tests with those of the simulation tests shows that the errors are very small. Moreover, using the discrete element method to explore the in-soil performance of the structures of the agricultural tillage machinery is feasible. Lastly, the proposed method can help select the future structural parameters of agricultural machinery.

Author Contributions: Conceptualization, S.J.; methodology, B.C.; software, H.J.; validation, P.G.; formal analysis, C.H.; investigation, X.W.; resources, W.G.; data curation, H.J.; writing—original draft preparation, S.J.; writing—review and editing, W.G.; visualization, P.G.; supervision, B.C.; project administration, S.J.; funding acquisition, W.G. All authors have read and agreed to the published version of the manuscript.

Funding: This research was funded by the National Natural Science Foundation of China (Project No. 32360440) and the Agricultural Science and Technology Public Relations Project of Xinjiang Production and Construction Corps (Project Name: Research and Development of Highly Efficient Residual Film Recycling and Resource Utilization Machinery).

Institutional Review Board Statement: Not applicable.

Data Availability Statement: The data presented in this study are available upon request from the authors.

Acknowledgments: The authors would like to thank their schools and colleges, as well as the funding providers of the project. All support and assistance are sincerely appreciated.

Conflicts of Interest: The authors declare no conflicts of interest.

References

1. Yan, C.; Mei, X.; He, W.; Zheng, S. Present situation of residue pollution of mulching plastic film and controlling measures. *Trans. Chin. Soc. Agric. Eng.* **2006**, *22*, 269–272.
2. Liang, R.Q.; Chen, X.G.; Zhang, B.C.; Meng, H.W.; Jiang, P.; Peng, X.B.; Kan, Z.; Li, W.M. Problems and countermeasures of recycling method sand resource reuse of residual film in cotton fields of Xinjiang. *Trans. Chin. Soc. Agric. Eng.* **2019**, *35*, 1–13.
3. Gao, H.; Yan, C.; Liu, Q.; Ding, W.; Chen, B.; Li, Z. Effects of plastic mulching and plastic residue on agricultural production: A meta-analysis. *Sci. Total Environ.* **2019**, *651*, 484–492. [PubMed]
4. Wang, H.; Wang, M.; Wang, J.; Lu, Y.; Ying, Y.; He, Y. Control technology and control strategy of residual film pollution in Xinjiang farmland. *J. Chin. Agric. Mech.* **2022**, *43*, 166.
5. Hu, C.; Wang, X.F.; Chen, X.G.; Tang, X.Y.; Zhao, Y.; Yan, C.R. Current situation and control strategies of residual film pollution in Xinjiang. *Trans. Chin. Soc. Agric. Eng.* **2019**, *35*, 223–234.
6. Hou, S.; Hu, S.; Kong, J.; Zhang, H.; Na, M.; Dong, X. Present situation of research on plastic film residue collector in China. *Trans. Chin. Soc. Agric Eng.* **2002**, *18*, 186–190.
7. Sun, J.; Liu, Z.; Yang, F.; Sun, Q.; Liu, Q.; Luo, P. Research Review of Agricultural Equipment and Slope Operation Key Technologies in Hilly and Mountains Region. *Trans. Chin. Soc. Agric. Mach.* **2023**, *54*, 1–18.
8. Yu, J.; Fu, H.; Li, H.; Shen, Y. Application of discrete element method to research and design of working parts of agricultural machines. *Trans. Chin. Soc. Agric. Eng.* **2005**, *21*, 1–6.
9. Tong, J.; Mohammad, M.A.; Zhang, J.; Ma, Y.; Rong, B.; Chen, D.; Menon, C. DEM Numerical Simulation of Abrasive Wear Characteristics of a Bioinspired Ridged Surface. *J. Bionic Eng.* **2010**, *7*, 175–181.
10. Ding, Q.; Ding, W.; Meng, W.; Han, Y. Characteristics and quantifications of soil structure for soil tillage research. *Trans. Chin. Soc. Agric. Mach.* **2007**, *38*, 62–66.
11. Fang, H.; Ji, C.; Chandio, F.A.; Guo, J.; Zhang, Q.; Arslan, C. Analysis of soil dynamic behavior during rotary tillage based on distinct element method. *Trans. Chin. Soc. Agric. Mach.* **2016**, *47*, 22–28.
12. Guo, Z.; Tong, J.; Ren, L.; Zhou, Z. Research method on the interacting problem of tillage tool soil. *Trans. Chin. Soc. Agric. Mach.* **2001**, *32*, 102–104.
13. Aikins, K.A.; Ucgul, M.; Barr, J.B.; Awuah, E.; Antille, D.L.; Jensen, T.A.; Desbiolles, J.M. Review of discrete element method simulations of soil tillage and furrow opening. *Agriculture* **2023**, *13*, 541. [CrossRef]
14. Kesner, A.; Chotěborský, R.; Linda, M.; Hromasová, M.; Katinas, E.; Sutanto, H. Stress distribution on a soil tillage machine frame segment with a chisel shank simulated using discrete element and finite element methods and validate by experiment. *Biosyst. Eng.* **2021**, *209*, 125–138. [CrossRef]
15. Ucgul, M.; Fielke, J.M.; Saunders, C. Three-dimensional discrete element modeling (DEM) of tillage: Accounting for soil cohesion and adhesion. *Biosyst. Eng.* **2015**, *129*, 298–306. [CrossRef]
16. Zheng, K.; He, J.; Li, H.; Diao, P.; Wang, Q.; Zhao, H. Research on polyline soil breaking blade subsoiler based on subsoiling soil model using Discrete Element Method. *Trans. Chin. Soc. Agric. Mach.* **2016**, *47*, 11.
17. Hu, J.; Zhou, C.; Hou, C.; Wang, J. Simulation analysis of seed-filling performance of magnetic plate seed-metering device by discrete element method. *Trans. Chin. Soc. Agric. Mach.* **2014**, *45*, 94–98.
18. Wu, J.; Wang, G.; Bi, Q.; Hall, R. Digging force and power consumption during robotic excavation of cable shovel: Experimental study and DEM simulation. *Int. J. Min. Reclam. Environ.* **2020**, *35*, 12–33. [CrossRef]
19. Han, S.; Li, Y.; Zhou, X.; Kan, Z.; Meng, H.; Qi, J. DEM Simulation and Experimental Validation of the Performance of an Orchard Deep Applicator for Manure. *Appl. Sci.* **2023**, *13*, 3709. [CrossRef]
20. Zhang, C.; Fan, X.; Li, M.; Li, G.; Zhao, C.; Sun, W. Simulation Analysis and Experiment of Soil Disturbance by Chisel Plow Based on EDEM. *Trans. Chin. Soc. Agric. Mach.* **2022**, *53*, 52–59.
21. Zeng, Z.; Ma, X.; Cao, X.; Li, Z.; Wang, X. Critical review of applications of discrete element method in agricultural engineering. *Trans. Chin. Soc. Agric. Mach.* **2021**, *52*, 1–20.
22. Guo, W.; He, X.; Wang, L.; Zhao, P.; Hu, C.; Hou, S.; Wang, X. Development of a comb tooth loosening and pneumatic stripping plough layer residual film recovery machine. *Trans. Chin. Soc. Agric. Eng.* **2020**, *36*, 1–10.
23. Mindlin, R.D. Compliance of elastic bodies in contact. *J. Appl. Mech.* **1949**, *16*, 259–268. [CrossRef]
24. Mindlin, R.D.; Deresiewicz, H. Elastic spheres in contact under varying oblique forces. *Am. Soc. Mech. Eng.* **1953**, *20*, 327–344. [CrossRef]
25. Zhang, R.; Han, D.; Ji, Q.; He, Y.; Li, J. Calibration methods of sandy soil parameters in simulation of discrete element method. *Trans. Chin. Soc. Agric. Mach.* **2017**, *48*, 49–56.
26. Li, J.; Tong, J.; Hu, B.; Wang, H.; Mao, C.; Ma, Y. Calibration of parameters of interaction between clayey black soil with different moisture content and soil-engaging component in northeast China. *Trans. Chin. Soc. Agric. Eng.* **2019**, *35*, 130–140.
27. Qi, Y.; Beriot, N.; Gort, G.; Lwanga, E.H.; Gooren, H.; Yang, X.; Geissen, V. Impact of plastic mulch film debris on soil physicochemical and hydrological properties. *Environ. Pollut.* **2020**, *266*, 115097. [CrossRef] [PubMed]

28. Lü, J.; Sun, H.; Dui, H.; Peng, M.M.; Yu, J.Y. Design and experiment on conveyor separation device of potato digger under heavy soil condition. *Trans. Chin. Soc. Agric. Mach.* **2017**, *48*, 146–155.
29. Qian, D.; Zhang, J. A summary of study of adhesion and friction between soil and metals. *Trans. Chin. Soc. Agric. Mach.* **1984**, *1*, 69–78.

agriculture

MDPI

Article

Development of a Crawler-Type Self-Propelled Machine with Trenching, Fertilizing, and Soil-Covering Components for Hilly Orchard

Jun Li [1,2,3], Chaodong Mai [1,2], Ye Zeng [1,2], Zhao Li [1,2], Runpeng Jiang [1,2], Qinglin Weng [1,2], Jiamin Cai [1,2], Qian Wang [1,2] and Can Li [1,2,*]

1 College of Engineering, South China Agricultural University, Guangzhou 510642, China; autojunli@scau.edu.cn (J.L.); mcd@stu.scau.edu.cn (C.M.); zengye_leo@outlook.com (Y.Z.); leez2018@163.com (Z.L.); runpeng@stu.scau.edu.cn (R.J.); qlweng@stu.scau.edu.cn (Q.W.); c13533895601@stu.scau.edu.cn (J.C.); eiq1021@stu.scau.edu.cn (Q.W.)
2 Guangdong Laboratory for Lingnan Modern Agriculture, Guangzhou 510642, China
3 State Key Laboratory of Agricultural Equipment Technology, Beijing 100083, China
* Correspondence: canll@scau.edu.cn

Abstract: In response to the issues of high energy consumption, limited functionality, and uneven soil–fertilizer mixing in mechanical operations for trenching and fertilizing in hilly orchards, this study proposes the design of a crawler-type self-propelled machine, integrating three main functions: trenching, fertilizing, and soil covering. The key components of the trenching device, fertilizing device, and soil-covering device were designed. Three fertilizing simulation models (pre-plant, mid-plant, and post-plant) were established using EDEM discrete element software. The soil–fertilizer mixing effects under each mode were analyzed, with results indicating that the post-plant fertilizing mode better meets the soil–fertilizer mixing requirements for deep organic fertilizer application. Using trenching speed, forward speed, and bending angle of the trenching knife as experimental factors, with operating power consumption and soil–fertilizer mixing uniformity as evaluation indicators, a Box–Behnken experiment was conducted to optimize the parameters of the trenching and fertilizing components. A regression model was established to analyze the interaction between experimental factors and indicators. The optimal operational parameter combination was determined as follows: trenching speed of 265.03 r/min, forward speed of 0.40 m/s, and bending angle of trenching knife of 130°. Under these parameters, the trenching power consumption and soil–fertilizer mixing uniformity were 1.74 kW and 77.15%, respectively. Orchard verification tests on the machine showed that under the optimal parameters, the relative errors in trenching power consumption and soil–fertilizer mixing uniformity between the field tests and simulations were 7.40% and 4.50%, respectively. These results meet the agronomic requirements for trenching and fertilizing, and the study provides valuable references for the application of related technologies in orchard trenching and fertilizing operations.

Keywords: agricultural machinery; trenching; fertilizing; soil covering; EDEM simulation; parameter optimization

Academic Editor: Mustafa Ucgul

Received: 25 January 2025
Revised: 13 February 2025
Accepted: 17 February 2025
Published: 19 February 2025

Citation: Li, J.; Mai, C.; Zeng, Y.; Li, Z.; Jiang, R.; Weng, Q.; Cai, J.; Wang, Q.; Li, C. Development of a Crawler-Type Self-Propelled Machine with Trenching, Fertilizing, and Soil-Covering Components for Hilly Orchard. *Agriculture* **2025**, *15*, 430. https://doi.org/10.3390/agriculture15040430

1. Introduction

Fertilization is an essential component of orchard management, playing a pivotal role in promoting the growth and development of fruit trees. Efficient and rational fertilization is fundamental to ensuring fruit yield, quality, and economic benefits [1,2]. Currently, issues such as over-fertilization, nutrient imbalances in the soil, and environmental pollution arise in the daily management and use of fertilizers in orchards. To achieve precise fertilization and optimize soil properties in orchard environments, the use of mechanical systems to replace traditional manual methods for deep fertilizer application and soil mixing has gained considerable attention in recent years [3]. Research indicates that compared to surface broadcasting, deep fertilization can increase rice yields by 10% [4]. The use of mechanical side-deep fertilization not only saves labor but also reduces the loss of essential elements such as nitrogen, phosphorus, and potassium, while improving fertilizer utilization efficiency. Developed countries have long been engaged in the research of orchard fertilization machinery, integrating agricultural machinery design with agronomic requirements. These nations have established high standards in orchard construction and management. Their trench fertilization machinery is now more large-scale, specialized, and standardized [3]. The development of orchard trench fertilization machinery in China started later and primarily uses rotary tillage fertilizers for operations. Additionally, many orchards are located in hilly and mountainous areas where problems such as rugged terrain and poor soil conditions are prominent. Therefore, soil loosening and fertilization are especially important [5]. Given the aging rural population, labor shortages, and high labor costs, particularly in hilly orchards, the development of mechanized and intelligent trench fertilization machinery is critical to addressing these issues. This represents a key trend in the future development of Chinese orchards and is of significant importance for the advancement of the orchard industry.

In recent years, the development of the equipment and the evaluation of the operation effect of the prototype have been the subject of extensive research among scholars in Table 1 [6,7]. To analyze the interaction between tools and soil, researchers have established interaction models. Li et al. [8] proposed a dynamic vibration model for self-excited trenching tools, effectively reducing power consumption during trenching. Zeng et al. [9] analyzed the trenching and cutting process and established a dynamic model for soil particles on the thrower blade, with field trials showing a power consumption of 25.96 kW and a furrow depth stability factor of 88.31%, meeting national technical standards. In addition, trenching is a crucial aspect of orchard management, playing a vital role in optimizing the structure of trenching equipment to reduce trench resistance during the operation. Scholars have conducted in-depth research on optimizing the structural parameters of trenching machinery to reduce operational power consumption, improve agricultural productivity, and enhance the quality of trenching. However, existing trenching and fertilizing systems are relatively complex, and prototype machines often serve a single function [10–13]. Furthermore, soil–fertilizer mixing uniformity is another key indicator of fertilization effectiveness [14–16]. Niu et al. [17] designed a soil–fertilizer collision mixing and covering device that simultaneously mixes and covers soil and fertilizer, achieving a uniformity of 87.02% for soil–fertilizer mixing and 94.37% for coverage. Zhang et al. [18] designed a spiral fertilizer mixing device for furrowing fertilization with a Lacey index of 0.76, meeting the operational requirements for trench fertilization. Yuan et al. [19] developed a layered backfilling device for soil–fertilizer mixing, with an organic fertilizer ratio relative standard deviation of 0.11% in bench tests, meeting the requirements for deep organic fertilizer application. Despite these efforts made by scholars, trenching and fertilizing machines still face challenges in practice, such as high power consumption, large volume, and limited functionality. Orchards in China are mostly located in hilly areas, requiring smaller, more adaptable machines that are not affected by the terrain. Reducing trenching power consumption, improving fertilizer utilization, and achieving mechanized fertilization are urgent issues that need to be addressed.

Table 1. Several trenching, fertilizing, and soil-covering machines at home and abroad.

Representative Equipment	Developer	Type of Fertilizer	Performance	Characteristics
 Self-propelled intelligent orchard trenching and fertilizing machine	Sun et al., China	Organic and chemical fertilizers	Trenching depth: 0~400 mm Trenching width: 200~350 mm Trenching distance: 2~2.5 m	Self-propelled; automatic driving; double-row trenching
 Self-propelled multifunctional fertilizing machine	Yifeng Machinery Co. LTD., China	Organic or chemical fertilizers	Trenching depth: 200~350 mm	Self-propelled
 Chain reversal trenching and fertilization machine	Tan et al., China	Organic fertilizer	Trenching depth: 0~300 mm Trenching width: 0~600 mm	Self-propelled
 Chain reversal trenching machine	Vermeer, America	-	Trenching depth: 760~1220 mm Trenching width: 100~200 mm	Self-propelled
 Chain reversal trenching machine	Kawabe, Japan	-	Trenching depth: 1000~1200 mm Trenching width: 0~120 mm	Self-propelled

With the widespread application of computer-based virtual simulation technology in the agricultural sector, simulation techniques have provided an efficient and feasible method for studying the soil disturbance process during deep tillage. The discrete element method (DEM) simulation technology has become an effective tool in trench fertilization research. On the one hand, it can effectively reveal the impact of the machinery on soil fragmentation, soil–fertilizer mixing, and other processes during the trenching operation, as well as visualize the internal interactions between soil particles and the machine surface. On the other hand, it effectively assists designers in the design, testing, and optimization of various discrete material-handling equipment [20–22]. Chen et al. [23] designed a small vertical spiral trenching–fertilizing machine suitable for hilly terrains with a 25° slope, integrating three functions—trenching, fertilizing, and soil covering—and optimized the structure of the trenching tool based on an EDEM simulation model. Zhu et al. [24] developed an organic fertilizer application machine with three integrated functions: fertilizing, soil–fertilizer mixing, and soil covering, and optimized the tool speed ratio based on EDEM simulations. Tan et al. [25] designed a chain-type reversible trenching–fertilizing machine for integrated trenching, fertilizing, and backfilling, optimizing the structure of the scraper blade using EDEM simulation. The application of virtual simulation technology in agricultural machinery not only shortens the trial production cycle but also reduces market response time and development costs, providing substantial practical value.

In summary, this study designed a crawler-type self-propelled machine that integrates trenching, fertilizing, and soil covering operations in a single cycle, which was suitable to the hilly orchards. Through theoretical analysis, various key mechanisms such as trenching, fertilizing, and soil covering devices were designed. Subsequently, simulation models for three fertilization modes (pre-plant, mid-plant, and post-plant) were proposed and established using EDEM software, and the effects of soil–fertilizer mixing uniformity under these three modes were analyzed. Furthermore, the study also used EDEM to simulate the trenching and fertilizing process and employed a Box–Behnken experimental design to optimize component parameters and identify the optimal parameter combination. Definitively, the efficacy of the structural parameters was verified through field trials. This research provides a theoretical foundation for the mechanization of orchard operations.

2. Materials and Methods

2.1. Structure and Working Principle of the Whole Machine

Figure 1 illustrates the design of the crawler-type self-propelled trenching, fertilizing, and soil-covering machine, with its key technical parameters summarized in Table 2. The machine consists primarily of a crawler chassis, trenching device, fertilizing device, soil-covering device, fertilizer box, and control system.

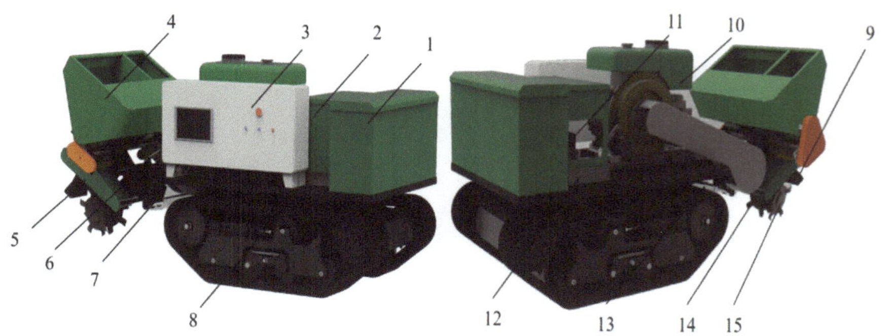

Figure 1. Structure of the crawler-type self-propelled orchard trenching, fertilizing, and soil-covering machine. (1) Counterweight box; (2) battery case; (3) control box; (4) fertilizer box; (5) soil covering plate; (6) trenching drive box; (7) lifting mechanism; (8) crawler chassis; (9) fertilizer conveyor channel; (10) engine; (11) power supply; (12) connection frame; (13) lifting mechanism; (14) trenching cutter disc; (15) trenching curved blade.

Table 2. Key technical parameters of the trenching and fertilizing machine.

Main Parameters	Numerical Value
Dimension (L × W × H) (mm)	2000 × 1000 × 1000
Matching power (kw)	18.8
Maximum driving speed (m/s)	1.0
Trenching depth (cm)	0~35
Width of trenching application (cm)	0~30
Fertilizer tank capacity (m^3)	0.2
Unicyclic fertilizer application (g)	164
Maximum operating time (h)	8

Figure 2 shows the working principle of the crawler-type self-propelled trenching, fertilizing, and soil-covering machine in the orchard. Before the trenching and fertilizing operation, the trenching cutter is activated and begins to rotate. The depth of trenching is adjusted by modifying the extension and retraction of the electric push rod on the trenching device. The trenching device is then lowered to the preset trenching position. The trenching cutter enters the soil, cutting and throwing the soil upward. Simultaneously, the fertilizer spiral conveyor transports organic fertilizer from the fertilizer box to the fertilizer discharge channel above the trenching mechanism. The discharge fertilizer spiral conveyor then moves the fertilizer from the discharge channel into the fertilizing tray above the trenching device, where it falls into the trench created by the trenching device. After trenching and fertilizing, the soil thrown up by the trenching cutter is blocked by the soil-covering device and falls back into the trench, thus completing the integrated operation of self-propelled forward movement, single-line trenching, deep organic fertilizer application, and soil covering.

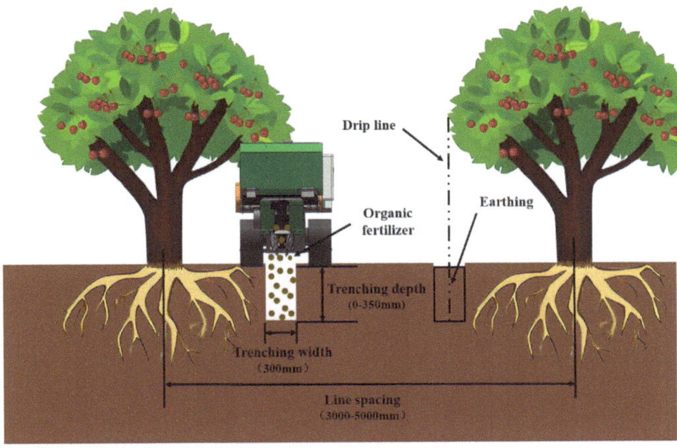

Figure 2. Principle diagram of the whole machine.

2.2. *Key Component Design of Whole Machine*

2.2.1. Design of Trenching Device

According to the agronomic standards for trenching and fertilizing in orchards [17], the required trench depth is generally between 20 and 35 cm, with a trench width of 20 to 30 cm. Considering the operational requirements for soil cutting and soil throwing during the trenching process, this study employed a dual-disc trenching device. The trenching knives are cup-shaped curved blades, which are symmetrically and alternately installed on both sides of the trenching disc by bolts, with the two discs symmetrically mounted on both sides of the trenching drive box at a 14° inclination angle. In order to ensure the strength and hardness, the material of the trenching knives is 65 Mn. Each side is equipped with four trenching knives arranged at 90° intervals along the circumference of the disc, as shown in Figure 3. The trenching disc drives the blades in a reverse rotation manner, ultimately forming a trapezoidal trench with cross-sectional dimensions ($a \times b \times h$) of $30 \times 35 \times 30$ cm, as shown in Figure 4.

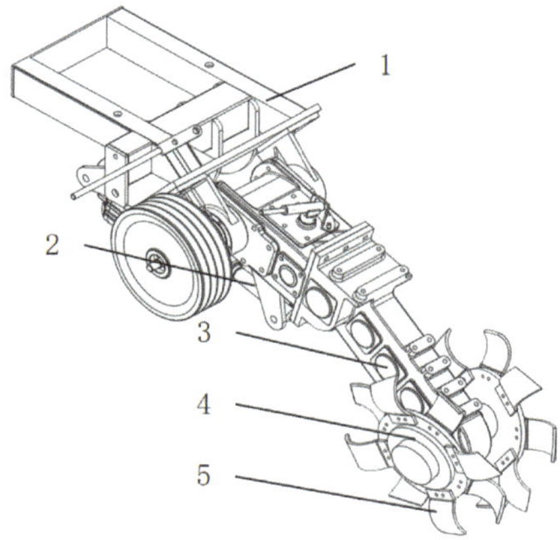

Figure 3. Structural schematic of the trenching device. (1) Connection frame; (2) lifting mechanism; (3) trenching drive box; (4) trenching disc; (5) cup-shaped curved blade.

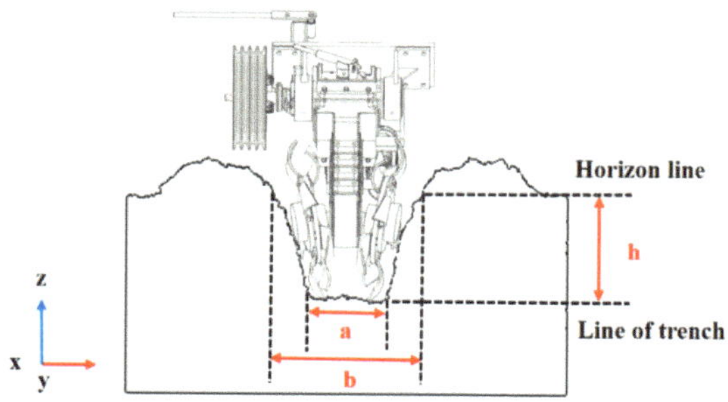

Figure 4. Trenching effect schematic.

The trenching disc, as the main working component of the disc-type trenching machine, is critical for achieving the desired trench depth while optimizing energy consumption and improving operational efficiency. The diameter of the disc has a decisive effect on the cutting arc length of the trenching knives, which in turn affects both the soil cutting power consumption and the milling power consumption. Based on the above analysis, the influence of increasing disc diameter D on power consumption is not unique. Therefore, selecting an appropriate disc diameter is crucial to balancing earth-cutting power consumption and milling power consumption, and an empirical formula needs to be introduced [26]. The relationship between the disc diameter and the cutting arc length can be expressed as:

$$B = \theta(R_1 + \delta_1) = \frac{1}{2}\theta(D_1 + 2\delta_1) \tag{1}$$

where B—the cutting arc length, mm; θ—the tearing line angle of the soil, °; R_1—the radius of the disc, mm; D_1—the diameter of the disc, mm; and δ_1—the working length of the trenching knife, mm.

From Formula (1), as the disc diameter increases, the cutting arc length also increases, which reduces the curvature of the soil-cutting profile, thereby lowering the power consumption during the cutting process. The relationship between the disc diameter and the theoretical trench depth can be expressed as:

$$D_1 + 2\delta_1 = (1.2 \sim 1.4)H \tag{2}$$

where H—the theoretical trench depth, mm.

Considering the impact of ground flatness on machine operations, we initially set the trench depth to 400 mm, with the working length of the trenching knife set to 140 mm. Using Formula (2), we determined that the disc diameter should range between 200 and 280 mm. Considering the comprehensive effects of disc diameter on torque distribution and structural dimensions, we increased the disc diameter within the reasonable range. To ensure that the trenching disc can cut soil evenly and complete the trenching operation with the minimum power consumption, the optimal disc diameter was determined to be 240 mm [18].

2.2.2. Design of Fertilizing Device

Figure 5 shows the structural schematic of the fertilizing device. The internal volume of the fertilizer box is approximately 0.2 m^3, and the angle between the two wall surfaces of the fertilizer box and the horizontal plane is 45°, ensuring that the fertilizer flows smoothly down. The fertilizer discharge spiral is the main working component of the discharge device, and its size directly determines the total fertilizer discharge rate, discharge uniformity, and discharge stability. Taking into account the installation of the fertilizer discharge spiral conveyor and the fertilizer box in Figure 6, as well as the size of the chassis, the effective length of the fertilizer discharge spiral is initially designed to be 760 mm [17].

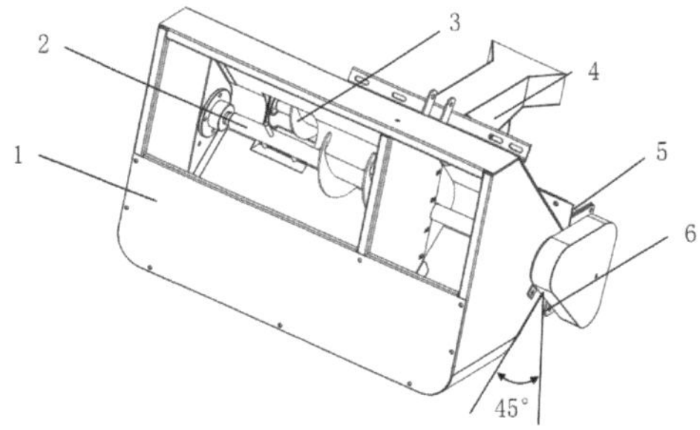

Figure 5. Structural schematic of the fertilizing device. (1) Fertilizer box; (2) fertilizer conveyor spiral; (3) fertilizer discharge spiral conveyor; (4) fertilizing tray; (5) discharge tray; (6) fertilizer drive motor.

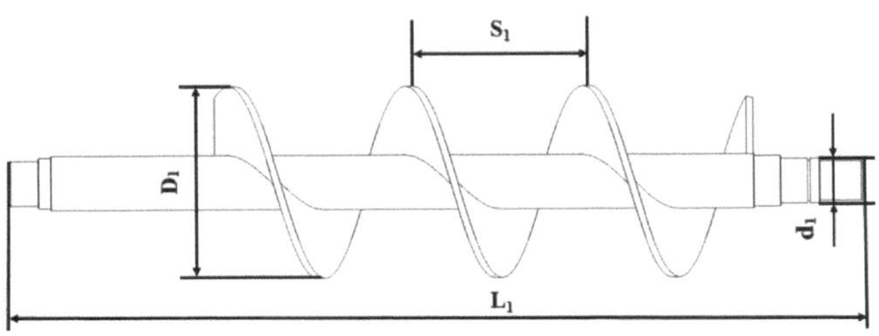

Figure 6. Structural schematic of the fertilizer discharge spiral conveyor.

The nominal diameter and pitch of the fertilizer discharge spiral are key parameters influencing the fertilizing rate. The nominal diameter of the spiral is related to the fertilizing speed, fill coefficient, and fertilizer bulk density. The calculation for the diameter is given by:

$$D_1 = K\left(\frac{Q_C}{\varphi\lambda\varepsilon}\right)^{\frac{2}{5}} \tag{3}$$

where D_1—the nominal diameter of the fertilizer discharge spiral, mm; φ—the material's comprehensive characteristic coefficient; K—the material's comprehensive coefficient; c—

the ratio of pitch to diameter; λ—the material's unit volume mass; and ε—the inclined conveyor coefficient.

For the orchard trenching–fertilizing machine, under continuous fertilization operation, the fertilizing rate can be expressed as:

$$Q = v_m g / s \tag{4}$$

where Q—the fertilizing rate, t/h; g—the amount of fertilizer per distance traveled, t; and s—the forward travel distance, m.

According to the agronomic requirements for fertilizing fruit trees [18], the fertilizing amount per tree on one side is 3 kg/m. The forward speed of the fertilizing machine is taken as 0.3 m/s. From Formula (4), the fertilizing amount is 3 t/h. Combining this with the characteristics of organic fertilizer, the fill coefficient for compound fertilizer is 0.45 and the material coefficient is 0.05. For a horizontal conveyor, the inclined conveyor coefficient is taken as 1.0 and the unit volume mass is 0.8 t/m^3. Substituting these values into Formula (3), the diameter of the spiral blades is found to be 116 mm. According to the mechanical industry standards, the spiral conveyor diameter is rounded to 100 mm.

The pitch and shaft diameter of the spiral conveyor directly affected the fertilizer discharge efficiency. The pitch and shaft diameter can be expressed as:

$$\begin{cases} S_1 = c_1 D_1 \\ d_1 = c_2 D_1 \end{cases} \tag{5}$$

where c_1—the pitch coefficient; c_2—the shaft diameter coefficient; d_1—the inner diameter of fertilizer discharge spiral conveyor, mm; and S_1—pitch, mm.

Based on the Manual of Agricultural Machinery Design [27], the pitch coefficient is taken as 0.8, the shaft diameter coefficient is 0.3, the pitch S_1 is equal to 80 mm, and the internal diameter of the fertilizer spiral d_1 is equal to 30 mm, which meet the required specifications.

The theoretical single-turn discharge amount is a key indicator for evaluating the discharge capacity of the fertilizer spiral conveyor. It is calculated as:

$$q = \left[\frac{\pi ({D_1}^2 - {d_1}^2) S_1}{4} - b_t h_t L_t \right] p_c \varphi_1 \tag{6}$$

where q—the single-turn discharge amount, g; D_2—the outer diameter of the discharge spiral, mm; d_1—the inner diameter of the discharge spiral, mm; S_1—the pitch, mm; b_t—the average thickness of the spiral blades, mm; h_t—the depth of the spiral blades, mm; L_t—the average length of the spiral blades, mm; p_c—the bulk density of compound fertilizer, g/mm^3; and φ_1—the fill coefficient of the discharge spiral.

Substituting the values of the discharge spiral's diameter, shaft diameter, and pitch into Formula (6), the theoretical single-turn discharge amount of the fertilizer spiral conveyor is approximately 164 g.

2.2.3. Design of Soil-Covering Device

The soil-covering device mainly consists of a soil-covering plate and side plates, as shown in Figure 7. The purpose of the soil-covering plate is to intercept the soil thrown up by the trenching device, guiding it to fall into the trench. The side plates play a crucial role in preventing the soil from being thrown outside the trench, thus preventing soil loss. To match the symmetric distribution of the trenching blade in the trenching device, the side plates of the soil-covering device are designed as single-side plates, and the soil-covering device is symmetrically installed on the symmetric trenching blade disc. This symmetric layout ensures uniformity of the soil-covering effect, making the entire soil-covering process more efficient and stable, providing strong support for fertilization operations in the orchard.

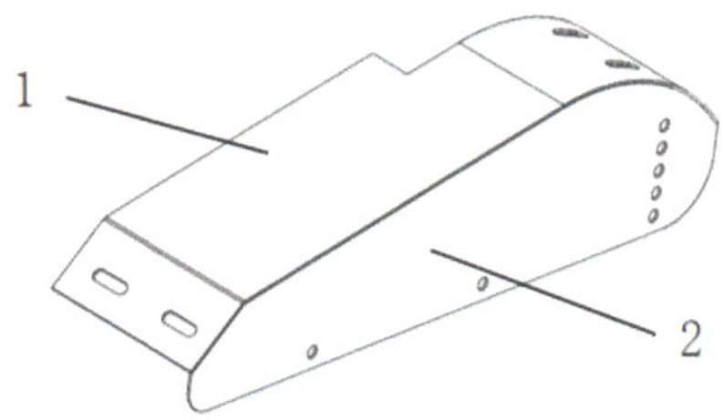

Figure 7. Structural schematic of the soil-covering device. (1) Soil-covering plate; (2) side plate.

2.3. Discrete Element Method Simulation Experiments

2.3.1. Discrete Element Simulation Modeling

In this study, EDEM 2022 software was used to construct a DEM simulation model to simulate the trenching, fertilizing, and soil-covering operations in the orchard. To improve simulation efficiency, a simplified model of the trenching, fertilizing, and soil-covering device was created in SolidWorks 2020, omitting components such as the gearbox, bolts, and nuts. The model was then saved in ".IGS" format and imported into the soil discrete element model to study the dynamic interaction between the machine and soil.

To better approximate the real field conditions, a soil bed with dimensions of 2000 mm in length, 1000 mm in width, and 450 mm in height was established, as shown in Figure 8, generating a total of 300,000 soil particles. In order to ensure the accuracy of the simulation test and a high efficiency of calculation, the radius of the soil particles was set to 4 mm, with the particles modeled as spherical [28]. A particle factory was added above the soil bed, and the simulation model ran for 0 to 5 s to fill the bed with soil and allow it to stabilize. After the soil reached a stable state, a compression plate was applied at a constant speed of 0.2 m/s for 5 to 6 s to compact the soil. After 6 s, the compression plate was reversed at the same speed until it was fully unloaded, thus completing the simulation of the 2000 mm-long, 1000 mm-wide, and 450 mm-high soil bed, which was then exported as the soil discrete element model.

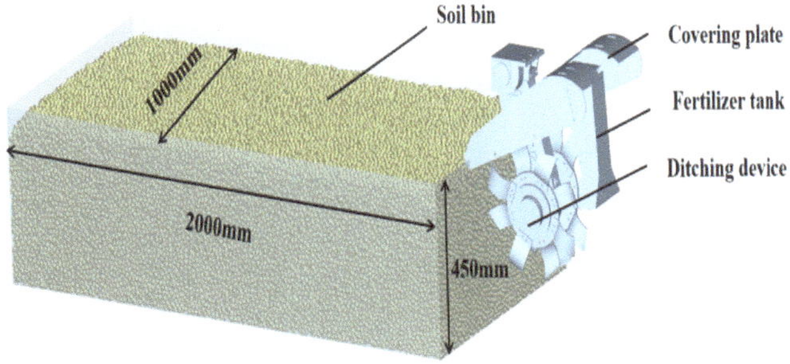

Figure 8. EDEM simulation model.

2.3.2. Setting of Simulation Parameters

In response to the high soil viscosity and moisture content characteristics of litchi plantation areas in the hilly regions of South China, where elastic–plastic deformation occurs under the action of the blades, this study employed the Hertz–Mindlin with JKR contact model for the interactions between particles. The geometry material was selected as 65 Mn, and the Hertz–Mindlin no-slip contact model was used between the particles and the geometry. In the EDEM pre-processing module, appropriate contact mechanics models were set, and the simulation parameters for the material and contact properties are provided in Tables 3 and 4, based on previous studies [8,29].

Table 3. Material property parameters.

Materials	Poisson's Ratio	Density (kg/m^3)	Young's Modulus (Pa)
Soil	0.3	2600	1×10^6
65 Mn	0.2	7800	7×10^9
Fertilizer	0.25	1320	1×10^7

Table 4. Contact attribute parameters.

Parameter	Restitution Coefficient	Coefficient of Static Friction	Dynamic Friction Coefficient
Soil and soil	0.5	0.4	0.15
Soil and 65 Mn	0.6	0.6	0.13
Soil and fertilizer	0.25	0.6	0.02
Soil and steel	0.5	0.5	0.05
Fertilizer and fertilizer	0.11	0.3	0.1

After importing the model, the device was positioned at the front of the soil bed, with the lower end of the device set 300 mm from the soil surface. The trenching forward speed was set at 0.3 m/s, and the trenching rotation speed was set at 300 r/min. A particle factory was added above the fertilizer discharge slot, with a total of 40,000 fertilizer particles. Based on the size of the organic fertilizer, the particle radius was set to 3 mm as spherical particles. During the simulation, the generated fertilizer particles fell into the trench via the fertilizing slot. The time step for the simulation was set to 10%, with a total simulation duration of 8 s. The data write-out interval was set to 0.01 s, and the grid size was set to 2 Rmin to ensure that the number of grids was no fewer than 1×10^5. To improve the simulation computational speed, the simulation area was minimized as much as possible.

2.3.3. Simulation Test of Fertilization Scheme

If unmixed fertilizer is placed directly into a trench and immediately covered with soil, many advantages of organic fertilizer—such as its buffering properties, soil loosening, moisture retention, and aeration capabilities—cannot be fully leveraged. This approach also fails to improve the internal soil environment as intended. Therefore, thoroughly mixing granular organic fertilizer with the soil is a crucial step in realizing its full functional benefits.

Based on existing organic fertilizer application methods, fertilization can be classified into three modes, as shown in Figure 9: pre-plant, mid-plant, and post-plant fertilization [24]. To select the most appropriate method, an analysis of the fertilization patterns of these three modes is necessary:

(1) Pre-Plant Fertilization: The fertilizing trough is aligned parallel to the machine's forward direction, with the trenching device positioned directly behind the fertilizer trough. Fertilizer is evenly spread along the outer edge of the tree root zone, forming a strip or band of fertilizer. The trenching device then cuts into the soil, simultaneously cutting and throwing both soil and fertilizer upward. The covering device directs the thrown soil and fertilizer back into the trench. This process achieves the mixing of soil and fertilizer without actual trench formation during the fertilization step, but may result in less uniform mixing compared to other methods.

(2) Mid-Plant Fertilization: The fertilizing trough remains parallel to the forward direction, but the trenching device is positioned directly ahead of the fertilizer trough. Organic fertilizer is discharged from the fertilizing trough onto the surface of the soil. The trenching device then cuts into the soil, throwing both soil and the recently discharged fertilizer upward. Concurrently, the covering device guides the thrown mixture back into the trench. This method mixes soil and fertilizer by combining them during the cutting process, but some fertilizer may remain on the soil surface, affecting uniformity.

(3) Post-Plant Fertilization: The trenching device cuts into the soil and throws soil upward, while the fertilizing trough is positioned behind the trenching device. As the trenching device cuts into the soil, organic fertilizer is discharged from the fertilizing trough into the open trench. The soil thrown and discharged fertilizer are then guided by the covering device to fall back into the trench, achieving a thorough mix of soil and fertilizer within the trench. This method tends to result in better soil–fertilizer mixing, as the fertilizer is directly introduced into the trench, where it is immediately combined with the displaced soil.

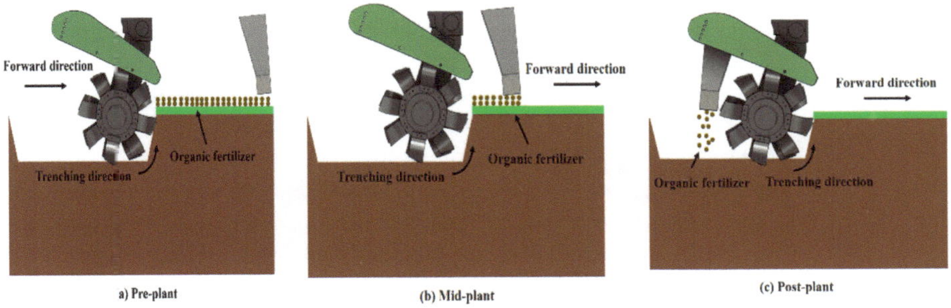

Figure 9. Fertilization models of three different modes.

To choose the most suitable fertilization approach for enhancing soil–fertilizer mixture uniformity, it is essential to analyze and compare the fertilization distribution patterns (fertilizer discharge behavior) of these three modes. By examining how each method distributes fertilizer and mixes it with soil—using metrics such as surface fertilizer leakage rate and soil–fertilizer mixture uniformity—a more informed decision can be made about which fertilization model best meets the goals of efficient and effective organic fertilizer application.

Three-dimensional models for the pre-plant, mid-plant, and post-plant fertilization methods were created and are shown in Figure 10. After importing the models, the device was positioned at the front of the soil bed, with its lower end set 300 mm above the soil surface. The trenching forward speed was set to 0.3 m/s and the trenching blade speed to 300 r/min, consistent with actual operating parameters. The simulation was set to run for 12 s, allowing the entire trenching, fertilizing, and covering device to pass completely over to the other side of the virtual soil bed. The time step was set at 10% of the simulation duration, with data saved at 0.01 s intervals. To facilitate quantitative evaluation of the soil–fertilizer mixing design requirements and to accurately assess the effectiveness of the soil–fertilizer mixing, this study referenced relevant research in the field and selected the surface fertilizer leakage rate (K) and soil–fertilizer mixture uniformity (M) as the primary evaluation metrics. These metrics were used to determine the optimal fertilization method during the operation.

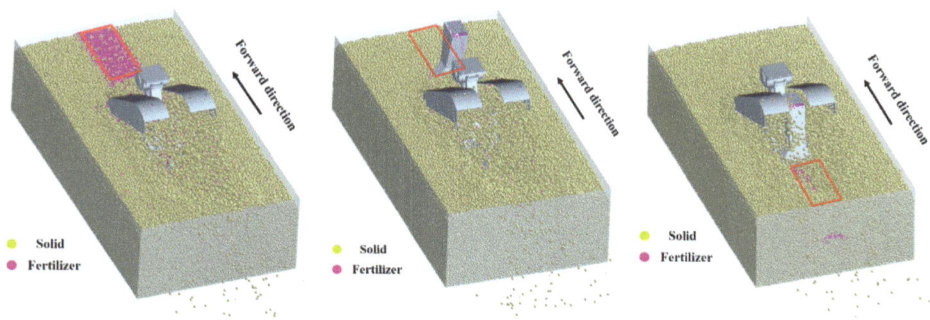

(**a**) Pre-plant (**b**) Mid-plant (**c**) Post-plant

Figure 10. Fertilization simulation models of three different modes.

2.3.4. Scheme and Method of the Simulation Test

Based on the practical operational requirements of litchi orchards and preliminary field experiments, as well as relevant literature, it has been observed that parameters such as the forward speed of the entire machine, trenching speed, and bending angle of the trenching blade significantly influence operational power consumption and the uniformity of soil–fertilizer mixing [17,30].When the forward speed is less than 0.2 m/s, the trenching consistency is well maintained. However, the overall trenching efficiency is low, leading to poor economic performance, and this speed may cause fertilizer to accumulate in localized areas. Conversely, when the forward speed exceeds 0.5 m/s, the operational efficiency of the entire machine significantly improves, increasing the area worked per unit time. Nonetheless, excessively high speeds can lead to increased power consumption, operational instability, and insufficient soil coverage. Consequently, a forward speed range of 0.2 to 0.5 m/s is considered optimal.

Regarding trenching speed, when it is below 250 r/min, soil is not effectively ejected from the trench, resulting in inadequate mixing of soil and fertilizer. Moreover, the trenching components endure high torque. When the speed exceeds 350 rpm, the soil ejection speed increases, leading to greater soil throw distance and more uniform soil–fertilizer mixing. However, this also results in higher specific power consumption. Therefore, a trenching speed range of 250 to 350 rpm is deemed appropriate.

As for the bending angle of the trenching blade, when it is less than 130°, the trench width decreases, reducing cutting power consumption. However, the disturbed soil area after blade entry is smaller, resulting in poorer soil-throwing performance and less effective soil–fertilizer mixing. When the bending angle exceeds 140°, the operational profile enlarges, the disturbed soil area increases, and the soil-throwing performance improves, leading to better mixing. Nonetheless, this also increases cutting power consumption. Therefore, a bending angle range of 130° to 140° is considered optimal.

To reduce operating power consumption and improve soil–fertilizer mixing performance, forward speed, disc cutter speed, and cutter deflection angle were selected as experimental factors. The evaluation indicators were trenching power consumption and soil–fertilizer mixing uniformity. A three-factor, three-level orthogonal experiment was designed using the Box–Behnken response surface method. The experimental factors and their levels are encoded in Table 5.

Table 5. Table of factor levels of the experimental test.

Code	Factors		
	Forward Speed x_1 (m/s)	Trenching Speed x_2 (r/min)	Bending Angle of Trenching Knife x_3 (°)
−1	0.2	250	130
0	0.3	300	135
1	0.4	350	140

2.3.5. Simulation Evaluation Method

1. Trenching Power Consumption

After the simulation run, the trenching resistance torque was exported as one of the experimental results in the EDEM software's post-processing module. The trenching power consumption is calculated using the following Formula (7):

$$P = \frac{Tn}{9550} \tag{7}$$

where P—the trenching power consumption, kW; n—cutter disc rotation speed, r/min; and T—cutter disc torque, N·m.

2. Soil–Fertilizer Mixing Uniformity Measurement

In the simulation, the soil-fertilizer mixing uniformity was measured as follows: In the EDEM post-processing module, under "Setup Selections", the "Grin Bin Group" was selected for the middle 1000 mm of the analysis area. This area was divided into a 20 × 20 × 20 mm grid, and the 10 evenly distributed measurement regions were analyzed. The number of soil and fertilizer particles in different regions was recorded, as shown in Figure 11.

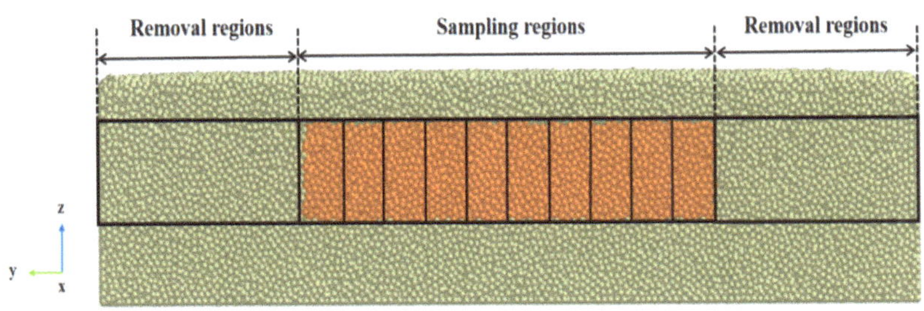

Figure 11. Measurement of soil–fertilizer mixing uniformity.

The average value of the soil–fertilizer mixture in each grid is denoted by Formula (8):

$$\overline{Q} = \frac{\Sigma Q_i}{n} \tag{8}$$

where $\overline{Q}$—The average value of the soil–fertilizer mixture, g; Q_i—the mass of soil–fertilizer mixture in the grid, g; and n—the total number of grids in the analysis area.

The standard deviation of the soil–fertilizer mixture in each grid is calculated by Formula (9) as follows:

$$s = \sqrt{\frac{\Sigma (Q_i - \overline{Q})^2}{n-1}} \tag{9}$$

The coefficient of variation (C_v) for the mixing uniformity is given by Formula (10):

$$C_v = \frac{s}{\overline{Q}} \times 100\% \tag{10}$$

where C_v—the coefficient of variation for soil–fertilizer mixing uniformity.

The uniformity of the mixture M is then calculated by Formula (11) as follows:

$$M = 1 - C_v \tag{11}$$

3. Fertilizer Loss Rate

In the EDEM post-processing module, under "Setup Selections", the "Grin Bin Group" was used to count the number of fertilizer particles distributed across different soil layers. The middle 1000 mm of the analysis area was selected and divided into three layers with a height of 10 cm each. The number of fertilizer particles in each layer was counted, as shown in Figure 12.

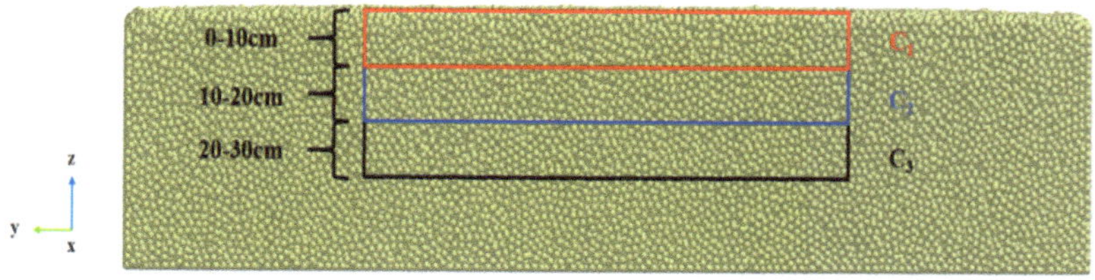

Figure 12. Measurement of fertilizer loss rate.

The fertilizer loss rate is calculated by Formula (12) as follows:

$$K = \frac{C_1}{C_1 + C_2 + C_3} \times 100\% \tag{12}$$

where C_1—the number of fertilizer particles in the first layer; C_2—the number of fertilizer particles in the second layer; C_3—the number of fertilizer particles in the third layer; and K—the fertilizer loss rate, %.

2.4. *Field Test Condition and Method*

2.4.1. Field Test Condition

To validate the accuracy of the simulation model and optimize the performance of the trenching, fertilization, and soil-covering system, a field experiment was conducted in 21 December 2024 at the Mingyuan Ecological Park in Gaozhou City, Guangdong Province, China. The weather was sunny, and the temperature was 20 degrees. The equipment used in the field test included (1) The trenching fertilization and soil-covering machine prototype, (2) a JRT-711 soil moisture content meter (range: 0–100%, accuracy: ±0.5%), (3) a JK-750-I soil compaction meter (range: 0–10 MPa, accuracy: ±1%), (4) a multimeter (range: 0–1000 V, accuracy: ±0.5%), (5) a ring knife (volume: 100 cm^3), (6) measuring tape (range: 50 m, accuracy: 0.001 m), (7) a steel ruler (range: 50 cm, accuracy: 0.1 cm), (8) a stopwatch (accuracy: 0.01 s), and (9) an electronic scale (capacity: 1000 g, accuracy: 0.01 g). The soil type was clay loam, the fertilizer was granular organic fertilizer, the moisture content was 14%, the density was 1.29 g/cm^3, and the soil compaction from 0 to 30 cm was 300 N/cm^2.

The field test aimed to measure the actual trenching depth, fertilizer application uniformity, soil-covering efficiency, and power consumption, and to compare the results with the simulation data for validation and further optimization of the machine's operational parameters.

2.4.2. Test Methods and Evaluation Indicators

1. Power Consumption Test

According to the power consumption evaluation test method for trenching machinery, as specified in Chinese Mechanical Industry Standard JB/T 11908-2014 "Agricultural disc ditcher" [31], the following procedure was followed: consistent with the simulation parameters, the machine's trenching blade rotational speed was set to 260 r/min, the forward speed was set to 0.4 m/s, and the trenching component was lowered to the target depth of 300 mm. The length of the test area was 300 cm; to ensure data accuracy, the middle 100 cm were designated as the data collection zone. Within the data collection zone, five points at equal intervals were selected. At each point, a multimeter was used to measure the current and voltage at the engine's power supply terminals, from which the trenching power consumption was calculated. The test was repeated over three passes, yielding a total of 15 measurement points for current and voltage readings.

2. Fertilization Uniformity Test

The soil–fertilizer uniformity test was conducted in accordance with the fertilization machinery evaluation test method specified in Chinese National Industry Standard GB/T 20346.2-2022 "Equipment for distributing fertilizer—Part 2: Fertilizer distributor in lines" [32]. The following procedure was followed: consistent with the simulation parameters, the machine's trenching blade rotational speed was set to 260 r/min, the forward speed was set to 0.4 m/s, and the length of the test area was 300 cm. After fertilization, within the middle 100 cm of the area, five equidistant measurement points were selected. At each measurement point, a 10 cm-long section of soil and fertilizer was collected and weighed separately. The test was repeated for three passes, resulting in a total of 15 measurement points for fertilizer mass, which were then used to calculate the soil–fertilizer mixture uniformity.

3. Fertilizer Leakage Rate

Taking the soil surface as the horizontal plane, after mixing and covering the soil with fertilizer, the soil–fertilizer mixture area was vertically divided into three depth layers: surface layer (0–10 cm), upper layer (10–20 cm), and lower layer (20–30 cm). The fertilizer leakage rate was defined as the ratio of the amount of fertilizer in the upper and lower layers to that in the surface layer.

3. Test Results and Discussion

3.1. EDEM Simulation Results and Discussion

During the operation process, the spatial positional relationships between the trenching, fertilizing, and soil-covering devices, as well as the synergistic coordination among the components, have a significant impact on the mixing effect of soil and fertilizer. To analyze the soil and fertilizer transport mechanisms of each device at different operational stages, this study employed EDEM post-processing technology to evaluate the kinetic energy and motion trends of simulated soil and fertilizer particles. As shown in Figure 13, taking operational conditions with a forward speed of 0.3 m/s and a trenching speed of 350 r/min as an example, the kinetic energy and motion trends of soil and fertilizer particles were observed. During the trenching process, the trenching device cuts through the soil via high-speed rotation, causing soil particles to be thrown upward and dispersed. Although the soil dispersion process is chaotic and irregular, most soil particles exhibit similar motion trends, eventually landing within a relatively fixed dispersion range. At this stage, the soil-covering device deflects the soil in a predetermined direction, guiding the motion trajectory of the soil particles. The detached soil is thrown backward along the leading surface of the soil-covering plate under the action of the high-speed rotating trenching cutter, with its kinetic energy gradually increasing. Due to inertia and internal forces, the soil continues to move and reaches the discharge surface, no longer influenced by the trenching cutter. Under the guiding effect of the discharge surface, the soil particles descend along a specific trajectory and collide with the fertilizer particles discharged from the fertilizing chute. At the moment of collision, the kinetic energy of the soil particles decreases, and the trajectories of some soil particles are altered, thereby affecting the landing point of the soil–fertilizer mixture. Ultimately, the soil–fertilizer mixture precisely falls into the trench created by the trenching device. Subsequently, the soil continues to move downward and backward, covering the soil–fertilizer mixture and completing the entire process of trenching, fertilizing, and soil covering.

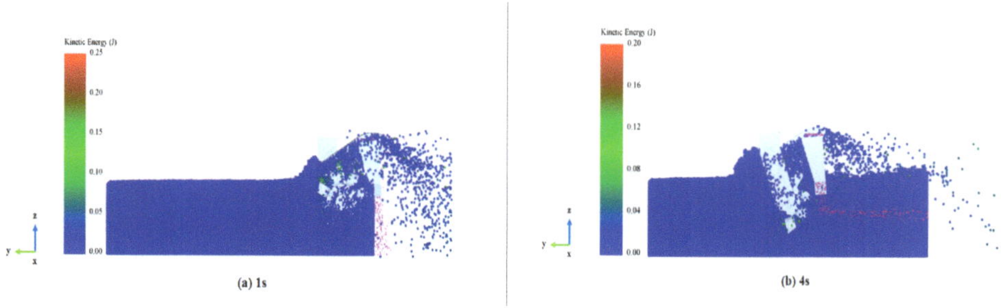

Figure 13. Process of soil–fertilizer mixing and covering. (**a**) At 1 s; (**b**) at 4 s.

3.1.1. Simulation Analysis of Three Different Fertilization Schemes

After the simulation, post-processing analysis of the experimental results was conducted using the Grid Bin Group method for layer-by-layer statistical analysis of the working area. The area was divided into three depth layers: 0~10 cm (surface layer), 10~20 cm (upper layer), and 20~30 cm (lower layer). The number of fertilizer particles in each depth layer was counted, and the leakage fertilizer rate was then calculated according to Formula (12).

In this study, the uniformity of soil–fertilizer mixing was calculated using a sampling method [25]. The Grid Bin Group method was used to select the central 1000 mm of the analysis area in EDEM's setup selections, dividing it into 20 × 20 × 20 mm cells to count the soil and fertilizer particle numbers in different regions. A group of 20 samples were randomly selected for the experiment, each group of samples was selected three times and the average value was obtained, the number of soil particles and fertilizer particles in the samples was counted. Finally, the uniformity of soil–fertilizer mixing was calculated according to Formula (11). The leakage fertilizer rate and soil–fertilizer mixture uniformity were used as evaluation metrics to calculate and assess the three fertilization schemes, as shown in Figure 14.

After a simulation comparison of pre-plant fertilization, mid-plant fertilization, and post-plant fertilization, the results were as follows: the pre-plant fertilization was mainly concentrated in the soil surface of 0~10 cm, resulting in a high leakage rate and lower soil–fertilizer mixture uniformity; the mid-plant fertilization was mainly concentrated in the soil surface of 0~10 cm below the surface, resulting in a high leakage rate but better soil–fertilizer mixture uniformity; and the post-plant fertilization was mainly concentrated in the upper and lower layers of the surface of 10~30 cm, and the fertilizer leakage rate was low. Both the soil and fertilizer mixture uniformity and the fertilizer leakage rate could better meet and suit the fertilization trend requirements of deep application of organic fertilizer in orchards. Thus, based on the scheme comparison and analysis, post-plant fertilization was found to be the most advantageous method for ensuring better soil–fertilizer mixture uniformity and lower leakage rate.

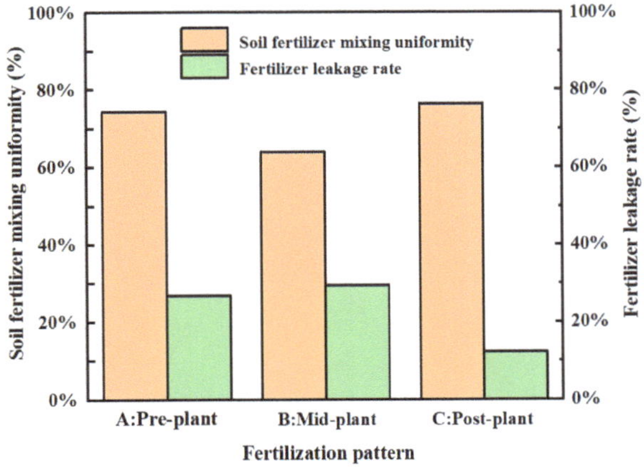

Figure 14. Fertilization results of three different schemes.

3.1.2. Response Surface Analysis

A Box–Behnken response surface test was carried out, and the test scheme and results are shown in Table 6.

Table 6. Response surface test scheme and results.

Test Number	Trenching Speed x_1/(r/min)	Forward Speed x_2/(m/s)	Bending Angle of Trenching Knife x_3/(°)	Trenching Power Consumption y_1/kW	Soil–Fertilizer Mixing Uniformity y_2/%
1	300	0.4	130	2.19	78.01
2	300	0.3	135	2.07	76.13
3	300	0.3	135	2.04	76.47
4	350	0.2	135	2.19	77.73
5	300	0.3	135	2.12	76.59
6	300	0.2	130	1.50	79.15
7	350	0.3	140	2.02	76.12
8	250	0.3	140	1.54	77.02
9	300	0.3	135	2.14	76.20
10	350	0.4	135	2.88	77.34
11	300	0.4	140	1.99	76.90
12	300	0.2	140	1.36	78.07
13	300	0.3	135	2.05	76.37
14	250	0.2	135	1.55	78.89
15	250	0.4	135	1.95	76.34
16	350	0.3	130	2.51	77.49
17	250	0.3	130	1.50	78.46

1. Regression Analysis of Trenching Power Consumption

A regression fitting analysis was conducted on the test results to derive a quadratic regression model for trenching power consumption with respect to the experimental factors. The analysis of variance (ANOVA) for the tillage power consumption model is presented in Table 7.

The regression analysis based on the experimental results led to a quadratic regression model for trenching power consumption in relation to the experimental factors. The ANOVA results showed that $p < 0.01$, indicating that the model was highly significant. The model's lack-of-fit term $p = 0.2263$, suggesting no significant lack of fit, and the regression formula was well fitted and suitable for predictive analysis. The factors x_1 (trenching speed), x_2 (forward speed), and x_3 (bending angle of trenching knife) had a highly significant impact on y_1 (trenching power consumption). Based on the p-values for the factors, the influence on tillage power consumption was ranked from largest to smallest as follows: x_1, x_2, x_3. The interaction terms x_1x_2 and x_2x_3 had significant effects on power consumption, while other interactions were not significant. Among the quadratic terms, the regression formula for each objective function, after excluding non-significant terms, are as follows:

$$y_1 = 2.08 + 0.3825x_1 + 0.3013x_2 - 0.0987x_3 + 0.0725x_1x_2 - 0.1325x_1x_3 + 0.0955{x_1}^2 - 0.2870{x_3}^2 \tag{13}$$

Table 7. ANOVA for the trenching power consumption model.

Variance Source	Sum of Squares	Degrees of Freedom	Mean Square	F	p	Significance
Model	2.46	9	0.2738	134.52	<0.0001	**
x_1	1.17	1	1.17	575.16	<0.0001	**
x_2	0.7260	1	0.7260	356.76	<0.0001	**
x_3	0.0780	1	0.0780	38.34	0.0004	**
x_1x_2	0.0210	1	0.0210	10.33	0.0148	*
x_1x_3	0.0702	1	0.0702	34.51	0.0006	**
x_2x_3	0.0009	1	0.0009	0.4423	0.5273	-
${x_1}^2$	0.0337	1	0.0337	16.57	0.0047	**
${x_2}^2$	0.0078	1	0.0078	3.83	0.0914	-
${x_3}^2$	0.3615	1	0.3615	177.63	<0.0001	**
Residual	0.0142	7	0.0020			
Lack of fit	0.0089	3	0.0030	2.24	0.2263	-
Pure error	0.0053	4	0.0013			
Total	2.48	16				

Note: $p \leq 0.01$ is highly significant, marked as **; $0.01 < p \leq 0.05$ is significant, marked as *; $p > 0.05$ is not significant, marked as -.

The influence surface of the interaction between variables on the trenching power consumption is shown in Figure 15. Figure 15a shows the interaction of trenching speed and forward speed on trenching power consumption. When the forward speed and the bending angle of the trenching knife were constant, the trenching power consumption increased with the increase in the trenching speed. The trenching power consumption curve along the direction of forward speed changed faster than the speed along the direction of trenching speed, indicating that at the test level, the trenching power consumption was more significantly affected by the forward speed than by the trenching speed. Figure 15b shows the interaction of trenching speed and bending angle of the trenching knife on trenching power consumption. When the forward speed and the trenching speed were constant, the trenching power consumption increased first and then decreased gradually. This was because when the bending angle of the trenching knife exceeded 130°, the torque force required by the trenching knife during rotation became larger, and when the bending angle of the trenching knife exceeded 135°, the torque force required by the trenching knife during rotation became smaller, but the trenching process had significant blade wear and a poor trenching effect on the soil. Therefore, the appropriate bending angle was beneficial for reducing the trenching torque and improving the quality of trenching.

2. Soil–Fertilizer Mixture Uniformity Model Variance and Response Surface Analysis

A regression fitting analysis was performed on the test results to derive a quadratic regression model for the experimental factors and soil–fertilizer mixture uniformity. The variance analysis for the soil–fertilizer mixture uniformity model is shown in Table 8.

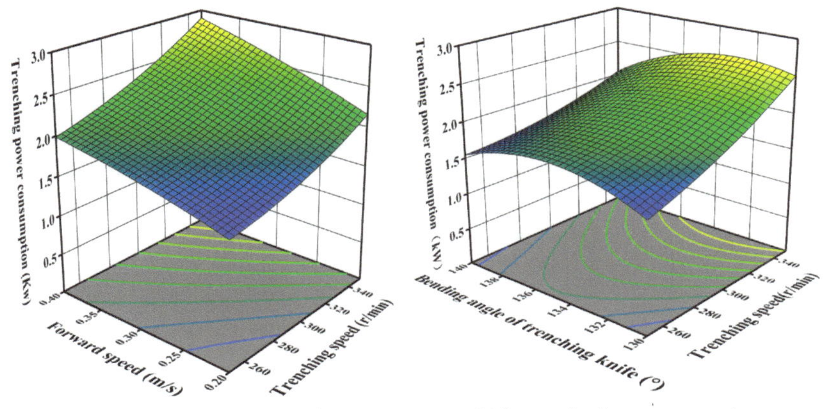

(**a**) Interaction between x_1 and x_2. (**b**) Interaction between x_1 and x_3.

Figure 15. Interaction effects of variables on power consumption.

Table 8. Variance analysis for soil–fertilizer mixture uniformity model.

Variance Source	Sum of Squares	Degrees of Freedom	Mean Square	F	p	Significance
Model	15.16	9	1.68	19.44	0.0004	**
x_1	0.5151	1	0.5151	5.95	0.0449	-
x_2	3.45	1	3.45	39.77	0.0004	**
x_3	3.13	1	3.13	36.07	0.0005	**
x_1x_2	1.17	1	1.17	13.46	0.0080	**
x_1x_3	0.0012	1	0.0012	0.0141	0.9087	-
x_2x_3	0.0002	1	0.0002	0.0026	0.9608	-
$x_1{}^2$	0.2257	1	0.2257	2.60	0.1506	-
$x_2{}^2$	4.14	1	4.14	47.78	0.0002	**
$x_3{}^2$	2.00	1	2.00	23.07	0.0020	**
Residual	0.6065	7	0.0866			
Lack of fit	0.4632	3	0.1544	4.31	0.0960	-
Pure error	0.1433	4	0.0358			
Total	15.77	16				

Note: $p \leq 0.01$ is highly significant, marked as **; $p > 0.05$ is not significant, marked as -.

Based on the regression fitting analysis, a quadratic regression model was obtained for experimental factors and soil–fertilizer mixture uniformity. The variance analysis results indicated that $p < 0.01$, showing that the model was highly significant. The lack-of-fit term $p = 0.0960$ suggests that there were no significant lack-of-fit factors, and the regression formula was well fitted and suitable for predictive analysis. The factors x_2 and x_3 had a highly significant impact on y_2 (soil–fertilizer mixture uniformity). The factor x_1 also had a significant effect on soil–fertilizer mixture uniformity. The p-value analysis showed that the influence on soil–fertilizer mixture uniformity decreased in the following order: x_3, x_2, x_1. The interaction term x_1x_2 had an extremely significant effect on soil–fertilizer mixture uniformity, with a p-value of less than 0.001, indicating a highly significant interaction between tillage speed and cutting speed on mixture uniformity. Among the quadratic terms, except for $x_1{}^2$, the remaining terms significantly influenced soil–fertilizer mixture uniformity. After excluding non-significant terms, the regression formula for each objective function is as follows:

$$\begin{aligned} y_2 &= 76.35 - 0.2537x_1 - 0.6526x_2 - 0.6250x_3 \\ &+0.5400x_1x_2 + 0.9915x_2{}^2 - 0.6890x_3{}^2 \end{aligned} \tag{14}$$

The influence surface of the interaction between variables on the soil–fertilizer mixture uniformity is shown in Figure 16. Figure 16 shows the interaction of trenching speed and forward speed on the soil–fertilizer mixture uniformity. When the forward speed and the bending angle of the trenching knife were constant, the soil–fertilizer mixture uniformity decreased slowly with the increase in the trenching rotation speed, because in the process of increasing the trenching rotation speed, the scattering speed of soil particles also increased, which weakened the effect of collision and stirring between soil particles and fertilizer particles. When the trenching speed and bending angle of the trenching knife were constant, the soil–fertilizer mixture uniformity decreased first and then increased slowly with the increase in the advancing speed. The variation in the soil–fertilizer mixture uniformity curve along the direction of forward speed was faster than that along the direction of trenching speed, indicating that at the test level, the influence of forward speed on soil–fertilizer mixture uniformity was more significant than that under the influence of trenching speed.

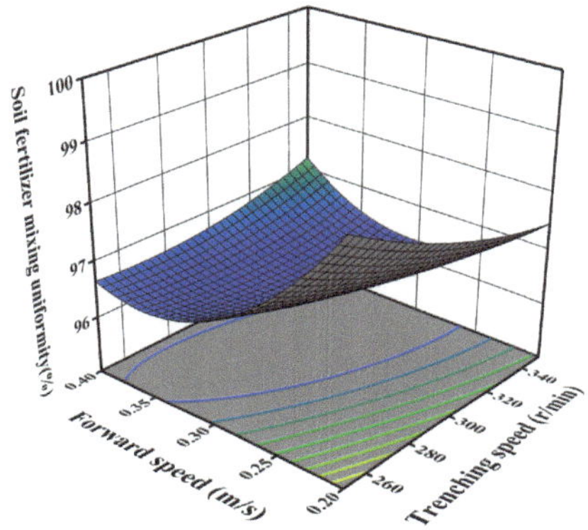

Figure 16. Interaction effects of variables on soil–fertilizer mixture uniformity.

3.1.3. Optimization and Validation of Seedling Extraction Parameters

Due to the inconsistent effects of various factors on trenching power consumption and soil–fertilizer mixture uniformity, global multi-objective optimization was required. The objective functions were to minimize the trenching power consumption y_1 and maximize the soil–fertilizer mixture uniformity y_2. The variables were the trenching speed x_1, forward speed x_2, and bending angle of the trenching knife x_3. The optimization was performed using quadratic models between the experimental factors and indices in the Design-Expert 12.0 software's optimization module, with equal weights assigned to both performance indicators. The optimization model was:

$$\begin{cases} Miny_1 \\ Maxy_2 \\ 250 \le x_1 \le 350 \\ 0.2 \le x_2 \le 0.4 \\ 130^\circ \le x_3 \le 140^\circ \end{cases} \tag{15}$$

The optimization results were as follows: the trenching speed was 265.03 r/min, the forward speed was 0.40 m/s, and the bending angle of trenching knife was 130°. According to the optimization results, the trenching component model was established to simulate the trenching power consumption, and the soil–fertilizer mixture uniformity were 1.74 kW and 77.15%, respectively.

3.2. Field Test Results and Discussion

3.2.1. Trenching Power Consumption Test

The real-time operating power consumption in the data collection zone is shown in Table 9, and the trenching test is shown in Figure 17. The average values for the simulation and field trials were 1.79 kW and 1.93 kW, respectively, with a relative error of 7.44%. This indicates that the mechanical characteristics of the simulated soil model are in good agreement with those of the actual soil. The simulation model calculated a lower operational power consumption than the field trial measurements, likely due to the presence of residual roots and gravel in the actual trenching environment. This will be the focus of future model optimizations. The optimized model can be used to analyze the interaction between the trenching tool set and residual roots or gravel until they are either smoothly cut or slide along the blade edge. This can provide design guidance for optimizing parameters in litchi orchard trenching operations.

Table 9. Results of trenching power consumption test.

Number of Tests	Simulation Test (kw)	Orchard Test (kw)	Relative Error (%)
1	1.62	1.75	7.42
2	1.71	1.86	8.06
3	2.04	2.19	6.84
Mean value	1.79	1.93	7.44

Figure 17. Trenching test.

3.2.2. Soil–Fertilizer Mixture Uniformity

After the trenching, fertilizing, and covering process in Figure 18, the soil–fertilizer mixture uniformity at various sampling points was determined and is shown in Table 10. The field trial soil–fertilizer mixture uniformity was 76.16%, and the average relative error between the simulation and field trial mixture uniformity was 4.50%, indicating that the established simulation model can accurately predict the optimization results. Based on the technical specifications for soil–fertilizer mixture uniformity, the field trial results meet the mixing quality evaluation requirements.

Figure 18. Soil–fertilizer mixture uniformity test.

Table 10. Results of soil–fertilizer mixture uniformity test.

Number of Tests	Simulation Test (%)	Orchard Test (%)	Relative Error (%)
1	78.68	75.74	3.73
2	79.64	75.93	4.65
3	80.96	76.81	5.12
Mean value	79.74	76.16	4.50

According to the relevant evaluation test method for trenching machinery, as specified in "JB/T 11908-2014 Agricultural disc ditcher" [31] and "GB/T 20346.2-2022 Equipment for distributing fertilizer—Part 2: Fertilizer distributor in lines" [32], the results strongly confirmed the reliability of the regression model, as demonstrated by the relative errors between the actual and predicted values, all of which were below the threshold of 10%, and show harmonious agreement between the actual and expected values. The research results can provide reference for the structural transformation of orchard trenching machines and the management of trenching operation parameters.

In orchard trenching and fertilization operations, soil and fertilizer mix uniformity is crucial to ensuring full utilization of the fertilizer, thereby enhancing the uptake of nutrients by the root system, which is a key factor. Mechanized fertilization overcomes the problem of uneven nutrient distribution and concentration caused by the single fertilization method, greatly reduces labor intensity, and saves time and labor cost. Objective factors such as soil type, wind speed, and the physical properties of the fertilizer may have a slight impact on the trenching and fertilization performance of the equipment, but the machine's performance far exceeds industry standards. From an innovative standpoint, the integration of trenching, fertilizing, and soil covering reflects the characteristics of automation and intelligence to a certain extent. This crawler-type self-propelled orchard trenching and fertilizing machine has been tested only in the experimental orchards of South China Agricultural University and the Mingyuan Ecological Garden in Gaozhou. However, to further enhance its environmental adaptability and technological advancement, systematic testing should be conducted in more typical orchard environments, particularly under the dwarf rootstock high-density planting mode, to verify the machine's passability and operational stability.

In order to promote the development of unmanned orchards, the control system of the crawler-type self-propelled orchard trenching and fertilizing machine can continue to be improved in the next plan, the control system will be upgraded to an intelligent navigation system based on multi-modal perception to achieve centimeter-level positioning accuracy and dynamic obstacle avoidance functions, and the adaptive operation module will be developed by combining a deep reinforcement learning algorithm. The equipment can optimize the depth of trenching and control the amount of fertilizer in real time according to soil hardness, root distribution, and other parameters to realize automatic driving and adaptive trenching fertilization and soil-covering operations in different orchard environments, and thus reduce the manual participation and improve the production efficiency of the orchard.

4. Conclusions

A crawler-type self-propelled orchard trenching and fertilizing machine has been designed that is capable of trenching, fertilizing, and soil-covering operations in a single cycle. It exhibits certain levels of semi-automation, intelligence, and stability. Through theoretical analysis, various mechanisms have been devised, including the key components of the trenching device, fertilizing device, and soil-covering device. Three fertilizing simulation models (pre-plant, mid-plant, and post-plant) were established using EDEM discrete element software, with results indicating that the post-plant fertilizing mode better meets the soil–fertilizer mixing requirements for deep organic fertilizer application. Using trenching speed, forward speed, and bending angle of the trenching knife as experimental factors, with operating power consumption and soil–fertilizer mixing uniformity as evaluation indicators, a Box–Behnken experiment was conducted to optimize the parameters of the trenching and fertilizing components. A regression model was established to analyze the interaction between the experimental factors and indicators. The optimal operational parameter combination was determined as follows: trenching speed of 265.03 r/min, forward speed of 0.40 m/s, and bending angle of the trenching knife of 130°. Under these parameters, the trenching power consumption and soil–fertilizer mixing uniformity were 1.74 kW and 77.15%, respectively.

Orchard verification tests on the machine demonstrate that the machine operated with stability across all its components, exhibiting high consistency in the trenching depth, with trenching depth stability and soil–fertilizer mixing uniformity. The relative errors in trenching power consumption and soil–fertilizer mixing uniformity between the field tests and simulations were 7.40% and 4.50%. These results meet the agronomic requirements for trenching and fertilizing. The distribution of fertilizer particles was uniform, and the machine's operational performance meets the industry standards stipulated by national regulations.

Author Contributions: Conceptualization, C.L. and C.M.; methodology, C.L. and Y.Z.; software, C.M.; validation, C.L., C.M. and Z.L.; formal analysis, C.L., C.M. and R.J.; investigation, C.L. and J.C.; resources, C.L. and Q.W. (Qinglin Weng); data curation, C.L., C.M. and Q.W. (Qian Wang); writing—original draft preparation, C.L. and C.M.; writing—review and editing, C.L. and C.M.; visualization, C.L. and C.M.; supervision, J.L.; project administration, J.L.; funding acquisition, J.L. All authors have read and agreed to the published version of the manuscript.

Funding: This research was supported by the earmarked fund for the Guangdong Laboratory for Lingnan Modern Agriculture (Grant No. NZ2021040 NT2021009), the Guangdong Province Rural Revitalization Strategy Special Project (Grant No. 2023-440000-60010000-9818), the 2022 Provincial Science and Technology Project of Agricultural High-Tech Industry Demonstration in Jinggangshan (Grant No. 20222-051256), the Special Project of Rural Vitalization Strategy of Guangdong Academy of Agricultural Sciences (Grant No. TS-1-4), the China Agriculture Research System (Grant No. CARS-32), and the Specific University Discipline Construction Project of South China Agricultural University in 2023 (Grant 2023B10564002).

Institutional Review Board Statement: Not applicable.

Data Availability Statement: The data presented in this study are available on request from the corresponding author.

Acknowledgments: The authors thank the helpful feedback from reviewers, which contributed to improving the quality of this manuscript.

Conflicts of Interest: The authors declare no conflicts of interest.

References

1. Hou, Z.; Yang, H. Effects of soil fertilization periods and methods on the growth and fruiting of fruit trees. *Pract. Rural Technol.* **2019**, *11*, 48.
2. Heidarisoltanabadi, M.; Elhami, B.; Imanmehr, A.; Khadivi, A. Determination of the most appropriate fertilizing method for apple trees using multi-criteria decision-making (MCDM) approaches. *Food Sci. Nutr.* **2023**, *12*, 1158–1169. [CrossRef] [PubMed]
3. Sun, J.W.; Chen, Z.X.; Song, R.H.; Fan, S.; Han, X.; Zhang, C.F.; Wang, J.X.; Zhang, H. An intelligent self-propelled double-row orchard trenching and fertilizing machine: Modeling, evaluation, and application. *Comput. Electron. Agric.* **2025**, *229*, 109818. [CrossRef]
4. Li, L.; Wu, T.Y.; Li, Y.S.; Hu, X.; Wang, Z.X.; Liu, J.F.; Qin, W.; Ashraf, U. Deep fertilization improves rice productivity and reduces ammonia emissions from rice fields in China; a meta-analysis. *Field Crops Res.* **2022**, *289*, 108704. [CrossRef]
5. Liu, S.X.; Xu, C.B.; Zhang, H.J.; Jiang, H.; Quan, Z.K.; Wang, J.X. Research Status and Development Analysis of Base-fertilizer Application Equipment of Orchard. *Chin. Soc. Agric. Mach.* **2020**, *51*, 99–108.
6. Sugirbay, A.; Zhao, K.; Liu, G.; Hu, G.; Chen, J.; Mustafin, Z.; Iskakov, R.; Kakabayev, N.; Muratkhan, M.; Khan, V.; et al. Double Disc Colter for a Zero-Till Seeder Simultaneously Applying Granular Fertilizers and Wheat Seeds. *Agriculture* **2023**, *13*, 1102. [CrossRef]
7. Qin, K.; Lang, X.T.; Shen, Z.G.; Wu, Z.M.; Bi, H.J.; Cao, C.M.; Sun, Y.; Ge, J.; Fang, L.F. Design and Experiment of Reciprocating Furrowing and Loosening Machine for Tea Garden. *Trans. Chin. Soc. Agric. Mach.* **2024**, *55*, 29–39.
8. Li, J.; Li, H.C.; Chen, Y.Y.; Lin, P.Y.; Zhang, Q.Q.; Cheng, Y.; Yang, Z.; Huang, G.W. Research on Ditching Mechanism of Self-Excited Vibration Ditching Machine. *Agronomy* **2023**, *13*, 905. [CrossRef]
9. Zeng, Y.; Li, J.; Li, H.C.; Zhang, Q.Q.; Li, C.; Li, Z.; Jiang, R.P.; Mai, C.D.; Ma, Z.; He, H.W. Research on the ditching resistance reduction of self-excited vibrations ditching device based on MBD-DEM coupling simulation. *Front. Plant Sci.* **2024**, *15*, 1372585. [CrossRef] [PubMed]
10. Ye, R.; Ma, X.; Zhao, J.; Liao, J.; Liu, X.; Xi, L.; Su, G. Optimization and Design of Disc-Type Furrow Opener of No-Till Seeder for Green Manure Crops in South Xinjiang Orchards. *Agriculture* **2023**, *13*, 1474. [CrossRef]
11. Chen, W.X.; Ren, J.B.; Huang, W.L.; Chen, L.B.; Weng, W.X.; Chen, C.C.; Zheng, S.H. Design and Parameter Optimization of a Dual-Disc Trenching Device for Ecological Tea Plantations. *Agriculture* **2024**, *14*, 704. [CrossRef]
12. Zhang, G.; Zhang, Z.; Xiao, M.; Bartos, P.; Bohata, A. Soil-Cutting Simulation and Parameter Optimization of Rotary Blade's Three-Axis Resistances by Response Surface Method. *Comput. Electron. Agric.* **2019**, *164*, 104902. [CrossRef]
13. Matin, M.A.; Hossain, M.I.; Gathala, M.K.; Timsina, J.; Krupnik, T.J. Optimal Design and Setting of Rotary Strip-Tiller Blades to Intensify Dry Season Cropping in Asian Wet Clay Soil Conditions. *Soil. Tillage Res.* **2021**, *207*, 104854. [CrossRef]
14. Dai, F.; Guo, W.J.; Song, X.F.; Zhang, Y.; Shi, R.J.; Wang, F.; Zhao, W.Y. Optimization of mechanized soil covering path based on the agronomic mode of full-film double-ditch with double-width filming. *Int. J. Agr. Biol. Eng.* **2022**, *15*, 139–146. [CrossRef]
15. Guo, D.; Wen, L.; Jia, D.; Cheng, Z.; Ning, M.; Wen, J. Optimal Design and Experiment of Arc-groove Double-spiral Fertilizer Discharge Device. *Trans. Chin. Soc. Agric. Mach.* **2022**, *53*, 118–125.
16. Wang, X.Z.; Fang, W.Q.; Han, D.L.; Chen, X.G. Review of the Research on Soil Disturbance by Tools. *Appl. Sci.* **2023**, *13*, 338. [CrossRef]
17. Niu, Y.; Zhang, J.Y.; Qi, J.T.; Meng, H.W.; Peng, H.J.; Li, J.H. Design and Test of Soil–Fertilizer Collision Mixing and Mulching Device for Manure Deep Application Machine. *Agriculture* **2023**, *13*, 709. [CrossRef]
18. Zhang, H.J.; Xu, C.B.; Liu, S.X.; Jiang, H.; Zhang, C.F.; Wang, J.X. Design and Experiment of Orchard Double Row Ditching-fertilizer Machine with Automatic Depth Adjustment. *Trans. Chin. Soc. Agric. Mach.* **2021**, *52*, 62–72.
19. Yuan, Q.; Xu, L.; Niu, C.; Ma, S.; Yan, C.; Zhao, S.; Liu, F.; Wang, K. Development of organic fertilizer deep applicator soil-fertilizer mixing layered backfill device for orchards. *J. Agric. Eng.* **2021**, *37*, 11–19.
20. Wang, J.W.; Xu, Y.N.; Wang, C.Y.; Xiang, Y.S.; Tang, H. Design and simulation of a trenching device for rice straw burial and trenching based on MBD-DEM. *Comput. Electron. Agric.* **2023**, *207*, 107722. [CrossRef]
21. Zhang, Z. Development and Experiment of Simulating Orchard Trencher Based on the Simulation of Soil Cutting Process by Edged Tool. Master's Thesis, Shandong Agricultural University, Taian, China, 2020.
22. Wang, X.Y.; Tang, Y.R.; Lan, H.P.; Liu, Y.; Zeng, Y.; Tang, Z.H.; He, Y.C.; Zhang, Y. Performance Analysis and Testing of Spiral Quantitative Fertiliser Distributors in Orchards. *Appl. Sci.* **2023**, *13*, 8941. [CrossRef]

23. Chen, P.L.; Su, J.H.; Xu, J.; Liu, M.H. Design and Experiment of Hilly Orchard Vertical Spiral Ditching-fertilizing Machine. *Trans. Chin. Soc. Agric. Mach.* **2024**, *55*, 223–233+274.
24. Zhu, X.H.; Zhao, H.S.; Fu, S.K.; Li, X.D.; Guo, W.C.; Zhang, L.X. Design and Test of Applicator for Kiwifruit Orchards to Mix Organic Fertilizer into Soil without Furrowing. *Trans. Chin. Soc. Agric. Mach.* **2023**, *54*, 133–142.
25. Tan, H.C.; Ma, S.; Shen, S.C.; Ma, J.L.; Zhou, H.N.; Xu, L.M. Design and experimentation of a chain reversal trenching and fertilization device for vineyard organic fertilizer. *Trans. Chin. Soc. Agric. Eng.* **2024**, *40*, 12–23.
26. Kang, J.M.; Li, S.J.; Yang, X.J.; Liu, L.J.; Li, C.R. Experimental verification and simulation analysis on power consumption of dise type ditcher. *Trans. Chin. Soc. Agric. Eng.* **2016**, *32*, 8–15.
27. China Academy of Agricultural Mechanization Science. *Manual of Agricultural Machinery Design*; Machinery Industry Press: Beijing, China, 1998.
28. Ucgul, M.; Fielke, J.M.; Saunders, C. Three-dimensional discrete element modelling (DEM) of tillage: Accounting for soil cohesion and adhesion. *Biosyst. Eng.* **2015**, *129*, 298–306. [CrossRef]
29. Fotuhi, M.J.; Hazem, Z.D.; Bingül, Z. Adaptive Joint Friction Estimation Model For Laboratory 2 DOF Double Dual Twin Rotor Aerodynamical Helicopter System. In Proceedings of the 2018 6th International Conference on Control Engineering & Information Technology (CEIT), Istanbul, Turkey, 25–27 October 2018; pp. 1–6.
30. Ma, C.; Meng, H.W.; Zhang, J.; Zhang, C.; Zhao, Y.; Wang, L.H. Research and experiment on the trenching performance of orchard trenching device. *Sci. Rep.* **2023**, *13*, 18941. [CrossRef] [PubMed]
31. *JB/T 11908-2014*; Agricultural Disc Ditcher. National Standardization Administration: Beijing, China, 2014.
32. *GB/T 20346.2-2022*; Equipment for Distributing Fertilizer—Part 2: Fertilizer Distributor in Lines. National Standardization Administration: Beijing, China, 2022.

agriculture

MDPI

Article

Establishment of a Discrete Element Model for Wheat Particles Based on the Ellipsoidal Method and CFD–DEM Coupling

Boxuan Gu [1,2], Can Hu [1,2], Jianfei Xing [1,2,3], Xiaowei He [1,2], Xufeng Wang [1,2], Kai Ren [4] and Long Wang [1,2,*]

1 College of Mechanical and Electrical Engineering, Tarim University, Alar 843300, China; boxuangu71@gmail.com (B.G.); 120140004@taru.edu.cn (C.H.); 1202000012@taru.edu.cn (J.X.); 120140001@taru.edu.cn (X.H.); wxf@taru.edu.cn (X.W.)
2 Xinjiang Production and Construction Corps (XPCC) Key Laboratory of Utilization and Equipment of Special Agricultural and Forestry Products in Southern Xinjiang, Alar 843300, China
3 College of Engineering, China Agricultural University, Beijing 100000, China
4 Xinjiang First Division Agricultural Development Service Center, Alar 843300, China; rkai6827@gmail.com
* Correspondence: 120140002@taru.edu.cn

Abstract: The precision of simulation plays a pivotal role in determining the design parameters of the pressure pipe and distributor in a pneumatic centralized seeding system. This study adopted the discrete element method (DEM) to investigate wheat seed models and their motion characteristics within a pneumatic precision seed-metering device. Using Xinchun No. 6 wheat as the experimental subject, multi-sphere combination models (5, 7, 9, and 11 balls) were employed to describe the seed particle morphology. Moreover, by utilizing the coupling method of the Discrete Element Method (DEM) and Computational Fluid Dynamics (CFD), along with bench tests, the air suspension velocity of seeds and the motion characteristics of the seed-supplying device were analyzed under different particle models. The physical properties of the wheat seeds were measured during the experiments. The simulation results indicated that, as the seed supply rate increased, the airflow velocity distribution within the model became more uniform, enhancing the stability of the suspension velocity. Comparisons between experiments and simulations validated the reliability of the particle models, with the minimum relative error in the suspension velocity determined as 0.21% for the 9-balls model. In addition, compared to the other models, the 9- and 5-balls models more accurately simulated the dynamic behavior of seeds within the seed-supplying device. For the 9-balls model, the relative error of particle velocity in the seed-supplying device is 1.39%, and, in the simulation of displacements in the X and Y directions of the seed-supplying device, the average error is 9.51%. The effectiveness of the multi-sphere combination models was verified, indicating their ability to accurately reflect the dynamic behavior of wheat seeds and improve the design and optimization efficiency of pneumatic precision seed-metering devices.

Keywords: wheat granule model; suspension speed; seed supply device; discrete element method; CFD–DEM

Academic Editor: Andreas Gronauer

Received: 19 January 2025
Revised: 7 February 2025
Accepted: 7 February 2025
Published: 10 February 2025

Citation: Gu, B.; Hu, C.; Xing, J.; He, X.; Wang, X.; Ren, K.; Wang, L. Establishment of a Discrete Element Model for Wheat Particles Based on the Ellipsoidal Method and CFD–DEM Coupling. *Agriculture* **2025**, *15*, 369. https://doi.org/10.3390/agriculture15040369

1. Introduction

Wheat is a primary staple food for humans and plays a critical role in maintaining food security [1]. Thus, enhancing yield per unit area and achieving high and stable production are key priorities in wheat. In 2023, the wheat planting area in Xinjiang, China reached a new peak of 882,666.67 hectares. The growth status and yield of wheat are closely related to the quality of sowing at the time of planting. The seeder sowing quality plays a crucial role in ensuring high yields [2]. Due to its strong adaptability, wide applicability, the fact that it causes no damage to seeds, and its relaxed requirements for seed size, the pneumatic collection and discharge sowing method can greatly improve the frequency and efficiency of sowing. Pneumatic centralized-row seeding machines are mainly composed of a seed separation system, an air source, and a seed distribution system. The seed separation system quantitatively separates seeds from the seed box according to the seeding rate requirements, then forms a seed–air mixture with the airflow from the air source device, and, finally, transports the mixture to the furrow openers via the seed distribution system's pipeline network. In recent years, this method has gained widespread adoption in field sowing and fertilization machinery [3–6]. Traditional testing methods, such as using high-speed cameras to examine the interactions between mechanical components and seed particles, have been employed to gain insights into the motion posture and dynamics of wheat seeds within the pneumatic sorting seeder. However, such methods are costly, time-consuming, and cannot easily capture accurate boundary conditions, particle displacement, and contact forces. The discrete element method (DEM) has become a feasible tool for research on the behavior of large particles as it can accurately analyze the flow characteristics of particles and the interaction between mechanical components to determine the motion of seed particle groups [7–9].

Simulation accuracy reflects the ability of simulation results to truly reflect actual conditions. Numerous studies have reported that the representation of particle shape is one of the key challenges in the development of DEM simulations as it significantly influences the mechanical properties of granular materials [10–12]. The simulation accuracy of seeds can affect the structural parameter design of pneumatic precision seed-metering devices. The multi-sphere model is currently the principle method used for wheat seed simulations. Fan et al. validated their proposed wheat granule model using several evaluation criteria, including the wheat seed volume fraction, the static angle of repose, the hopper discharge, the rotating drum, and "free-flow screening", by comparing experimental and simulation results [13]. Their work demonstrates that the multi-ball model can effectively simulate the behavior of wheat seeds, yet their focus is confined to static properties and does not extend to the dynamic interactions within pneumatic systems. Sun et al. developed dual ellipsoid models based on seed shape analysis, using 13, 17, 25, and 33 balls. Their findings indicate that the 17- and 25-balls models based on dual ellipsoids are suitable for modeling wheat seeds with an aspect ratio greater than 2.05 [14]. However, their research did not explore the coupling of DEM with Computational Fluid Dynamics (CFD) for gas–solid two-phase flow simulations, which is crucial for pneumatic seeding systems. In contrast, Gao et al. employed discrete element simulation to study the two-dimensional discharge flow of elliptical particles with different aspect ratios in silos, systematically measuring the density and velocity distribution of particles at the outlet. Under the same outlet dimensions, the velocity of particles passing through the orifice decreases as the aspect ratio increases [15]. Although their work provides valuable insights into particle flow dynamics, it is limited to silo discharge and does not address the complexities of pneumatic conveying systems. Building upon previous work, this study goes a step further by combining the ellipsoid method with the multi-ball approximation method, rigorously validating the model through suspension velocity tests and seed-feeding device tests based on CFD–DEM coupled simulations. Unlike traditional high-speed camera-based methods, our approach enables precise tracking of particle trajectories and contact forces, addressing the shortcomings in simulating complex pneumatic seed metering systems.

With the advancement of computer-based digital simulation technology, DEM and computational fluid dynamics (CFD), which is based on finite element fluid dynamics simulations, have been widely applied in the field of agricultural machinery design [16–19]. The coupling simulation of CFD–DEM for gas–solid two-phase flow provides a new approach for the DEM modeling of wheat particles.

The research tasks and objectives of this study are as follows:

1. Select the Xinchun No. 6 wheat variety as the experimental object. Combine the ellipsoid method with the multi-ball method and the ellipsoid equation to establish multi-ball combination models with 5 balls, 7 balls, 9 balls, and 11 balls for Xinchun No. 6 wheat.
2. Conduct air suspension bench tests to measure the suspension velocities of wheat seeds under different seed supply amounts, and use the CFD–DEM coupling to carry out air suspension coupling simulation tests to simulate the suspension process of wheat seeds in the airflow. Additionally, build a bench for the seed-feeding device, observe the motion state of wheat seeds in it, analyze their velocities, and use the CFD–DEM coupling to conduct simulation tests on the seed-feeding device to simulate the motion of wheat particles under different seed supply amounts.
3. Compare and analyze the simulation accuracies of different multi-ball combination models (5-balls, 7-balls, 9-balls, and 11-balls) in simulating the air suspension velocities of wheat seeds, the displacements, and velocities of particles in the seed-feeding device. This aims to verify the feasibility of simulating the air motion characteristics of wheat based on the CFD–DEM coupling simulation method, so as to provide a theoretical reference for the design of key components of pneumatic collection and discharge seeders.

2. Materials and Methods

2.1. Shape Analysis

In this study, we chose the 'Xinchun No. 6' wheat variety as the experimental material. The shape and size of wheat seeds are crucial for establishing geometric models. Figure 1 illustrates the three-axis dimensions and reference coordinate system of wheat seeds, with the coordinate axes labeled in millimeters (mm). The figure depicts the size and configuration of wheat seed particles, characterized by three axial dimensions: the long axis (L), the middle axis (W), and the short axis (T).

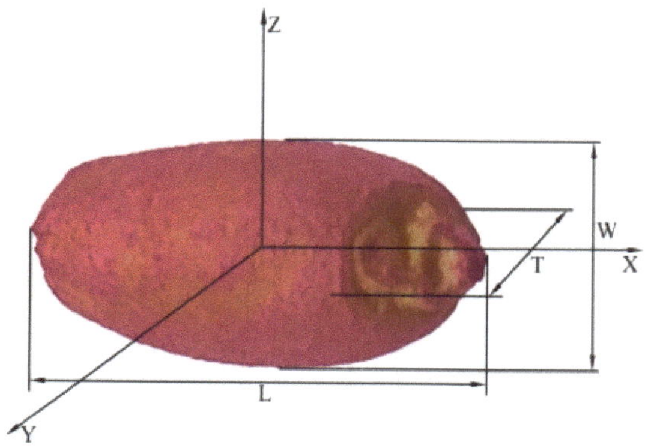

Figure 1. Geometric characteristics of wheat seeds for Xinchun No. 6.

2.2. Discrete Element Modeling of Wheat Grains

The shape and size of seeds directly influence the structural design parameters of the key components in pneumatic seeding systems and can be represented by the length (L), width (W), and thickness (T). A vernier caliper can be directly used to measure these axial dimensions for larger wheat seeds.

A toal of 50 wheat seeds were randomly selected as the measurement sample, and a vernier caliper (Biaokang BK-318, Shanghai Dingleng Industrial Development Co., Ltd. (Shanghai, China), with an accuracy of 0.01 mm) was used to measure their length (L), width (W), and thickness (T). Wheat seeds generally have an ellipsoidal shape. Consequently, relying solely on a single axial dimension to determine the size of the seeds lacks a comparable basis. Based on the measured three-axis dimensions, the corresponding equivalent particle diameter and sphericity can be calculated as follows:

$$D = \sqrt[3]{LWT}, \tag{1}$$

$$S_P = D/L, \tag{2}$$

where D represents the equivalent spherical diameter of the seed (mm); L is the length of the seed (mm); W is the width of the seed (mm); T is the thickness of the seed (mm); S_P is the sphericity, which indicates the degree of difference between the actual seed shape and a perfect sphere.

To provide a more intuitive representation of the frequency distribution of seed length, width, and thickness, histograms with overlaid normal distribution curves were plotted. As shown in Figure 2, the length of wheat seeds primarily concentrated within the range of from 6.4 mm to 6.6 mm, the width mainly fell between 3.2 mm and 3.4 mm, and the thickness was predominantly distributed between 2.9 mm and 3.0 mm. Statistical analysis of the seed measurement data revealed the mean values for seed length, width, thickness, equivalent diameter, and sphericity. The average length of the wheat seeds was 6.5 mm, the average width was 3.43 mm, the average thickness was 2.97 mm, the equivalent diameter was 4.02, and the sphericity was 0.6.

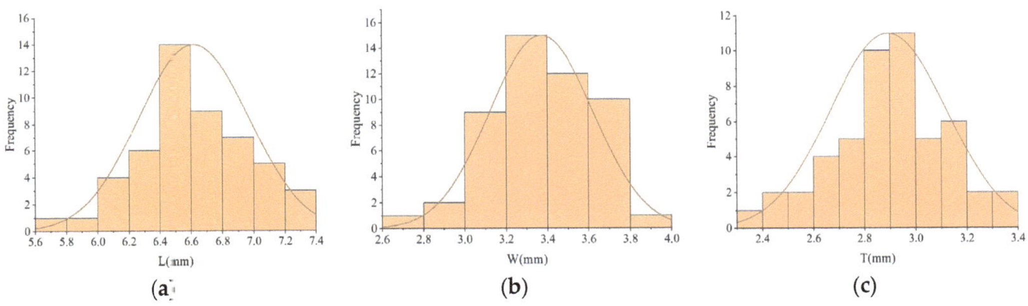

Figure 2. Three-axis size distribution of wheat seeds: (**a**) Distribution of length and size of wheat seeds; (**b**) Distribution of width and size of wheat seeds; (**c**) Distribution of thickness and size of wheat seeds.

By combining the multi-sphere method and ellipsoid equation, we established an ellipsoid model to describe the real shape of wheat particles. The experimental measurements of the three-axis dimensions of wheat revealed the width W (b) and thickness T (c) of the grains of New Year 6 wheat to be very close. Therefore, the ellipsoid model of wheat grains can be further simplified into an axisymmetric model with an equal width and thickness, with the average half-width (thickness) represented as $\overline{b} = \frac{b+c}{2}$, where b and c are the half width (thickness) of the ellipsoid in the Y and Z directions, respectively. The corresponding ellipsoid equation is described as follows:

$$f(x, y, z) = \frac{x^2}{a^2} + \frac{y^2 + z^2}{b^2} - 1 = 0, \tag{3}$$

where a represents the semi-major axis of the ellipsoid. To construct the ellipsoid model, a series of constituent spheres are used to fill the ellipsoid. The centers and radii of these constituent spheres are precisely calculated on the X–Y cross-sectional planes of the ellipsoid. Specifically, the center coordinates of the constituent spheres are set along the horizontal axis, namely, at the origin (0, 0, 0), to ensure that the contour of the combined spherical wheat grain model approximates the true ellipsoid.

To further refine the precision of the combined spherical model, multiple constituent spheres are positioned tangent to the contour of the ellipsoid. Leveraging the symmetry of the ellipsoid, it suffices to compute the coordinates for half of the contour. The semi-major axis a of the ellipsoid is divided into k equal segments, resulting in $k + 1$ nodes. However, to construct the combined spherical model, the boundary points located at both ends of the contour are omitted. Only the remaining k internal nodes are utilized to determine the constituent spheres tangent to the ellipsoid. As a result, the entire ellipsoid model encompasses $N = 2k - 1$ constituent spheres.

The horizontal axis of the i-th constituent ball ($i = 0, 1, \ldots, k - 1$) can be represented as $\frac{i}{k}a$ (Figure 3). To ensure that the composite sphere model approximates an ellipsoid, we require that the circular cross-section of each constituent sphere be tangent to the elliptical cross-section of the ellipsoid on the X–Y cross-section. The tangent equation when the i-th constituent sphere is tangent to the ellipse and is described as follows:

$$(x - x_i)^2 + y^2 = R^2, \tag{4}$$

By equating the first derivatives at the point of tangency (x_{ti}, y_{ti}), the coordinates of the tangency point x can be derived as follows:

$$x_{ti} = \frac{a^2 - \overline{b}^2}{a^2} x_i, \tag{5}$$

By solving the system of equations formed by the ellipse equation and the condition of equal tangent slope, the coordinate of y can be obtained as follows:

$$y_{ti} = \sqrt{\overline{b^2}[1 - (\frac{a^2}{a^2 - b^2})(\frac{x_i}{a})^2]}, \tag{6}$$

In practical modeling, only the absolute values are of concern.

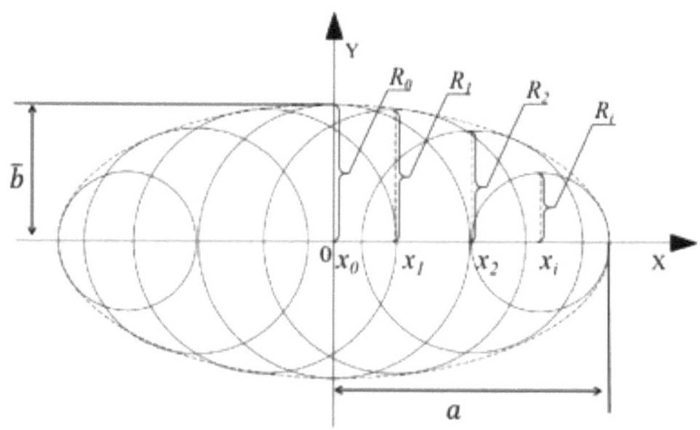

Figure 3. Calculation of the wheat particle multi-sphere model established in this study.

The radius R of the sub-sphere is calculated by substituting the tangent point coordinates into the elliptic equation as follows:

$$R_i = \sqrt{\overline{b}^2 - (a^2 - \overline{b}^2)\left(\frac{x_i}{a}\right)^2}, \tag{7}$$

Figure 4 presents the multi-sphere combination models with 5, 7, 9, and 11 balls established in this study for Xinchun No. 6 wheat.

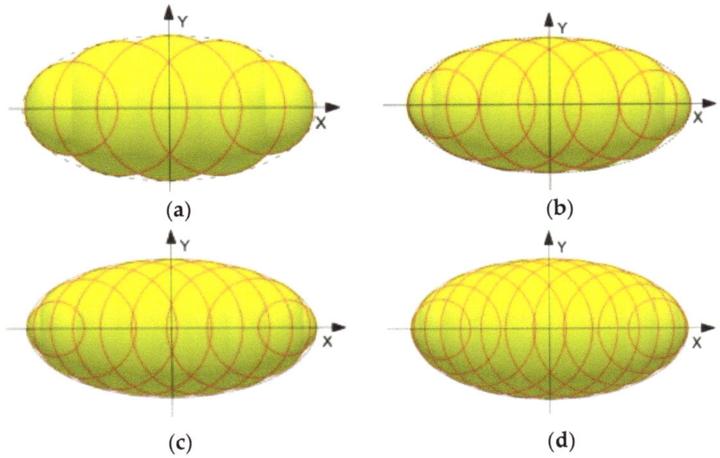

Figure 4. Multi-sphere model of wheat grains for Xinchun No. 6: (**a**) 5 balls; (**b**) 7 balls; (**c**) 9 balls; (**d**) 11 balls.

The wheat seeds are approximately ellipsoidal in shape. In this study, they are simplified into an axisymmetric model (assuming the width and thickness are equal) to construct an ellipsoid model. This simplification, to some extent, ignores the slight differences between the width and thickness but retains the general outline and main dimensional characteristics of the seeds. For the design and optimization of the pneumatic collective-discharging seeding system, the focus is on the macroscopic motion characteristics of the seeds in the airflow and the motion trend of the seeds under the action of the airflow.

Determination of Wheat Seed Physical Properties

The wheat seed density was determined by taking five samples of Xinchun No. 6 wheat seeds in appropriate amounts and measuring their masses. An appropriate amount of water was added to a graduated cylinder with a precision of 0.2 mL. After all the water droplets on the inner wall of the cylinder had slid down, the initial water volume was recorded. The wheat seeds were then added and the total volume after the seeds were immersed in water was recorded. This process was repeated for each of the five groups (Figure 5a).

Using a halogen moisture analyzer (model LHS-125A, manufactured by Shanghai Yixin Scientific Instrument Co., Ltd. (Shanghai, China), with a moisture display accuracy of 0.02%) to measure the moisture content of wheat. Approximately 20 g of wheat seeds were placed on the instrument to measure the sample weight. The internal halogen heating element rapidly heated the sample, causing the moisture to evaporate. During the drying process, the instrument continuously measured the sample weight and displayed the results. The drying program was set to fast, with a drying temperature of 120 °C, and the shutdown mode was set to automatic. The measurements were repeated five times and averaged (Figure 5b).

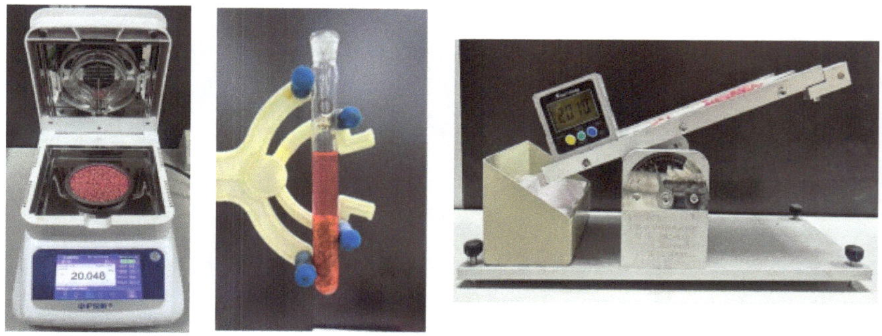

(**a**) (**b**) (**c**)

Figure 5. Determination of wheat seed physical properties: (**a**) moisture content measurements; (**b**) density measurements; (**c**) sliding friction angle measurements.

The sliding friction angle of wheat seeds characterizes the flow behavior of seed populations during movement within a seed bin. The sliding friction angle reflects the friction characteristics between the seed and the contact surface and has a direct impact on mechanized operations such as seed cleaning, transportation, and sowing. The EDEM simulation requires the friction coefficient as an input. Therefore, the sliding friction angle was measured to calculate the friction coefficient of the wheat seeds by placing an inclinometer on a stable horizontal surface and adjusting its inclination to the initial position. A total of 50 g of seeds were selected and spread flat on the inclined surface of the experimental device, forming a thin layer of seeds. The inclination of the inclinometer was then gradually increased while carefully observing the state of the seed layer. We slowly and uniformly increased the slope angle, and immediately stopped increasing the slope angle when most of the seeds began to fall. The slope angle at this time point was recorded using the scale or angle measuring device on the slope meter, and was taken as the sliding friction angle of the wheat seeds. The measurements were repeated for five groups and averaged (Figure 5c).

Due to the irregular shape of wheat seeds, the rolling friction coefficient between wheat–wheat and wheat–contact materials has low sensitivity and is difficult to measure directly. The Poisson's ratio and shear modulus of wheat are taken from the experimental results of Liu et al., who calibrated these values by comparing the angle-of-repose experiments with DEM angle-of-repose simulations. The parameters between materials used in all simulations are shown in Table 1, and the remaining parameters are determined according to the reference [20].

Table 1. Parameters required for EDEM simulations.

Parameter	Value for Xin Chun No. 6 Wheat Seed	Value for PVC Pipes
Density (kg/m^3)	1330	1400
Moisture content (%)	6.278	/
Poisson's ratio	0.29 [20]	0.4
Shear modulus (MPa)	501 [20]	160
Restitution coefficient	0.50 [20]	0.50 [20]
Static friction coefficient	0.58 [20]	0.783
Coefficient of rolling friction	0.08 [20]	0.05 [20]

To validate the accuracy of the parameters for the wheat particles and contact materials, we compared the results of cylinder lifting tests with corresponding simulations.

Specifically, the repose angle of the particles was measured using both physical tests and EDEM simulations. The relative error between the simulated and actual repose angle was calculated as follows:

$$N = \frac{|\sigma - \theta|}{\theta} \times 100\%, \tag{8}$$

where θ and σ are the stacking angles of the bench and simulation tests, respectively.

Wheat particles of random sizes were loaded into the cylinder with a diameter of 55 mm and a depth of 100 mm. The cylinder was lifted by the testing machine at a speed of 10 mm/s (Figure 6a). When the particles in the cylinder were completely discharged and a pile without particle movement was formed, the repose angle was photographed and recorded (Figure 6b).

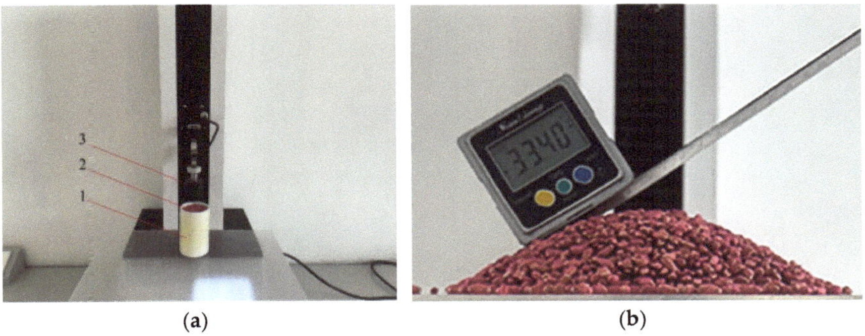

Figure 6. Stacking angle test: (**a**) Round lifting experimental device: 1—cylinder; 2—wheat seeds; 3—lifting device. (**b**) Round lifting pile angle.

Using the parameters above, model validation tests were performed on wheat grains, as shown in Figure 7. In the cylinder-lifting test, the measured angle of repose was 33.8°. The relative errors of the angles of repose for the 5 balls, 7 balls, 9 balls, and 11 balls were 3.69%, 0.83%, 1.36%, and 1.3% respectively. The rolling friction coefficient between wheat and acrylic was 0.58, which matched the bench-test results by Liu et al., ensuring the interaction between grains and the wall was realistic [20]. Finally, the rolling friction coefficients between wheat–wheat and wheat–contact materials were set as 0.08 and 0.05, respectively. These parameters of the wheat-grain model were proven effective.

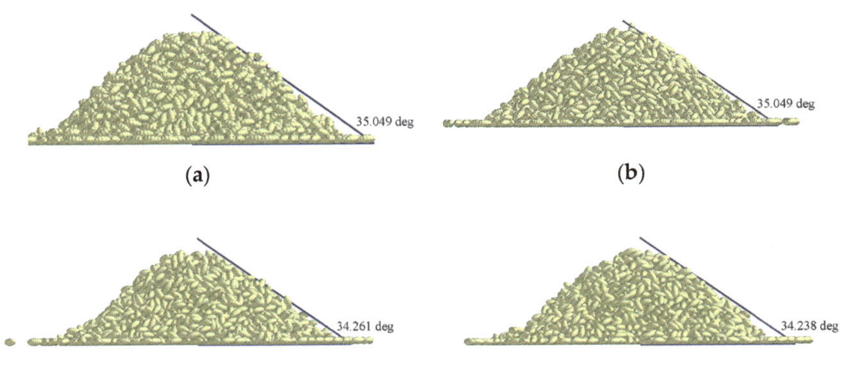

Figure 7. Simulated stacking angle test: (**a**) 5 balls; (**b**) 7 balls; (**c**) 9 balls; (**d**) 11 balls.

3. Validation of Wheat Particle Models

3.1. Air Suspension Velocity of Particle Models

The suspension velocity of materials refers to the airflow velocity at which particles suspend at a certain height in an airflow moving vertically upward. The suspension velocity characteristics are crucial in the design of agricultural machinery used for harvesting conveying, cleaning, and removing impurities. Numerous scholars have conducted suspension velocity measurement experiments and numerical simulation analyses on different crop particles. Ren et al. utilized a material suspension velocity test rig to measure the terminal velocities of hazelnuts and fallen leaves. They employed gas–solid two-phase flow theory and the coupling of computational methods to simulate and analyze the motion state, particle distribution, and airflow field distribution of hazelnuts from the inlet to the outlet of a pneumatic conveying device [21]. Dai et al., by measuring the suspension velocities of flax seeds, capsules, short stems, and capsule shells at different stages, derived the overall spacing distribution and separation characteristics of flax threshing material components. Furthermore, they employed a gas–solid coupling method to simulate the suspension velocities of each component of the flax threshing materials, verifying the feasibility of measuring suspension velocities based on CFD–DEM simulations [22].

Air Suspension Bench Test

In pneumatic conveying, a large number of particles move within a defined range, increasing the frequency of collisions between them, which may alter their trajectories and velocities. As the fluid viscosity increases, the drag force on the particles also increases, leading to a reduction in their suspension velocity. The free suspension velocity of ideally spherical particles in air is expressed as follows:

$$u_0 = \sqrt{\frac{4}{3}\frac{d_p\left(\rho_p - \rho_f\right)g}{C_D\rho_f}}, \tag{9}$$

where u_0 is the free floating velocity of the particle, m/s; d_p is the particle diameter, m; ρ_p is the particle density, kg/m^3; ρ_f is the air density, kg/m^3; g is the acceleration due to gravity; C_D is the drag coefficient.

where C_D is the drag coefficient calculated according to the method given by M Sommerfeld [23].

When $R_e \leq 1, C_D = \frac{24}{R_e}$

When $1 < R_e < 1000, C_D = \frac{24}{R_e}\left(1 + 0.15{R_e}^{0.687}\right)$

When $1000 \leq R_e \leq 2 \times 10^5, C_D = 0.44$

The air suspension velocity u'_0 of irregularly shaped particles can be corrected by the suspension velocity u_0 of an equivalent sphere as follows:

$$u'_0 = \frac{u_0}{\sqrt{K_s}}, \tag{10}$$

where K_s is the irregular shape correction coefficient for the suspension velocity of a single particle.

The suspension velocity u_n of a group of particles is lower than the suspension velocity u'_0 of a single particle. When the volume fraction φ_0 is within the range of 5–25%, it can be calculated as follows:

$$u_n = u'_0(1 - \varphi_0)^{\beta}, \tag{11}$$

According to the relationship between indicator β and the *Re* number described by Mills, the value in this study is determined as 2.3 [24].

The air suspension experimental setup consists of three key components: a wind regulation device, an airflow conveying pipeline, and a measurement apparatus. The structure of the setup is illustrated in Figure 8. The wind regulation device includes a frequency converter, electric motor, and centrifugal fan. The airflow pipeline has a lower stabilizer, feeding inlet, conical observation tube, and upper stabilizer. The measurement apparatus consists of a suspension height ruler, an HT-9829 wind speed meter (0.01 m/s precision), and a detection port. Auxiliary materials like protective, holding, blocking nets, and 20-mesh nylon nets are also used. The main technical parameters of the air suspension experimental setup are presented in Table 2.

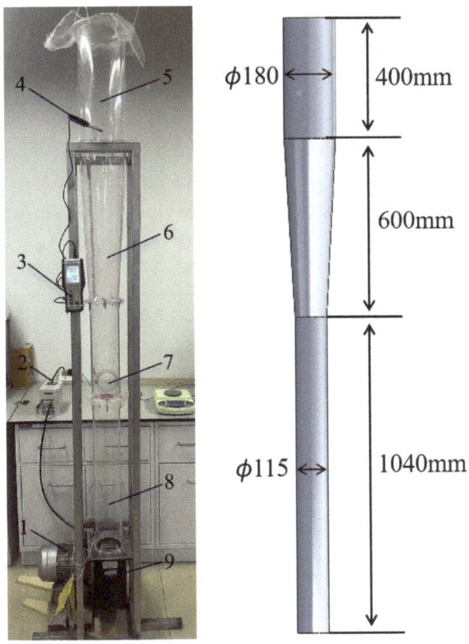

Figure 8. Structural diagram of the air suspension device: 1—Electric motor. 2—Frequency converter. 3—High-precision anemometer. 4—Wind speed detection port. 5—Upper flow stabilizer. 6—Conical observation tube. 7—Material feeding inlet. 8—Lower flow stabilizer. 9—Centrifugal fan.

Table 2. Parameters of the air suspension device test bench.

Parameter	Value
Centrifugal blower power (kw)	1.1
Centrifugal blower speed (r/min)	200–2800
Measurement range of anemometer (m/s)	0–30
Inverter power (kw)	1.5
Frequency converter frequency (Hz)	0–50

For the measurements, the material was placed on the material feeding net inside the lower flow stabilizer of the experimental setup. The power supply was then turned on to activate the centrifugal fan and the frequency converter was adjusted to control its rotational speed. The airflow velocity increased with the rotational speed of the centrifugal fan. When the material reached a stable suspension within a certain height range inside the conical observation tube, the high-precision anemometer read the airflow velocity value at the upper-end port of the conical tube. This value was utilized to calculate the suspension velocity of the material.

Wheat seeds weighing 10 g, 20 g, 30 g, 40 g, 50 g, 60 g, and 70 g were gently placed onto a permeable metal mesh within the test section of the pipeline through a paper chute at the material-feeding inlet. The experiment was maintained at a constant room temperature of 20 °C. The variable-frequency fan was then activated, and its rotational speed was adjusted to alter the air velocity within the air duct of the velocity measurement section.

The suspension state of the wheat seeds within the air duct was observed and the position of the seeds was recorded. An anemometer was used to measure the air velocity at this position. Due to the irregular shape of the wheat seeds, a stable suspension state could not be achieved within the velocity measurement section. Therefore, suspension was determined based on whether the material was in contact with the permeable metal mesh. The position of the wheat seeds and the corresponding air velocity at that position were recorded. The experiment was repeated five times for each wheat seed weight and the values were averaged.

The experimental results indicated that the air velocity fluctuated significantly when the seed supply ranged from 10 g to 20 g. However, when the seed supply was between 30 g and 70 g, the air velocity exhibited a stable upward trend. Table 3 presents the average suspension velocities under different seed supply amounts.

Table 3. Average suspension velocity under different seed supply rates.

Seed Supply (g)	Suspension Speed (m/s)
10	6.385
20	7.085
30	7.123
40	7.497
50	7.733
60	7.930
70	8.167

3.2. *Air Suspension Coupling Simulation Test*

3.2.1. Control Equations for Fluid and Particle Phases

The flow field provides the necessary energy for particle movement. In pneumatic conveying flow fields, momentum and energy exchanges occur between the flow field and particles. The Euler–Lagrange method is adopted to establish a coupling model for the solid and gas phases in pneumatic conveying flow fields. In the Euler–Lagrange method, the gas phase is treated as a continuous medium, while the solid phase is considered as a discrete phase, with the movement of solid particles following Newton's second law. During the CFD–DEM coupling process, the fluid terms are governed by the incompressible Navier–Stokes equations. The mass conservation equation is expressed as follows:

$$\frac{\partial(\alpha\rho)}{\partial t} + \nabla\cdot(\alpha\rho v) = 0, \tag{12}$$

where ρ is the fluid density, kg/m^3; α is the volume fraction of the fluid; v is the fluid velocity, m/s; t is time, s.

The momentum conservation equation is described as follows:

$$\frac{\partial(\alpha\rho)}{\partial t} + \nabla\cdot(\alpha\rho v) = -\alpha\nabla p + \alpha\nabla\cdot\tau - S_f + \alpha\rho g, \tag{13}$$

where p is the fluid pressure, Pa; τ is the viscous stress tensor; S_f is the force exerted by the particle phase on a unit volume of the fluid phase. S_f is obtained by calculating the fluid interaction forces of all particles in the unit and dividing them by the fluid volume of the unit. This generally involves summing the fluid interaction forces of each particle within the computing unit, taking into account the number M of particles and the volume V_{cell} of the fluid within the computing unit:

$$S_f = \frac{1}{V_{cell}} \sum_{i=1}^{M} F_{fluid,i}, \tag{14}$$

where $F_{fluid,i}$ is the fluid interaction force exerted on the i-th particle.

The motion of particles is driven by gravity, buoyancy, contact forces (e.g., interactions between particles and between particles and walls), and interaction forces (e.g., drag, shear lift, and rotational lift). Only when suspended particles pass through shock waves or nozzles do the pressure gradient force, virtual mass force, and Bassett force become significant and are usually not considered, while van der Waals forces only appear near the wall [25]. Therefore, we ignore these forces in our simulations, and the translational and rotational motion of particles is described by Newton's second law as follows:

$$m_p \frac{dv_p}{dt} = \sum_{i-1}^{k_c} F_{pc,i} + F_D + F_G + F_B, \tag{15}$$

$$I_p \frac{d\omega_p}{dt} = \sum_{i-1}^{k_c} T_{pt,i} + T_{pD}, \tag{16}$$

where m_p is the particle mass, g; v_p is the velocity vector of the particle, m/s; $\sum_{i-1}^{k_c} F_{pc,i}$ is the sum of all contact forces acting on particles; k_c is the number of particles; F_D is the fluid resistance, N; F_G is gravity, N; F_B is buoyancy, N; I_p is the moment of inertia of the particle, kg·m^2; ω_p is the angular velocity vector of the particle, rad/s; $\sum_{i-1}^{k_c} T_{pt,i}$ is the sum of torque vectors generated by all tangential and normal contact forces acting on the particles N·m; T_{pD} is the torque exerted by the fluid on the particles N·m.

3.2.2. CFD–DEM Coupling Simulation Experiment

A particle factory was initially established at the feeding network using the four different wheat particle models. The wheat seed particles were suspended in a conical observation tube through the CFD–DEM coupling. After the particles stabilized, the airflow velocity was detected at the wind speed detection port.

We used the 3D solid modeling software SolidWorks 2023 to model the suspension velocity measurement device, simplifying the 3D model to improve the calculation speed and only retaining the lower steady flow tube, conical observation tube, and upper steady flow tube. Meshing was conducted in ANSYS Fluent meshing 2020 R2 to partition the fluid domain model into grids. The grid size was approximately three times the diameter of the wheat particles. The maximum diameter of the long axis of the wheat model was 7.34 mm. As a consequence, a grid size of 22 mm was too large, resulting in calculation divergence. Therefore, the grid size was set to 16 mm to ensure a uniform size for the entire fluid domain grid. After completing the grid partition, a total of 26,880 cells were generated with a maximum skewness of 0.3526 and a cell mass of 0.25454. The results were imported into EDEM 2020 and ANSYS Fluent 2020 R2 for the coupling simulations, the simulation process is shown in Figure 9.

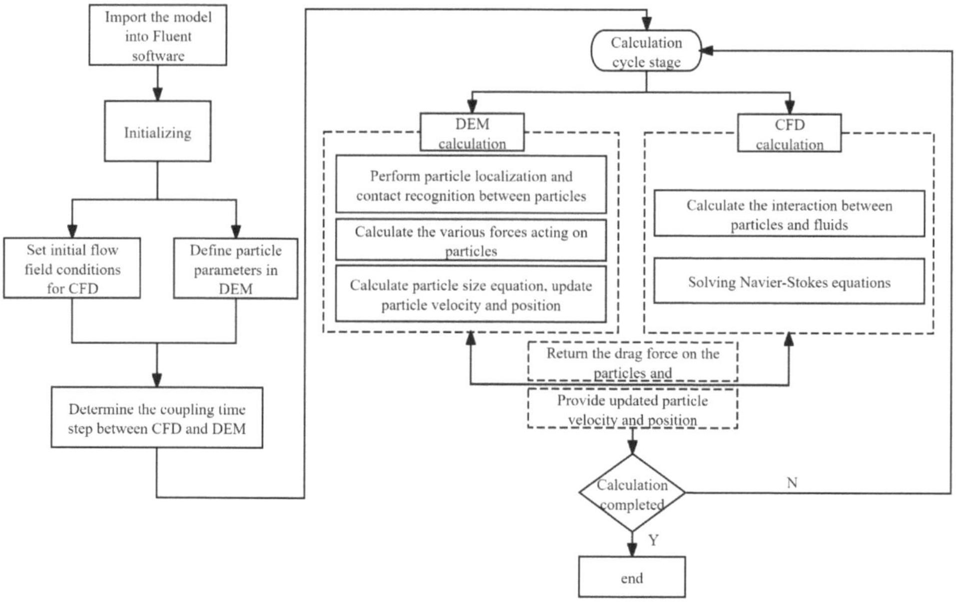

Figure 9. Coupled simulation calculation flowchart.

Table 1 reports the wheat seed parameters used in the EDEM simulations. A virtual plane was created in the model with a diameter of 90 mm and positioned identically to the discharge net inside the lower steady flow tube of the experimental device. This diameter was slightly smaller than that of the lower steady flow tube to ensure that particles were fully contained within the fluid region during generation. The particle generation was set to start 0.001 s after the simulation began, ensuring that particles were generated when the wind field was stable. The Rayleigh time step was set to 20–30% and the data saving interval to 0.01 s. In addition, the particle size variation of the particle model was configured to follow a standard normal distribution with a mean and standard deviation of 0.1 mm. The coupling switch was activated following the finalization of the settings.

The transient standard $k - \varepsilon$ turbulence model was used for the ANSYS Fluent finite element simulation. The dispersed multiphase model was also adopted, as well as the Euler–Lagrange coupling model to consider volume fractions. The inlet was configured as a velocity inlet and the outlet as a pressure outlet, with a pressure value of 0. The time step was set to 3×10^{-4} s, and each simulation run comprised 20,000 steps. The default number of iterations was maintained at 20, resulting in a total simulation duration of 6 s.

3.3. *Accuracy of Different Particle Model Seeding Devices*

The seed feeding device, which mixes air with wheat seeds, is a crucial component of the pneumatic centralized and row-seeding wheat planter. An excessively low airflow velocity in the seed metering system of pneumatic seeding machines can clog seeds within the seed conveying tube, while an excessively high airflow velocity can increase the seed breakage rate, energy consumption, and tube wear. Key performance indicators of the seed-feeding device include the seed movement speed and uniformity of the seeding rows, which also affect the seed distribution in the soil. In recent years, numerous scholars have conducted extensive research on the movement mechanisms within seed-feeding devices to determine their optimal design [26–28]. To further investigate the dynamic behavior of wheat particles, particularly within the seed feeding system, the seed feeding device of the pneumatic centralized and row-seeding wheat planter was selected for bench tests and simulations.

3.3.1. Test Bench of the Seed Supply Device

Figure 10 shows the test bench for the seed-feeding device of the pneumatic centralized-row seeding wheat planter and the seed dynamic observation system. On the test bench, an external-groove-wheel-type wheat metering device, a seed-feeding device, an RB-51D-3 centrifugal fan, and a high-speed camera (Fastec TS3-100S, Fastec Imaging) are installed. The parameters of the test bench for the seed-feeding device are listed in Table 4. During the experiment, wheat is placed in the seed box. The external-groove-wheel metering device is controlled by a drive motor to meter the seeds. The seeds enter the seed-feeding device by their own gravity and mix with air. The wind speed at the outlet of the centrifugal fan is 28 m/s. After the seed-feeding system operates stably, the high-speed camera is used to observe the movement state of the wheat seeds in the seed-feeding device. The high-speed camera has a frame rate of 2000 fps and a resolution of 1280 × 1024 pixels, capturing an image every 0.5 milliseconds. After the image acquisition, image-processing techniques are used to analyze the seed velocity. Three seeds are randomly selected, and their average velocity is obtained for analysis. Based on Newton's second law, combined with position and time information, the velocity of the wheat seeds 20 milliseconds after entering the seed-feeding device is 2.516 m/s. The calculation formula is as follows:

$$v_{11} = \sqrt{\left(\frac{x_t}{t}\right)^2 + \left(\frac{y_t}{t} + \frac{gt}{2}\right)^2}, \tag{17}$$

where speed v_{11} equals 20 ms after the selected seeds enter the seed supply device; x_t and y_t represent the horizontal and vertical positions of the seeds after entering the seed supply device for 20 ms, respectively; g is the gravitational acceleration, 9.8 m/s; t equals 20 ms.

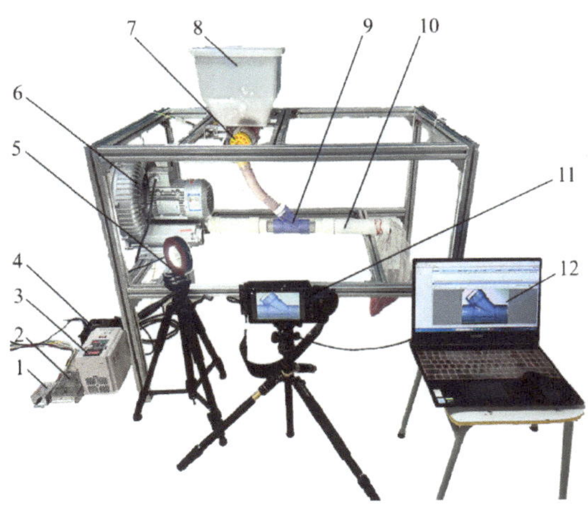

Figure 10. Wheat pneumatic conveying device test bench and seed movement state observation system: 1—motor speed controller; 2—24 V power supply; 3—frequency converter; 4—light source controller; 5—fill light; 6—centrifugal fan; 7—external groove wheel seeder; 8—types of boxes; 9—seed supply device; 10—seed delivery tube; 11—high-speed camera; 12—high-speed camera built-in software.

Table 4. Parameters of the seed feeding device test bench.

Parameter	Value
Centrifugal blower power (kw)	1.6
Centrifugal blower speed (r/min)	200–2800
Measurement range of anemometer (m/s)	0–30
Inverter power (kw)	2.2
Frequency converter frequency (Hz)	0–50

3.3.2. Coupling Simulation Testing of the Seed Supply Device

The simulation process is consistent with the bench test. The seed supply device was modeled in SolidWorks. Meshing was conducted in ANSYS Fluent 2020 R2 to partition the fluid domain model into grids. The grid size was set as 16 mm to ensure a uniform grid size throughout the fluid domain. EDEM software was employed to establish 3-, 5-, 7-, and 9-balls models based on wheat seeds. A particle factory was first established at the seed guide tube to generate 60 g of wheat particle seeds per second. The wheat seeds were mixed with air through CFD–DEM coupling and the particle model motion speed was obtained using EDEM post-processing.

To obtain quantitative data on seed movement, an analysis was conducted on the movement state of seeds with a seeding rate of 60 g per second. Specifically, three seeds were randomly selected and their average speed was determined. During the simulation, the seed was directly obtained through the speed command in the software. The timing commenced at the discharge of the seeds from the seed planter into the seed supply device and the seed position was captured after 20 ms.

4. Results and Discussion

4.1. Testing and Simulation of the Airflow Velocity for Different Particle Models

The wheat suspension simulation experiment reveals that wheat particles initially move upwards driven by the airflow. During this phase, the particles' trajectories exhibit a clear upward trend, accompanied by a deviation towards the tube wall. This phenomenon can be attributed to the force exerted by the airflow on the particles and the inter-particle collisions, which cause the particle group to collectively gather near the tube wall where resistance is lower. As the thrust of the airflow gradually weakens and becomes insufficient to overcome the total gravitational force of the wheat particle group, the particles reach their highest point and begin to descend. During this descent, the particle group is influenced by gravitational acceleration. As a consequence, it accelerates downwards and moves towards the center of the tube until it passes the bottom of the conical observation tube. When the wheat particle group descends below the conical observation tube, the thrust of the airflow exceeds the gravitational force of the particle group, and the particles once again receive an upward thrust, initiating a new cycle of ascending movements. This process is repeated and after multiple ascents and descents, the wheat particle group reaches a relatively stable state near the tube wall within the conical observation tube (Figure 11).

Figure 12 depicts the velocity contour maps determined from the simulation experiments of the air suspension speeds of wheat particle models comprising 5, 7, 9, and 11 spheres and a seeding quantity of 20 g. As the number of particles increases, the overall velocity distribution becomes more uniform upon entering the conical observation tube. The velocity distributions across different particle models are even and lack significant velocity fluctuations and abrupt changes.

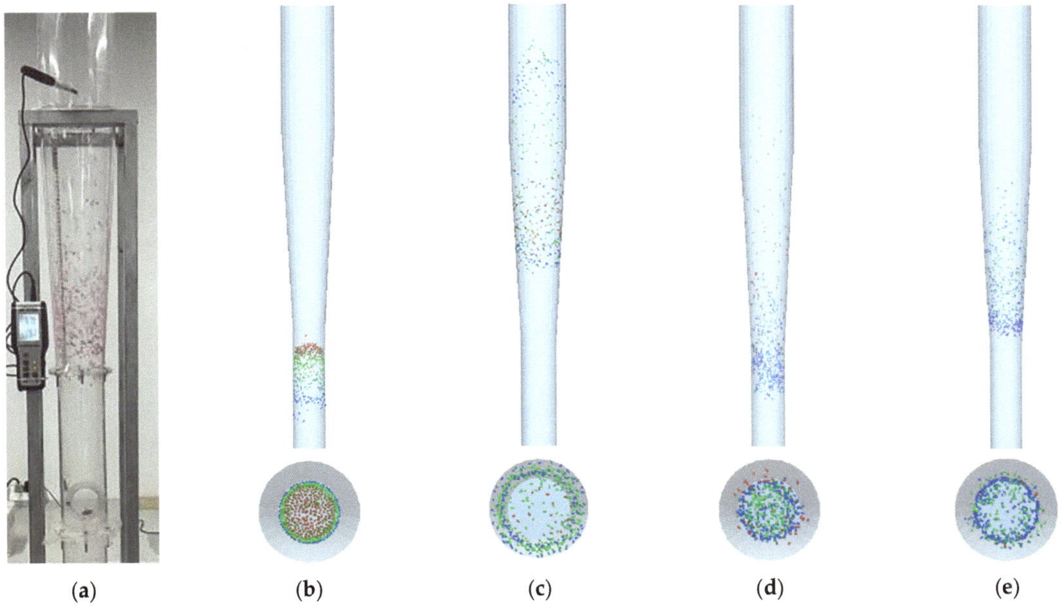

Figure 11. Suspension process of the air suspension experiment: (a) air suspension test of wheat suspension state; (b) simulation test of wheat particle ascent; (c) simulation test of wheat particle peak; (d) simulation test of wheat particle descent; (e) simulation test of wheat particle stable suspension.

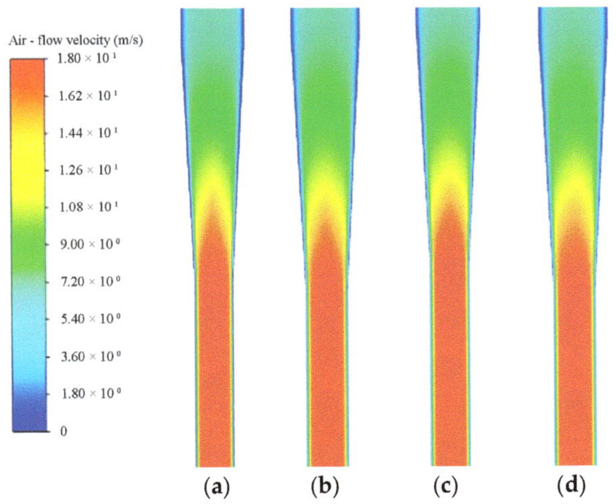

Figure 12. Velocity cloud map of the wheat particle air suspension simulation experiment: (a) 5 balls; (b) 7 balls; (c) 9 balls; (d) 11 balls.

Figure 13 presents the velocity contour maps within the conical observation tube from the air suspension simulation experiments for a wheat particle model with 9 spheres and seeding quantities ranging from 10 g to 70 g. The cross-sectional velocity distribution contour maps intuitively show the changes in the airflow velocity field within the tube. As the seeding quantity increases, the velocity distribution within the tube becomes more uneven, with lower airflow velocity regions emerging in localized areas, indicating that the presence of particles intensifies local turbulence and energy loss.

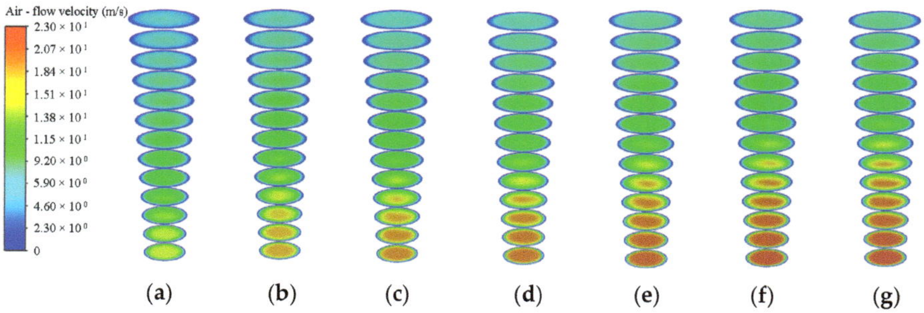

Figure 13. Velocity cloud map inside conical observation tube under different seed supply levels: (**a**) 10 g; (**b**) 20 g; (**c**) 30 g; (**d**) 40 g; (**e**) 50 g; (**f**) 60 g; (**g**) 70 g.

Wind speed measurements at different locations within the conical observation tube (Figure 14) reveal that the inlet air velocity increases with the wheat seeding quantity at from 10 g to 70 g, while the suspension velocities remain between 6 and 9 m/s. Comparing the velocity changes of the airflow along the tube under different seeding quantities indicates a decelerating trend in airflow velocity as the tube length increases at varying seeding quantities. The airflow velocity curves corresponding to different seeding quantities (10 g, 20 g, 30 g, 40 g, 50 g, 60 g, and 70 g) exhibit similar overall trends but differ in values. At lower seeding quantities (10 g and 20 g), the deceleration of the airflow is relatively small and the airflow velocity maintains a high level. This is because there are fewer particles in the tube, resulting in less resistance to the airflow and a lower energy loss in the gas–solid two-phase flow. In addition, most of the wheat particles are suspended stably near the wall, allowing the airflow to pass smoothly through the conical observation tube.

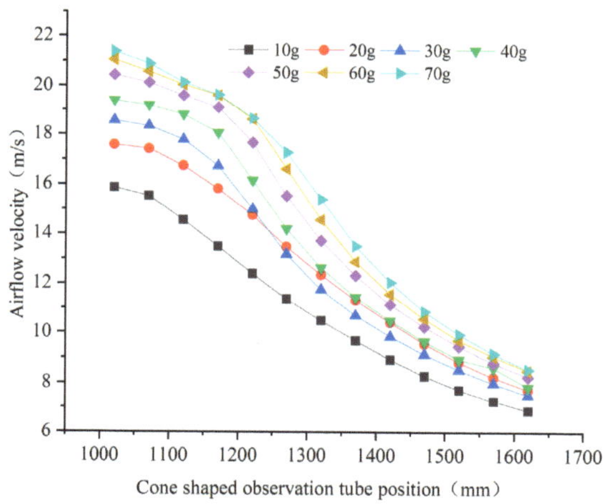

Figure 14. Wind speed at different positions inside the conical observation tube.

As the seeding quantity increases, the particle concentration within the tube rises, causing the particles to form more significant resistance and energy consumption relative to the airflow, which results in a more pronounced decrease in airflow velocity. This also reflects the influence of the particle density on the velocity distribution of the gas phase in a gas–solid two-phase flow. A more significant velocity drop is observed at the seeding quantities of from 30 g to 70 g. In particular, the decrease in wind speed between 1170 mm and 1420 mm at seeding quantities of 30 g and 40 g exceeds that at the seeding quantity of 20 g.

The number and density of particles suspended in the airflow increase with the seeding quantity, enhancing the collision, friction, and drag effects between the particles and the airflow. This leads to a more pronounced decrease in the airflow velocity along the tube. This conclusion provides a reference for selecting an appropriate seeding quantity in the design and regulation of seed conveying systems, ensuring stable and efficient gas–solid two-phase flow fields.

Figure 15 depicts the changing trend in airflow velocity under different seeding rates in simulation and air suspension device tests for scenarios involving 5, 7, 9, and 11 particles. The average relative errors of the air suspension velocity for 5, 7, 9, and 11 particles are 2.87%, 1.53%, 1.18%, and 1.39%, respectively. The errors for the 9- and 11-wheat particle models are relatively smaller. For the 9-particle model, the relative errors of the air suspension velocity at seeding rates ranging from 10 g to 70 g are 0.7%, 2.50%, 0.67%, 0.21%, 1.47%, 2.09%, and 0.60%, respectively, with a minimum relative error of 0.21%. As the seeding rate increases, the relative errors of the air suspension velocity for different particle numbers decrease. This may be due to the influence of fluid dynamics on the movement of suspended particles in the fluid. When the seeding rate increases, the fluid's effect on the particles becomes more uniform.

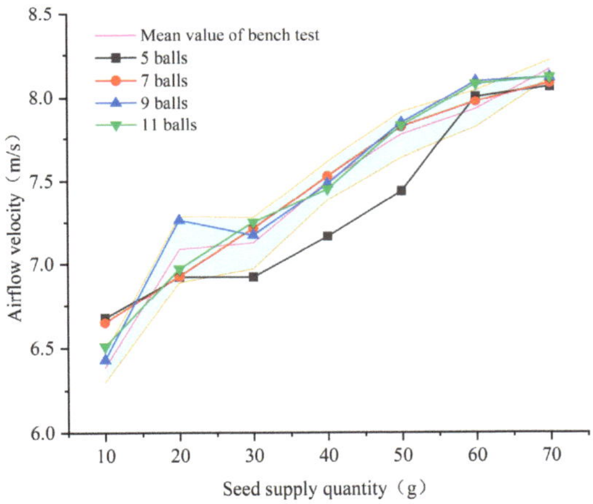

Figure 15. Air suspension experiment and simulation results.

4.2. Accuracy Test and Simulation of Seed Feeding Device

As shown in Figure 16, a 9-particle model was used to numerically simulate the particle motion inside the mixing chamber and seed delivery tube under different seed supply conditions (60 g, 70 g, 80 g, 90 g, 100 g, 110 g, and 120 g). By comparing the velocity variations of wheat grains moving along the length of the seed conveying tube at different seed supply rates, and under the condition of a constant inlet wind speed, we analyze the impact of variations in seed supply rate on the trend of grain velocity increase and the differences in eventual stable velocities. Across all tested seed supply rates, the grain velocities increase with the length of the tube. Initially, the velocities are low, but, as the grains are accelerated by the airflow within the conveying tube, their velocities gradually increase and eventually converge towards a stable value over a longer distance. Under the same inlet wind speed condition, although the overall trends of the grain velocity curves for different seed supply rates (ranging from 60 g to 120 g) are similar, there are still distinct differences. Specifically, when the seed supply rate is lower, the grains are more significantly influenced by the airflow, leading to a faster increase in velocity.

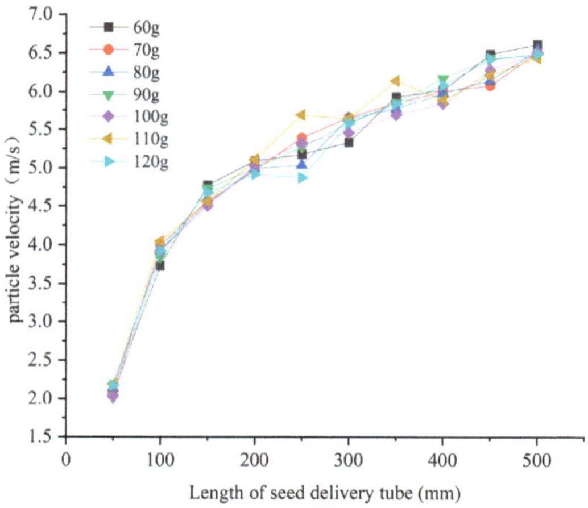

Figure 16. Particle velocity of seed feeding device under different seed feeding amounts.

As the seed supply increases, the particle density inside the tube also increases, and the collision, friction, and space occupation between wheat seeds slow down the overall acceleration of the particles, resulting in a slightly different final velocity growth curve. However, these differences are not very marked (Figure 17). Comparing the particle velocity variation curves and local flow field distribution under different seed supply conditions can aid in understanding the impact of seed supply changes on particle acceleration and the distribution characteristics in the mixing chamber and seed transport tube under the same inlet wind speed conditions. This has important reference value for the optimization design of seed delivery devices, the selection of seed supply parameters, and the improvement of seed delivery stability and uniformity in practical agricultural machinery applications.

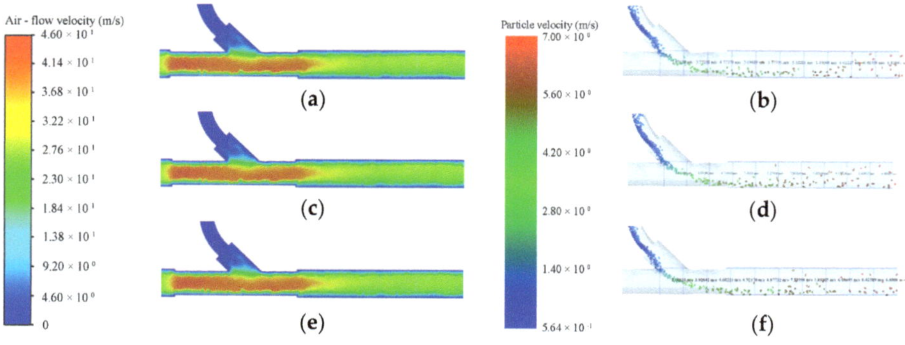

Figure 17. Cloud map of the seed-feeding device and particle velocity under different seed-feeding amounts; (**a**) velocity cloud map of the 60 g seed-feeding device; (**b**) velocity cloud map of the 60 g particle scenario; (**c**) velocity cloud map of the 90 g seed-feeding device; (**d**) velocity cloud map of the 90 g particle scenario; (**e**) velocity cloud map of the 120 g seed-feeding device; (**f**) velocity cloud map of the 120 g particle scenario.

In order to compare the accuracy of different particle models, the evaluation is carried out by comparing the displacement of wheat particles from entering the mixing chamber to 20 ms after entering the mixing chamber in the bench test and the simulation test, as shown in Figure 18. Since the high-speed camera can only capture the displacements in the X and Y directions, we only conduct an error analysis of the displacements in these two directions. Three wheat particles are randomly selected, and the displacement of each particle is calculated based on the change in its position in the time series. The displacement can be calculated by the distance between the initial and final positions of the particle. In the simulation, the same number of particles as in the bench test is selected, and their motion trajectories are recorded to calculate the displacement of the particles in the simulation. The displacements of the three particles in the bench test and the simulation test are compared. The displacement difference of each particle is calculated, and the average displacement error is calculated as shown in Table 5.

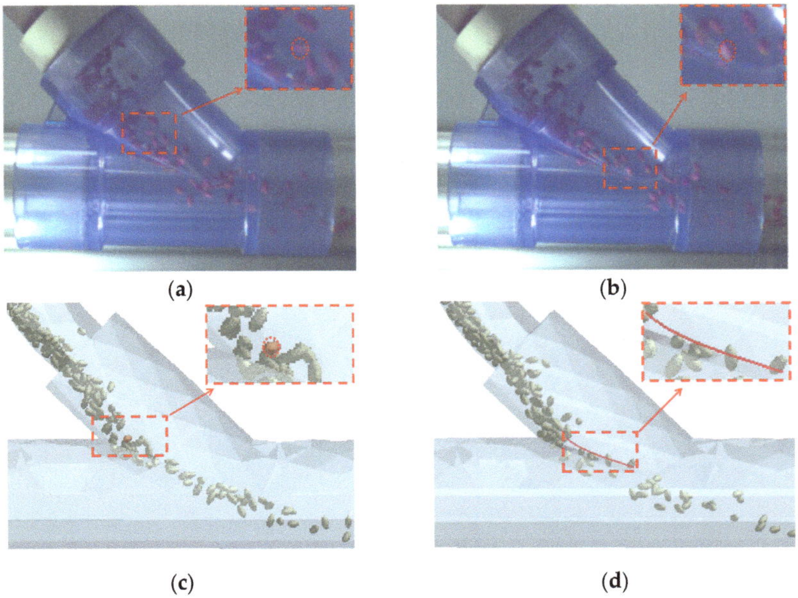

Figure 18. Tests and simulations of the pneumatic conveying device bench test: (**a**) initial position of test particles on the test bench; (**b**) position of test particles on the test bench 20 ms after entering the feeding device; (**c**) initial position of particles in the coupling simulation test; (**d**) position of particles in the coupling simulation test 20 ms after entering the feeding device.

Table 5. Displacement of different particle models.

Model	X-Direction Displacement (m)	Y-Direction Displacement (m)	X-Direction Error (%)	Y-Direction Error (%)	Average Error (%)
Bench test	0.047	0.016	/	/	/
5-balls model	0.035	0.019	25.53	18.75	22.14
7-balls model	0.042	0.019	10.64	18.75	14.70
9-balls model	0.041	0.015	12.77	6.25	9.51
11-balls model	0.030	0.019	36.17	18.75	27.46

The 9-balls model exhibits relatively low errors in both the X and Y directions. Notably, the error in the Y-direction is merely 6.25%, which is significantly lower than that of other models. The error of the 5-balls model in the X-direction is slightly higher than that of the 9-balls model, yet its error in the Y-direction is on par with those of other models. The 7-balls model has an X-direction error of 10.64%, demonstrating a relatively good simulation effect in the X-direction. The 9-balls model demonstrates the highest accuracy in simulating displacements in both the X and Y directions, with an average error of 9.51%. In particular, its error in the Y-direction is remarkably lower than that of the other models.

Consequently, the 9-balls model can better reflect the actual motion behavior of wheat particles in the pneumatic centralized-row seeding machine.

According to Newton's second law, combined with position and time information, the speed of wheat entering the feeding device after 20 milliseconds is obtained. The results of establishing particle models for four different particles and comparing them with bench tests are shown in Table 6. The errors of modeling particles with 5 balls, 7 balls, 9 balls, and 11 balls are 4.69%, 22.89%, 1.39%, and 6.319%, respectively. Among them, the modeling particle error of 9 balls is the lowest, with errors 3.3%, 21.5%, and 4.929% lower than those of 5 balls, 7 balls, and 11 balls, respectively.

Table 6. Wheat particle velocity in the coupled simulation experiment.

	5 Balls	7 Balls	9 Balls	11 Balls
Particle velocity (m/s)	2.634	3.092	2.551	2.675
Relative error (%)	4.690	22.890	1.390	6.319

5. Discussion

The pressure-boosting pipe and the distributor are key components of the pneumatic centralized-row seeding machine. Zhang et al. [29] introduced its working principle and structural design. Currently, most studies use CFD–DEM coupled simulation to optimize the parameters of these components. For example, Li et al. [30] designed an M-type seed distributor with this method. They studied the impacts of different outlet-pipe inclinations, top-cover cone angles, inlet diameters, fillet radii, and air-flow conveying velocities on the coefficient of variation of the M-type seed distributor's row-by-row discharge uniformity through distributor performance tests.

Simulation accuracy is a crucial factor in the design of the pressure-boosting pipe and distributor in the pneumatic centralized-row seeding system. Fan et al. validated the proposed wheat model by comparing experimental and simulation results, using several evaluation criteria including the volume fraction of wheat seeds, static angle of repose, hopper discharge, drum operation, and "gravity-flow screening" [13]. Sun et al. established double-ellipsoid models with 13, 17, 25, and 33 balls through the analysis of seed shapes. Meanwhile, they constructed an ellipsoid model using the ellipsoid equation for comparison with other models. The results showed that the 17-ball and 25-ball models based on the double-ellipsoid are suitable for modeling wheat seeds with a length-to-thickness ratio greater than 2.05 [14]. Previous research mainly focused on discrete-element modeling of the physical properties of wheat, with little attention given to the establishment of discrete-element models of wheat using the gas–solid two-phase flow coupling method. In this paper, a discrete-element model of wheat seeds was established through the ellipsoid method and the CFD–DEM coupled method to enhance the simulation accuracy during the design and optimization of the pneumatic centralized-row seeding system.

In the bench tests and simulation experiments of wheat suspension velocity, it was found that as the seed supply amount increased, the airflow velocity distribution of the model became more uniform, and the stability of the suspension velocity was enhanced. This phenomenon indicates that the increase in particle density makes the effect of the airflow on the particles more uniform, thereby reducing local turbulence and energy loss. This finding provides an important reference for the design of the pneumatic collective-discharging seeding system, especially in optimizing the seed supply amount and airflow velocity. However, there are certain uncertainties and error sources when conducting verification experiments related to the wheat suspension velocity. In terms of experimental equipment, the accuracy limitation of the wind speed measuring instrument may lead to errors in the measured airflow velocity.

In the simulations and experiments of the seed-feeding device, we discovered that, as the number of seeds supplied increases, the movement velocity of particles in the seed-conveying pipe gradually tends to stabilize. Although an increase in the number of seeds supplied leads to more collisions and frictions among particles, slightly slowing down the acceleration process of particles, the final movement velocities of particles tend to be consistent under different seed-supply amounts. This indicates that the design of the seed-feeding device can adapt to different amounts of seeds supplied within a wide range, ensuring the uniform distribution and stable transportation of seeds. In addition, the simulation results of the 9-balls model in the seed-feeding device are highly consistent with the experimental data, further validating the applicability of this model in complex gas–solid two-phase flow fields. This result provides a theoretical basis for the optimized design of the pneumatic centralized-row seeding machine, especially in terms of improving seeding uniformity and reducing the seed breakage rate. In the displacement test of wheat grains in the seed-supply device, due to the limitations of the experimental equipment, the camera can only capture the displacements of wheat grains in the X and Y directions, and cannot obtain the displacement data of the Z-axis. Although, in the current study, based on the existing displacement data in the X and Y directions, the 9-balls model has shown high accuracy, the lack of Z-axis displacement data makes the model validation incomplete. Future research needs to adopt more advanced measurement techniques to obtain the complete motion trajectory of seeds in three-dimensional space so as to verify the accuracy of the model more comprehensively and precisely.

In summary, predicting the movement behavior of wheat seeds in the seeding device and seed-conveying tube through CFD–DEM coupling simulation has important theoretical value. It helps analyze seed dynamic characteristics and design the seeding system. It also has significant practical application significance and can effectively improve R & D efficiency. The simulation and test results of the seed-supply device show that when the inlet wind speed remains constant, the seed speed changes very little as the seed-supply amount increases. This finding is of great reference value. It can be used for the optimized design of the seed-conveying device, the rational selection of the inlet wind speed, and improving the stability and uniformity of seed-conveying in practical agricultural machinery applications. In addition, the low relative error (only 1.39%) in the particle velocity simulation means it is expected to reduce field tests by 30–40%. This can significantly cut down the R & D costs of agricultural machinery manufacturing. The research results of this study provide key theoretical support for the design and optimization of the pneumatic collective-discharging seeder. In the future, the 9-balls model will be used. We will study the movement postures and laws of wheat seeds in the seed-conveying tube, explain the mechanisms affecting the uniformity of the gas–solid two-phase flow and distribution, further optimize the seed-conveying tube and distributor mechanism, and determine the optimal parameters. These are the main problems to be solved in our subsequent research.

Although this study has verified the effectiveness of the discrete element model of wheat seeds based on the CFD–DEM coupling method, there are still many problems to explore. First, the soil conditions in Xinjiang are complex and diverse. This brings many special challenges to the seeding process and requires further model optimization. In the actual seeding process, the feed rate and airflow speed may vary. This is due to different field conditions, seed types, and the working status of the seeder. Therefore, future research can consider more variable conditions to evaluate the model's performance more comprehensively.

6. Conclusions

In this study, the ellipsoid modeling method was adopted to establish the discrete element model of Xinchun No. 6 wheat seeds. Based on the DEM–CFD coupled simulation method, the kinematic characteristics of seeds with different particle models (5, 7, 9, and 11 balls) were simulated. Through the air suspension experiment, a comparative analysis of different particle models was carried out. The results show that the average relative errors of the air suspension velocities of the 5-balls, 7-balls, 9-balls, and 11-balls models are 2.87%, 1.53%, 1.18%, and 1.39% respectively. Among them, the errors of the 9 ball and 11 ball wheat particles are relatively small. Further research reveals that, when the seed mass ranges from 10 g to 70 g for the 9-balls model, the relative errors of the air suspension velocity are 0.7%, 2.50%, 0.67%, 0.21%, 1.47%, 2.09%, and 0.60%, respectively, with the minimum relative error being 0.21%.

By comparing the accuracy of displacement simulations of wheat particles in the seed-dropping device using different multi-ball models, a comparative analysis of bench tests and simulation tests was carried out. In the 20 ms period after the particles enter the mixing chamber, the 9-balls model is the most accurate in simulating displacements in both the X and Y directions, with an average error of 9.51%. The error in the Y-direction is only 6.25%. By comparing and analyzing the velocity of wheat 20 milliseconds after it enters the seed-feeding device, the errors of the 5-balls, 7-balls, 9-balls, and 11-balls modeled particles in the bench test and simulation test of the seed-feeding device are 4.69%, 22.89%, 1.39%, and 6.319%, respectively. Among them, the error of the 9-balls modeled particles is the lowest, being 3.3%, 21.5%, and 4.929% lower than those of the 5-balls, 7-balls, and 11-balls modeled particles, respectively.

Through air suspension and seed-feeding experiments, the feasibility of simulating the air-motion characteristics of wheat using the DEM–CFD coupled simulation method was verified. In the design of the pneumatic centralized-row seeding system, the motion state of wheat particles can be determined by the coupled simulation, thus providing a theoretical reference for the design of the key components of the pneumatic centralized-row seeding machine.

Author Contributions: Conceptualization, B.G. and L.W.; methodology, B.G.; software, B.G. and C.H.; validation, B.G., X.H. and J.X.; formal analysis, X.H.; investigation, C.H. and J.X.; resources, L.W.; data curation, K.R.; writing—original draft preparation, B.G.; writing—review and editing, B.G. and L.W.; visualization, B.G.; supervision, L.W. and X.W.; project administration, L.W.; funding acquisition, L.W. All authors have read and agreed to the published version of the manuscript.

Funding: This research was funded by the President Fund from Tarim University (TDZKYS202302) and Guiding Science and Technology Program Projects of Bingtuan (2023ZD094).

Institutional Review Board Statement: Not applicable.

Data Availability Statement: The data presented in this study are available upon request from the authors.

Acknowledgments: The authors would like to thank their schools and colleges, as well as the funding providers of the project. All support and assistance are sincerely appreciated.

Conflicts of Interest: The authors declare no conflicts of interest.

References

1. Wang, Y.B.; Li, H.W.; Wang, Q.J.; He, J.; Lu, C.; Liu, K. Design and Experiment of a Mechanical Wheat Shooting Seed Metering Device. *Trans. Chin. Soc. Agric. Mach.* **2020**, *51*, 73–84.
2. Ge, D.; Long, H.; Ma, L.; Zhang, Y.; Tu, S. Analysis Framework of China's Grain Production System: A Spatial Resilience Perspective. *Sustainability* **2017**, *9*, 2340. [CrossRef]

3. Klinzing, G.E. A review of pneumatic conveying status, advances and projections. *Powder Technol.* **2018**, *333*, 78–90. [CrossRef]
4. Li, J.W.; Yuan, W.S.; Zhai, H.L.; Zhu, S.S. Application of Pneumatic Conveying Technology in Agricultural Machinery. *J. Chin. Agric. Mech.* **2023**, *44*, 85–92, 115.
5. Zhu, Q.; Zhang, H.; Zhu, Z.; Gao, Y.; Chen, L. Structural Design and Simulation of Pneumatic Conveying Line for a Paddy Side-Deep Fertilisation System. *Agriculture* **2022**, *12*, 867. [CrossRef]
6. Hu, H.; Zhou, Z.; Wu, W.; Yang, W.; Li, T.; Chang, C.; Ren, W.; Lei, X. Distribution characteristics and parameter optimisation of an air-assisted centralised seed-metering device for rapeseed using a CFD–DEM coupled simulation. *Biosyst. Eng.* **2021**, *208*, 246–259. [CrossRef]
7. Zhao, J.L.; Zhao, H.N.; Tang, H.; Wang, X.G.; Yu, Y.J. Bionic threshing component optimized based on MBD-DEM coupling simulation significantly improves corn kernel harvesting rate. *Comput. Electron. Agric.* **2023**, *212*, 108075. [CrossRef]
8. Xu, N.N.; Tang, S.; Lin, D.D.; Geng, R.H.; Wang, X.W.; Liang, X.H. Complex granular flows of sticky-wet material on flip—Flow screens: Calibration of discrete element simulations. *Particuology* **2024**, *84*, 290–308. [CrossRef]
9. Zhou, L.; Yu, J.Q.; Wang, Y.; Yan, D.X.; Yu, Y.J. A study on the modelling method of maize-seed particles based on the discrete element method. *Powder Technol.* **2020**, *374*, 353–376. [CrossRef]
10. Cleary, P.W.; Sawley, M.L. DEM modelling of industrial granular flows: 3D case studies and the effect of particle shape on hopper discharge. *Appl. Math Model.* **2002**, *26*, 89–111. [CrossRef]
11. You, Y.; Zhao, Y. Discrete element modelling of ellipsoidal particles using super-ellipsoids and multi—Spheres: A comparative study. *Powder Technol.* **2018**, *331*, 179–191. [CrossRef]
12. Zhao, S.W.; Zhang, N.; Zhou, X.W.; Zhang, L. Particle shape effects on fabric of granular random packing. *Powder Technol.* **2017**, *310*, 175–186. [CrossRef]
13. Fan, J.H.; Wang, H.W.; Sun, K.; Zhang, L.; Wang, L.; Zhao, J.W.; Yu, J.Q. Experimental verification and simulation analysis of a multi-sphere modelling approach for wheat seed particles based on the discrete element method. *Biosyst. Eng.* **2024**, *245*, 135–151. [CrossRef]
14. Sun, K.; Yu, J.Q.; Liang, L.S.; Wang, Y.; Yan, D.X.; Zhou, L.; Yu, Y.J. A DEM—Based general modelling method and experimental verification for wheat seeds. *Powder Technol.* **2022**, *401*, 117353. [CrossRef]
15. Gao, Q.Q.; Chen, Y.C.; Zhao, C. Self—Similarity of density and velocity profiles in a 2D hopper flow of elliptical particles: Discrete element simulation. *Powder Technol.* **2022**, *402*, 117338. [CrossRef]
16. Liu, Z.D.; Wang, Q.J.; Li, H.W.; He, J.; Lu, C.Y.; Yu, C.C. Analysis and Experiment on Fertilizer Application Path of Pneumatic Seed and Fertilizer Hill-seeding Device Coupled with CFD–DEM. *Trans. Chin. Soc. Agric. Eng.* **2019**, *35*, 18–25.
17. Wang, C.; Li, H.W.; He, J.; Wang, Q.J.; Lu, C.Y.; Wang, J.X. Design and Experiment of Pneumatic Precision Wheat Seeding Device in Rice—Wheat Rotation Area. *Trans. Chin. Soc. Agric. Mach.* **2020**, *51*, 43–53.
18. Yang, Q.L.; Li, Z.H.; Li, H.W.; He, J.; Wang, Q.J.; Lu, C.Y. Numerical Analysis of Particle Motion in Centralized Fertilizer Distribution Device Based on CFD–DEM. *Trans. Chin. Soc. Agric. Mach.* **2019**, *50*, 81–89.
19. Li, L.W.; Meng, Z.J.; Wang, X.O.; An, X.F.; Wang, P.; Wu, G.W. Simulation Analysis of Gas-Solid Two-Phase Flow in Fertilizer Delivery Device of Pneumatic Rice Fertilizer Applicator. *Trans. Chin. Soc. Agric. Mach.* **2018**, *49*, 171–180.
20. Liu, F.Y.; Zhang, J.; Li, B.; Chen, J. Analysis and Calibration of Wheat Discrete Element Parameters Based on Packing Experiments. *Trans. Chin. Soc. Agric. Eng.* **2016**, *32*, 247–253.
21. Ren, D.; Yu, H.; Zhang, R.; Li, J.; Zhao, Y.; Liu, F.; Zhang, J.; Wang, W. Research and Experiments of Hazelnut Harvesting Machine Based on CFD–DEM Analysis. *Agriculture* **2022**, *12*, 2115. [CrossRef]
22. Dai, F.; Xu, P.; Yuan, Z.; Shi, R.; Zhao, Y.; Song, X.; Zhao, W. Simulation and Optimization Experiment: Working Process of a Cleaning Device for Flax Combine Harvester. *Agriculture* **2023**, *13*, 2123. [CrossRef]
23. Sommerfeld, M. Analysis of collision effects for turbulent gas–particle flow in a horizontal channel: Part I. Particle transport. *Int. J. Multiph. Flow* **2003**, *29*, 675–699. [CrossRef]
24. Mills, D.; Jones, M.G.; Agarwal, V.K. *Handbook of Pneumatic Conveying Engineering*, 1st ed.; CRC Press: Boca Raton, FL, USA, 2004. [CrossRef]
25. Song, J.; Wang, T.; Hu, G.; Zhang, Z.; Zhao, W.; Wang, Z.; Zhang, Y. Conveying characteristics of shrimp feed pellets in pneumatic conveying system and minimum power consumption dissipation factor. *Aquac. Eng.* **2023**, *102*, 102347. [CrossRef]
26. Lei, X.; Liao, Y.; Liao, Q. Simulation of seed motion in seed feeding device with DEM–CFD coupling approach for rapeseed and wheat. *Comput. Electron. Agric.* **2016**, *131*, 29–39. [CrossRef]
27. Wang, L.; Xiao, W.; Shi, B.; Li, X.; Liao, Q.; Liao, Y. Numerical and experimental study on the performance of a stabilising turbine inside a seed distribution device. *Biosyst. Eng.* **2024**, *246*, 219–232. [CrossRef]
28. Cao, X.; Wang, L.; Liao, Q.; Liao, Y. Simulation of the mixing component of a horizontal air-assisted centralised wheat metering device. *Biosyst. Eng.* **2024**, *243*, 13–26. [CrossRef]

29. Zhang, M.; Wu, C.Y. Analysis of the General Situation of Pneumatic Conveying Seeders and Their Application Prospects in China. *Chin. Agric. Mech.* **2008**, *2*, 70–72. [CrossRef]
30. Li, G.C.; Li, H.Y.; Yang, S.P.; Huang, Y.X.; Gao, X.J.; Fu, Z.L. Simulation and experiment of distributors of different outlet types of pneumatic collection and discharging systems based on CFD-DEM coupling. *J. Intell. Agric. Mech.* **2024**, *5*, 1–11.

agriculture

MDPI

Review

Application of Discrete Element Method to Potato Harvesting Machinery: A Review

Yuanman Yue, Qian Zhang, Boyang Dong and Jin Li *

College of Mechanical and Electronic Engineering, Northwest A & F University, Yangling 712100, China; yue2024@nwafu.edu.cn (Y.Y.); zhang2023@nwafu.edu.cn (Q.Z.); 2023110032@nwafu.edu.cn (B.D.)
* Correspondence: lijin2023@nwafu.edu.cn

Abstract: The Discrete Element Method (DEM) is an innovative numerical computational approach. This method is employed to study and resolve the motion patterns of particles within discrete systems, contact mechanics properties, mechanisms of separation processes, and the relationships between contact forces and energy. Agricultural machinery involves the interactions between machinery and soil, crops, and other systems. Designing agricultural machinery can be equivalent to solving problems in discrete systems. The DEM has been widely applied in research on agricultural machinery design and mechanized harvesting of crops. It has also provided an important theoretical research approach for the design and selection of operating parameters, as well as the structural optimization of potato harvesting machinery. This review first analyzes and summarizes the current global potato industry situation, planting scale, and yield. Subsequently, it analyzes the challenges facing the development of the potato industry. The results show that breeding is the key to improving potato varieties, harvesting is the main stage where potato damage occurs, and reprocessing is the main process associated with potato waste. Second, an overview of the basic principles of DEM, contact models, and mechanical parameters is provided, along with an introduction to the simulation process using the EDEM software. Third, the application of the DEM to mechanized digging, transportation, collection, and separation of potatoes from the soil is reviewed. The accuracy of constructing potato and soil particle models and the rationality of the contact model selection are found to be the main factors affecting the results of discrete element simulations. Finally, the challenges of using the DEM for research on potato harvesting machinery are presented, and a summary and outlook for the future development of the DEM are provided.

Keywords: potato harvesting machine; numerical simulation; DEM; damage avoidance

Academic Editor: Bruno Bernardi

Received: 29 December 2024
Revised: 23 January 2025
Accepted: 29 January 2025
Published: 31 January 2025

Citation: Yue, Y.; Zhang, Q.; Dong, B.; Li, J. Application of Discrete Element Method to Potato Harvesting Machinery: A Review. *Agriculture* **2025**, *15*, 315. https://doi.org/10.3390/agriculture15030315

1. Introduction

Potato is the fourth largest agricultural crop in the world [1] and one of the most globally consumed staple foods [2]. It is cultivated in over 100 countries and regions [3], playing a significant role in ensuring global food security and economic development [4–6]. According to data released by the *Food and Agriculture Organization of the United Nations* (FAO), global potato-cultivating areas are divided into six major growing environments, as shown in Figure 1 [7], with two each in temperate, subtropical, and tropical environments [8,9]. Global potato production reached 368.8 million tons in 2019, 371.1 million tons in 2020, and 376.1 million tons in 2021. It is expected that annual production will increase with the growth of the global population. Ensuring the efficient and stable development of the potato industry has become a key issue in the agricultural field [10]. The current mechanized harvesting rate of potatoes is low, and potato harvesters face problems such as high mechanical damage rates [11], low operational efficiency, high operational costs, and soil loss in the field during mechanized harvesting [7]. As a result, there is no potato harvester that can currently function in all operational environments, under all soil conditions, and with all potato varieties. Research and development of potato harvesters largely depends on the mechanical operational environment [12]. There are various methods of harvesting potatoes, such as traditional manual methods [13], animal-drawn methods [14], and combine harvesters [15]. Due to the labor-intensive nature of manual potato harvesting and the advancements in agricultural technology, there is an urgent need for transformation from primarily manual methods to mechanized harvesting modes; however, the biomechanical properties of potatoes make it difficult to achieve fully mechanized harvesting throughout the entire process [16]. As one of the main tuber crops and a significant source of starch [17], potatoes are susceptible to mechanical damage during the mechanized harvesting process, which can lead to rot, resulting in significant potato losses each year during the harvesting process [18,19]. Furthermore, sprouting during storage also contributes to potato losses [20]. To ensure sustainable development of the potato industry and global food security, there is a pressing need to develop a potato harvester that can adapt to various operational environments. However, although domestic potato combine harvesters have been developed to a certain extent, gaps exist between them and those developed in foreign countries. There are fewer applications in automatic navigation, intelligent identification, and precise operation. Operational precision and efficiency need to be improved, and their adaptability to complex terrains and different soils is limited. When working in hilly and mountainous areas, there are problems, such as poor passability and difficulty separating potatoes and soil. On the whole, there are shortcomings such as low intelligence level, insufficient adaptability, and insufficient reliability and durability. In contrast, foreign potato combine harvester technology is complex and expensive, increasing farmers' purchasing and operating costs; thus, it is not conducive for use in some economically underdeveloped areas. Additionally, for some special planting patterns or minority varieties of potatoes, the lack of targeted design and optimization means that the technology cannot adequately meet the diverse planting needs.

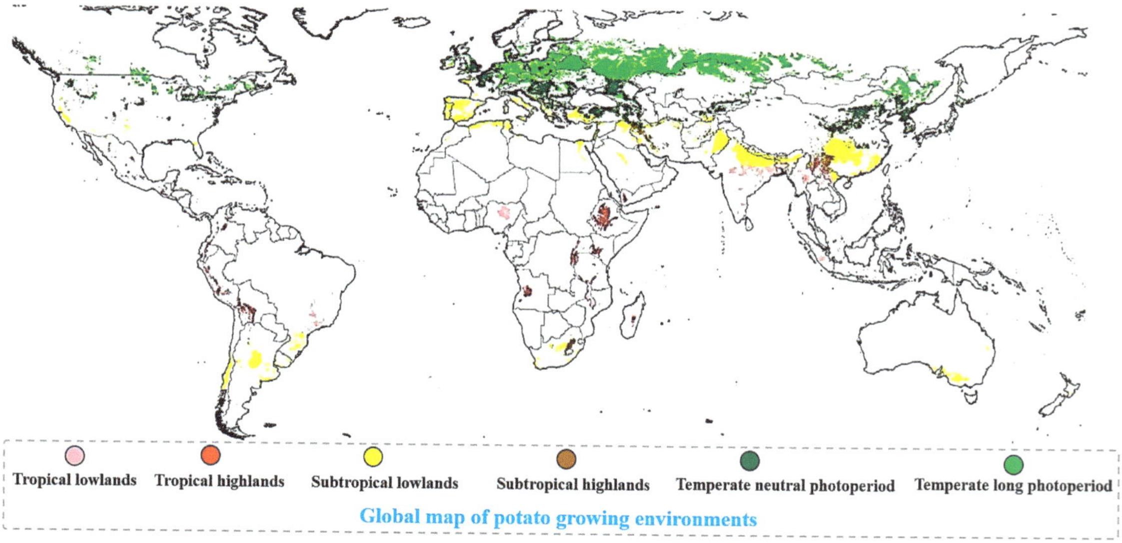

Figure 1. Global map of the distribution of potato growing environments [7].

With the rapid development of computer technologies [21] and new materials science [22], digital design and software simulation have become some of the main methods of engineering design [23–25] and have been widely applied in the field of engineering [26,27]. The application of machine learning and supercomputers has made it possible to simulate large-scale engineering projects [28–30]. The Discrete Element Method (DEM) is one of the primary methods for solving dynamic problems in discrete systems. This method involves numerical calculation and analysis of discrete systems under the premise of considering interactions such as contacts and separations, mutual movements, contact forces between particles, and energy transfers. Therefore, the DEM has been widely applied in agricultural machinery design, operation process simulation, soil flow characteristics, and plant emergence research [31–34].

The data included in this article are sourced from the *Food and Agriculture Organization of the United Nations*, *Web of Science*, etc., and the content and references cited in this article are from sources such as academic conferences, scientific books, and academic journals. The primary academic journals include "*Computers and Electronics in Agriculture*", "*Agriculture*", "*Transactions of the Chinese Society for Agricultural Machinery*", "*Transactions of the Chinese Society of Agricultural Engineering*", etc. Some content is also sourced from academic conference proceedings. The structure of this review is as follows: Part I provides an introduction to potato production and the global scale of the potato industry, as well as the challenges faced during mechanized potato harvesting. Statistical analysis of potato research papers from 2013 to 2014 shows an overall increasing trend, as illustrated in Figure 2a. Using "potato machinery", "agricultural machinery", and "discrete element analysis" as search terms, relevant research literature was statistically analyzed, as shown in Figure 2b. It can be observed that there is a linear correlation between the literature on discrete element analysis of agricultural machinery and research on potato machinery. Part II summarizes the basic principle, contact model, and mechanical parameters of DEM and introduces the simulation process based on the EDEM software. Part III reviews the application of the DEM in the mechanized digging, transportation, collection, and soil–potato separation processes during potato harvesting. Part IV summarizes and outlines the application and shortcomings of DEM in potato harvester research.

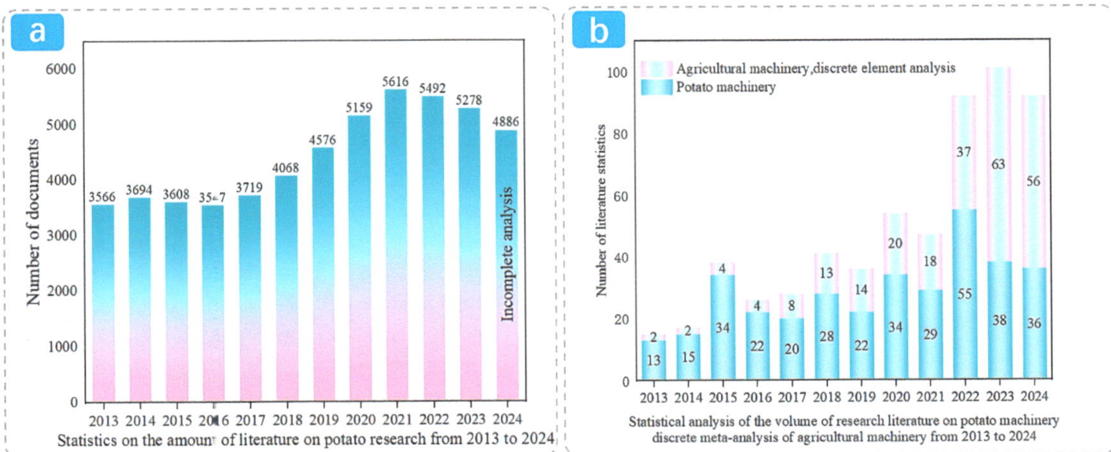

Figure 2. (**a**) Statistical analysis of the research literature on potatoes. (**b**) Statistical analysis of potato machinery, agricultural machinery, and discrete element analysis.

2. Basic Principles of the Discrete Element Method

The DEM is a numerical simulation and calculation method for analyzing and solving the motion laws and mechanical properties of complex discrete systems [35]. The method considers the entire medium as a series of discrete, independently moving particles, with changes in the medium described by the movement characteristics of each unit [36–39]. Initially applied in geotechnical mechanics research, DEM has been used to describe the movement of discontinuous media such as soils and rocks [40,41]. It assumes that there are contacts between particles, and particles in the discrete system can move, rotate, or deform. According to Newton's second law, the acceleration of each particle is obtained, and integration over time is used to calculate the velocity and displacement of the discrete system [42,43]. With the development of computer technology, the DEM has been widely applied in engineering research [44]. From the verification of simple 2D simulation results to complex 3D processes, and from single DEM modeling to the combination of DEM with other technologies, the superiority of DEM in engineering design has been demonstrated [45,46]. Simultaneously, DEM has also been applied in the analysis of soil dynamics. It assesses the importance of contact model particle parameters in soil modeling and obtains parameters for selecting Hertz–Mindlin contact model particles based on various moisture levels [47]. The development of DEM involves the refinement of contact models and particle shape models [48]. The selection of contact models and the determination of parameters directly affect the efficiency and accuracy of DEM simulations [49].

2.1. Contact Model

The contact model describes the contact behavior between particles [50]. In the early days of the DEM, particle interface action was thought to occur through either soft contact or hard contact, leading to important conclusions for the numerical model of discontinuous bodies [51,52]. In DEM, each particle is considered an independent rigid body, and the contact model is used to calculate the forces and torques between these bodies. This section provides a brief introduction to the eight common contact models in the DEM analysis software EDEM.

2.1.1. Hertz–Mindlin No-Slip Contact Model

The Hertz–Mindlin no-slip contact model is a model used to describe the contact behavior between granular materials and is widely applied in DEM simulations. This model combines the normal contact force from the Hertz model and the tangential friction force from the Mindlin model, forming the Hertz–Mindlin no-slip contact model [53,54]. This model can comprehensively describe the contact behavior between particles, including the normal and tangential force–displacement relationships and the frictional effects between particles. It is commonly used to simulate the interactions between particles and geometric shapes. Chen et al. [55] conducted an analysis of the DEM prediction sensitivity for sliding wear of single iron ore particles, which accelerated the calibration of DEM for predicting sliding wear. Li et al. [56] established a mathematical model for the "corn–machine" interaction system based on the Hertz–Mindlin no-slip model. They derived the kinematic and dynamic formulas for the movement of corn ears under the action of the sheller and established a force analysis model for corn–corn interactions, as shown in Figure 3a. Additionally, they analyzed the contact characteristics of "corn–corn" based on the Hertz–Mindlin and bonded models.

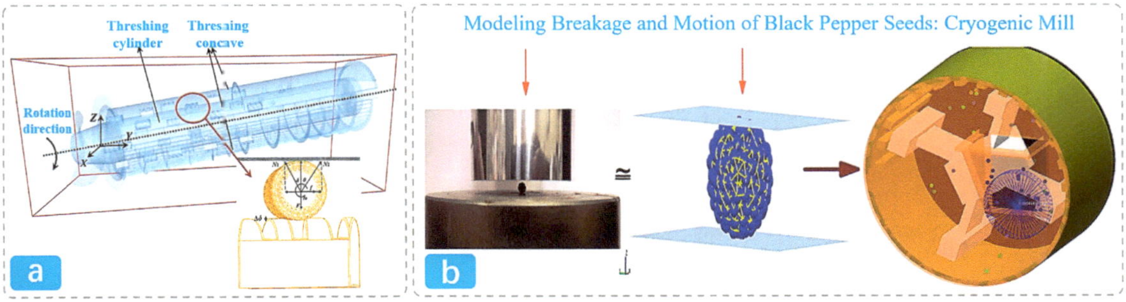

Figure 3. (**a**) Mechanical analysis model of maize [56]. (**b**) Simulation of the process of crushing black pepper seeds [57].

In summary, during DEM numerical calculations, the Hertz–Mindlin model requires calibration of parameters using experimental data to ensure the accuracy of the model. For example, parameters such as Young's modulus, Poisson's ratio, and friction coefficient of particles should be measured experimentally to verify the consistency between the model's predicted results and the experimental data.

2.1.2. Hertz–Mindlin Bonding Contact Model

The Hertz–Mindlin bonding contact model is specifically designed to simulate the adhesion and fracture behavior between particles in the EDEM software [58,59]. A layer of "binder" is designed between particles to reproduce how small particles adhere and form larger agglomerates. When subjected to external forces, if the adhesive force between these adhered particles exceeds the maximum limit, they will fracture and produce effects of particle breakage in the simulation. Ahmad et al. [60] studied the relationship between soil moisture and cohesive particles using the Hertz–Mindlin bonding model, validating the applicability of the model. Wang et al. [61] integrated the Hertz–Mindlin model with a bonding contact model. The dynamic analysis results were used to replicate actual drilling scenarios while minimizing computational demands. Ghodki et al. [57] coupled the cohesive particle model with the Hertz–Mindlin contact model to analyze the breakage process of black pepper seeds, as shown in Figure 3b. DEM modeling was found to better explain particle breakage and fluidity in grinders.

In summary, the Hertz–Mindlin bonding model allows particles to maintain a bonded state within a certain range of normal and tangential displacements until the maximum shear stress of the bonding force is reached, at which point the bonding breaks and the particles revert to the interaction of hard spheres. This model is particularly suitable for simulating materials that are prone to breakage and fracture under external forces, such as rocks and concrete.

2.1.3. Hertz–Mindlin Thermal Conductivity Model

The Hertz–Mindlin thermal conductivity model is a type of contact model in the EDEM software that considers the thermal conduction effects when simulating the contact behavior between particles [62]. As a variant of the Hertz–Mindlin model, it adds the calculation of thermal conduction on the basis of the Hertz–Mindlin model [63]. The Hertz–Mindlin model is based on the Hertz contact theory and the Mindlin–Deresiewicz tangential force model, allowing for accurate and efficient calculation of normal and tangential forces between particles. Gui et al. [64] analyzed particle mixing and thermal conduction based on the DEM, obtaining particle temperature distribution maps, as shown in Figure 4a. The study demonstrated an effective method for thermal conduction between particles. The Hertz–Mindlin thermal conduction model, on this basis, adds a temperature update mechanism, enabling thermal conduction between particles to be transmitted through contact, which is suitable for occasions requiring temperature analysis. Thermal conduction occurs after particle contact due to temperature differences. This model is particularly suitable for applications where the effect of heat transfer needs to be considered during particle contact [65,66].

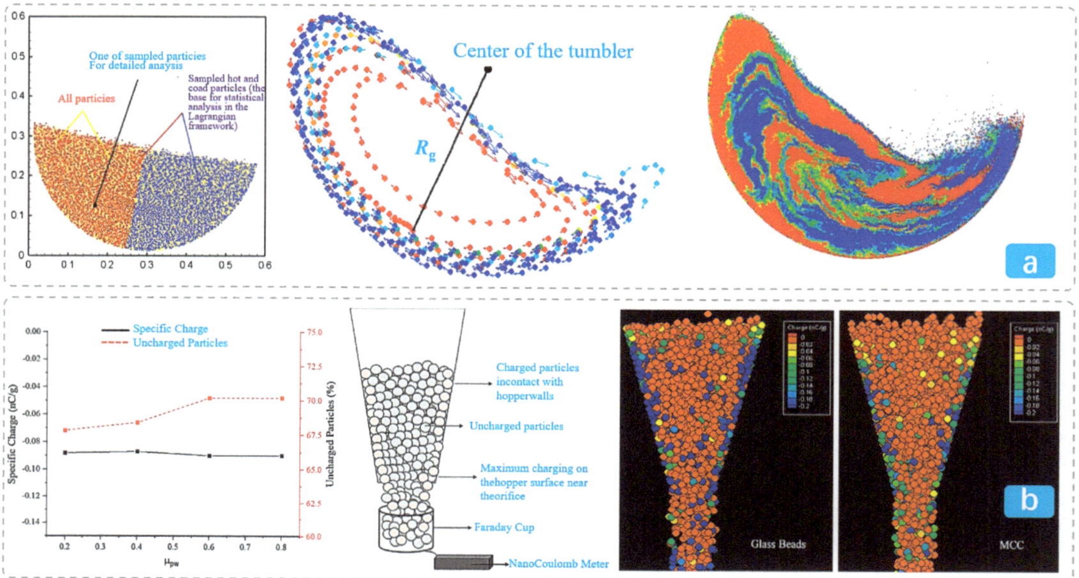

Figure 4. (**a**) Temperature simulation of heat-transferring particles [64]. (**b**) Specific charge analysis of frictionally charged particles [67].

2.1.4. Temperature Update Model

The temperature update model was proposed based on heat flux [68]. It is a physical model in the EDEM software used to simulate temperature changes during particle contact. This model must be used in conjunction with the Heat Conduction model to ensure the correct execution of the thermal conduction algorithm. It is particularly suitable for those application scenarios where heat transfer effects need to be considered during the process of particle contact, such as heat transfer simulation in particle flow in a rotating container.

2.1.5. Linear Cohesion Model

The linear cohesion contact model is a model used in the EDEM software to simulate bonding between particles. This model is suitable for rapid calculations of general cohesive particles and can also be used for moist materials. Unlike the JKR Cohesion model [69,70], which calculates cohesive forces that exist in both the normal and tangential directions of particle contact, the cohesive forces in this model only exist in the normal direction. The linear cohesion model considers the cohesive forces between particles, which are typically caused by intermolecular forces, electrostatic forces, or moisture content. During the simulation, when two particles come into contact and their surfaces are sufficiently close, the cohesive force will bond them together. This model is particularly suitable for applications that require the simulation of bonding behavior between particles, such as the stacking, mixing, or molding processes of powders.

2.1.6. Linear Spring Contact Model

The linear spring contact model is a commonly used contact model in DEM that describes the mechanical behavior of particles during contact. This model is based on the concepts of springs and dampers, where springs simulate the elastic forces between particles and dampers simulate the viscous behavior. The linear spring model typically includes linear elastic frictional behavior and viscous behavior, which act on a very small area, thus transmitting only one force. The linear spring contact model shares similarity with the nonlinear spring contact model. Caserta et al. [71] studied the relationship between the damping coefficient and contact duration in the continuous nonlinear spring–damper contact model within the DEM. Comparing this model to nonlinear contact force models shows that the proposed approximate relationship can effectively reduce the computation time.

In summary, this model is highly useful in the simulation of granular materials, especially when simulating particle interactions without requiring overly complex contact behaviors. For example, it can be used to simulate granular flows, powder stacking, and mixing processes. Due to its simplicity and computational efficiency, the linear spring model is very popular in industrial applications.

2.1.7. Motion Plane Contact Model

The motion plane contact model simulates the linear motion of geometries, with the entire geometry moving at the same velocity [72]. Only when using the moving plane contact model for contact force calculation can a linear velocity be added to the geometry [73]. When setting up the motion plane contact model in EDEM, it is necessary to select appropriate models and parameters based on the specific application scenario and physical characteristics. By adjusting these parameters, more realistic simulation results can be obtained.

2.1.8. Frictional Electrification Contact Model

The frictional electrification contact model is a computational model used to simulate the friction and electrostatic behavior of granular materials during contact [74,75]. This model is typically used to study the physical properties of granular materials, such as powder handling, mixing, and separation in industrial processes. The DEM model considers the contact forces between particles, including elastic forces, frictional forces, and electrostatic forces, as well as how these forces affect the motion and interactions of the particles. Mukherjee et al. [67] predicted frictional charging phenomena during hopper discharge using a simulation model based on the DEM. Specific charge information under different particle–wall frictions was obtained, as shown in Figure 4b. The study concluded that during the simulation of hopper discharge of monodisperse particles, approximately one-third of the particles are expected to remain unchanged, resulting in a relatively large distribution of charge concentrated in a small volume of material. Rasera et al. [76] modeled and analyzed the frictional charging process in both 2D and 3D using the DEM. The method for determining the DEM frictional charging model parameters was found to be effective, resulting in good agreement between the simulation data and experimental data. This approach will improve and simplify the DEM modeling of frictional charging during the handling of dry materials.

2.2. Simulation Process

EDEM is a DEM simulation analysis software widely used for simulating granular materials. The general steps of the EDEM simulation process are as follows: (1) Model building: Establish the model, defining the shape, size distribution, and material properties of the particles (e.g., density, friction coefficient, elastic modulus, etc.). Set the geometric shape and material properties of the container or equipment. Determine the particle generation method, such as through hoppers, airflow, etc. (2) Contact model settings: Choose an appropriate contact model, such as a linear spring–damper model, nonlinear contact model, etc. (3) Interaction characteristics definition: Define the friction, rolling, and torsional characteristics between the particles. If charged particles are involved, it is also necessary to configure the frictional charging model. (4) Simulation initialization: Set the initial conditions, such as the initial positions, velocities, and angles of the particles. Determine the time step and total simulation duration. (5) Boundary conditions and driving forces: Define boundary conditions, such as the interaction of particles with container walls. Add external forces, such as gravity, vibrations, airflow, etc. (6) Running simulation: Start the simulation and perform the calculations based on the specified parameters and models. Monitor the simulation process to ensure stability and that the calculations meet expectations. (7) Analyzing results: Collect and analyze simulation data, such as particle velocities, positions, temperature, pressure distribution, etc. Utilize EDEM's built-in tools or export data to other software for in-depth analysis. (8) Model validation: Compare the simulation results with experimental data or theoretical predictions to validate the accuracy of the model. Adjust model parameters as needed to enhance simulation accuracy. (9) Optimization and iteration: Optimize the process or equipment based on the simulation results. Multiple iterations of the simulation may be needed to find the best design solution. (10) Post-processing and reporting: Create charts, animations, and other visual materials to present the simulation results. The case is shown in Figure 5.

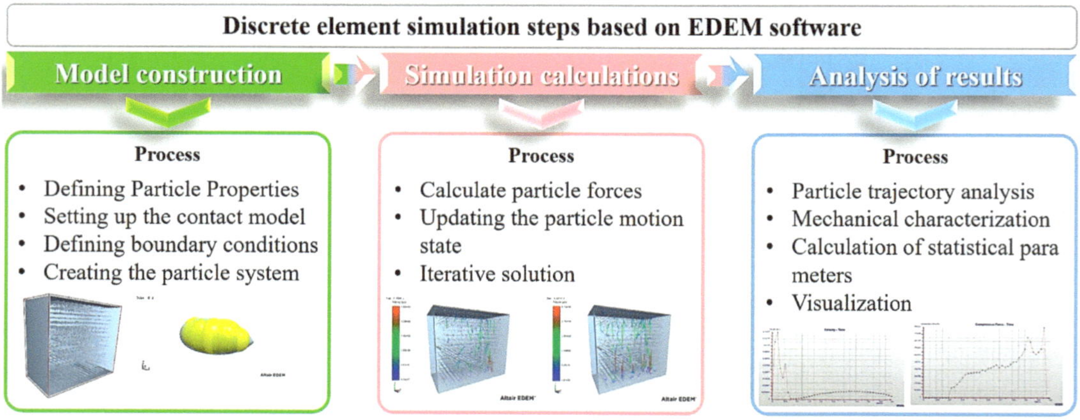

Figure 5. Discrete element simulation steps based on the EDEM software.

The EDEM software provides a user-friendly interface and a comprehensive set of tools to help users efficiently complete the aforementioned process. In addition, EDEM also supports interfaces with other engineering software, such as CAD, CFD, etc., to enable multiscale and Multiphysics-coupled simulations.

2.3. *DEM Interaction Design with Potato Harvesters*

2.3.1. Key Steps

Inputting mechanical parameters: Key design parameters of potato harvesting machinery, such as the shape, size, and entry angle of the digging shovel, the structure of the separating device, the size of the sieve hole, and the size and rotational speed of the conveying parts, are inputted into the DEM. At the same time, the operational parameters of the harvesting machinery, such as forward speed, digging depth, etc., are set; these serve as the initial conditions of the DEM model simulation and directly affect the accuracy of the simulation results.

Granular material simulation: In the DEM model, potatoes and soil are accurately simulated. Granular materials are endowed with corresponding physical and mechanical properties, including particle density, elastic modulus, Poisson's ratio, internal friction angle, cohesion, etc., in order to reflect the real physical properties of potatoes and soil. The mechanical behavior of potatoes during harvesting was studied by simulating the interaction of granular materials, as well as interactions between particles and harvesting mechanical components, such as collision, friction, extrusion, and so on. Simulation results output and feedback: After running the DEM model for simulation calculation, a series of simulation results data related to the harvesting process are output. These data include the trajectory, speed, and acceleration of potatoes and soil particles; interaction forces between particles and between particles and surfaces of mechanical components, such as friction, crowding force, collision force, etc.; they also include statistical data on damage such as skin abrasion and internal rupture of potatoes during harvesting, and information such as the degree and distribution of soil disturbance during digging and separation. These simulation results are fed back to the operator to provide an important basis for evaluating the performance of the harvesting machinery, identifying potential problems, optimizing machinery design, and adjusting operating parameters.

2.3.2. Key Technologies

Modeling of mechanical parts: Based on the actual structure and size of potato harvesting machinery, the three-dimensional model of each mechanical part is accurately constructed using the CAD software. During the modeling process, the shape and geometric features of the components are defined in detail, as well as the assembly relationship and connection between the components. In this way, the physical structure of the harvester is accurately presented in a digital form, which lays the foundation for the subsequent use of EDEM for simulation analysis.

Granular material modeling: In the DEM model, appropriate granular modeling methods are employed to simulate granular materials such as potatoes and soil. Common granular modeling methods include the spherical particle model, polyhedral particle model, aggregate particle model, etc. Based on the actual situation and research objectives, suitable granular modeling methods are selected, and parameters such as the shape, size distribution, and density of the particles are reasonably set. For example, for soil particles, the proportion of particles in different grain size ranges can be set according to the texture and gradation, and corresponding physical and mechanical properties are assigned to them; for potatoes, they can be simplified into spherical or ellipsoidal particles with certain shapes and dimensions, and their mechanical properties such as elasticity and plasticity are taken into account. Through accurate, granular material modeling, the behavior and interactions of granular materials in the potato harvesting process can be more realistically reflected.

Contact algorithms: In DEM (Discrete Element Method) simulations, accurately calculating the contact forces between particles and between particles and mechanical component surfaces is crucial. To achieve this, efficient and reliable contact algorithms are employed to simulate the contact behavior between particles. Common contact algorithms include the linear spring–damper model, the Hertz–Mindlin contact model, and contact algorithms that couple the discrete element method with the finite element method. These contact algorithms determine the magnitude and direction of contact forces based on different mechanical principles and assumptions by calculating parameters such as relative displacement and velocity between particles. In practical applications, it is essential to select the appropriate contact algorithm based on the characteristics of the particulate material and the requirements of the simulation problem. Additionally, parameters in the algorithm should be calibrated appropriately to ensure the accuracy and reliability of contact force calculations, thereby accurately reflecting the interactions between particles during the potato harvesting process.

Parallel computing: The DEM simulation of the potato harvesting process involves the movement and interaction calculations of a large number of particulate materials, resulting in a substantial computational workload and requiring a long computation time.

Data extraction and analysis: After completion of the DEM simulation, key data relevant to the potato harvesting process are extracted from the simulation results data files. These data include time series information such as the positions, velocities, accelerations, and force conditions of the particles, as well as statistical data on potato damage and soil disturbance levels. Using data analysis methods and tools, the extracted data undergo in-depth analysis and exploration. Statistical analysis methods can be employed to calculate indicators such as the rate of damage to potatoes and the distribution of damage types at different harvesting stages, allowing for assessment of the damage inflicted by the harvesting machinery on the potatoes. Additionally, time-domain and frequency-domain analysis methods can be used to study the variations in the movement speed and acceleration of the particulate materials over time, as well as the frequency components and energy distribution of the interaction forces between particles. This helps to reveal the dynamic characteristics and interaction mechanisms during the potato harvesting process. By extracting and analyzing the simulation results data, detailed information and valuable conclusions about the potato harvesting process can be obtained, providing a scientific basis for performance evaluation, optimization design, and adjustment of operational parameters of harvesting machinery.

Data visualization: Using data visualization technologies, the DEM simulation results are displayed in the form of intuitive and vivid graphics, images, animations, etc., so that researchers and engineers can observe, understand, and analyze the mechanical behavior and the movement of granular materials during potato harvesting. Common data visualization methods and tools include visualization software based on computer graphics (such as OpenGL, VTK, etc.), professional data analysis and visualization platforms (such as MATLAB, Python's related visualization libraries, etc.), and some commercial engineering simulation and visualization software.

Despite the many advantages of using DEM in the simulation of agricultural machinery, there are some limitations to large-scale simulations. Large-scale simulation involves large amounts of particles, with the amount of computation increasing exponentially. This requires high-performance hardware support, and the long time required for computation increases the costs significantly. As the number of particles increases and the model complexity increases, the simulation time increases significantly. This makes the iterative optimization design less efficient and unable to respond to real-world needs in a timely manner.

3. Global Analysis of the Current Status of Potato Harvesting Machinery

Mechanized potato harvesting technology integrates the processes of haulm destruction, digging, separation, collection, and loading. Germany, the United States, Italy, Poland, and other developed countries had already achieved integrated mechanized harvesting of potatoes by the 1990s; however, due to limitations in machine size, damage when harvesting fresh potatoes, and high prices, the requirements for mechanized potato harvesting have not been met in China. The potato combine harvester consists of different components—haulm destruction devices, digging devices, conveying devices, soil–potato separation devices, power devices, and control systems—that enable it to harvest mature potatoes. This reduces the cost of potato harvesting and promotes the development of the potato industry.

3.1. Small and Medium-Sized Potato Harvesters

Researchers have developed various types of small and medium-sized potato harvesters designed for hilly and mountainous terrains. These harvesters are characterized by strong maneuverability, low operational costs, and wide applicability. They can address challenges in mechanized potato harvesting on hilly terrains or small plots. This section provides a summary of current typical small and medium-sized potato harvesters, as shown in Table 1. Research on small and medium-sized potato harvesters started relatively late in China, but has developed rapidly in recent years. Enterprises and scientific research institutions have continuously innovated by introducing and absorbing foreign technologies and have developed a variety of products suitable for the nation's conditions. These have performed well when it comes to harvesting small plots in hilly and mountainous clay soil environments; however, overall performance, reliability, intelligence, and other aspects are still lagging behind those of foreign countries. International research on small and medium-sized potato harvesters started early, and the technology has matured. Potatoes in Italy, Poland, Japan, and South Korea are mainly planted in small and scattered mountainous areas, and they are mostly harvested by digging pickup loaders with fixed sorting devices. In countries with large-scale potato planting and relatively flat terrain, combine harvesters are widely used; these countries have made significant progress in intelligence and automation, with stable product performance and high reliability.

Table 1. Structure and technical features of small and medium-sized potato harvesters.

Name	Technical Characteristics	Appearance	Place of Production
WM4000 Potato Harvester	1. This machine is a single-row potato harvester; 2. Overall weight: 4450 kg; 3. Operating speed: 3~5 km/h; 4. Collection box capacity: 4000 kg; 5. Drawbar length: 1250 mm; 6. Its double-row embedded tooth cleaning device enables a more thorough separation of soil and stones [77].		French
1710A Potato Harvester	1. This machine is a trailed potato harvester; 2. Overall weight: 5600 kg; 3. Working efficiency: 8~10 acres/h; 4. Operating depth: 0~30 cm; 5. Floating disc cutter design reduces digging resistance while cutting weeds [78].		China
4TS-490 Potato Harvester	1. This machine can harvest 4 rows of potatoes in one operation; 2. The reduction in operational resistance reaches over 50%; 3. Structural simplicity; 4. Highly economical [77].		China
4U-90LH Potato Harvester	1. This machine is a trailed potato harvester; 2. Overall weight: 1700 kg; 3. Row spacing: 1050 mm; 4. Working efficiency: 1.5 acres/h; 5. This model is highly adaptable and can perform well in different soil conditions [79].		China

Table 1. *Cont.*

Name	Technical Characteristics	Appearance	Place of Production
4U-170B Potato Harvester	1. The machine adopts a double lift chain structure; 2. Overall weight: 1610 kg; 3. Row spacing: 1650 mm; 4. Working efficiency: 7~1.2 hectare/h 5. The unique lower side of the roller can compact the soil after digging and improve the rate of clearing potatoes, which is convenient for manual picking to improve operational efficiency [80].	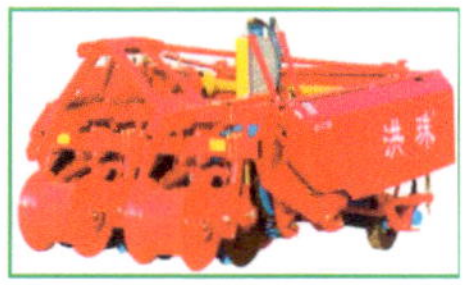	China
4UFD-1400 Potato Harvester	1. This machine is a trailed potato harvester; 2. Matching power: 44~58.8 kW; 3. Working width: 1400 mm; 4. Operating depth: 0~300 mm 5. It completes the tasks of digging up the potatoes, soil–potato separation, and screening in one go, significantly increasing the rate of harvesting [81].		China
4U-2-900 Potato Harvester	1. This machine is a small traction potato harvester; 2. Operating depth: 20~35 cm: 3. Working efficiency: 2~4 acres/h; 4. Breakage rate: $\leq$2; 5. Developed for mechanized potato harvesting in small plots in hilly and mountainous areas. It can complete digging and separating operations in one go [82].		China
TPH179 Potato Harvester	1. This machine is a crawler-type potato harvester; 2. Improves potato harvester maneuverability using tracks as the mode of movement; 3. The whole machine has higher ground clearance [82].		Japan
SHI-1500 Potato Harvester	1. The machine adopts a potato lifting chain conveyor; 2. Separation of potatoes and soil can be accomplished during transportation [83].		Korea
4UL-170C Potato Harvester	1. This machine is a trailed potato harvester; 2. Overall dimensions: 6670 $\times$ 2680 $\times$ 2890 mm (length $\times$ width $\times$ height); 3. Overall weight: 3270 kg; 4. Row spacing: 1700 mm; 5. Working efficiency: $\geq$6 acres/h; 6. Harvests 2 rows at a time [84].		China

3.2. Large Potato Combine Harvester

International research on large potato harvesters began early, and the technology has reached a mature stage. In regions with significant potato cultivation, such as Europe and North America, the use of combine harvesters is widespread. For example, large combine harvesters in countries like the United States and the Netherlands are equipped with efficient functions for digging, separating, cleaning, and packaging, allowing them to complete harvesting operations in one go. These machines are also highly automated and intelligent, capable of features like autonomous driving, smart recognition, and precise operations, which reduce the need for manual intervention and enhance operational efficiency and quality. Furthermore, advanced technologies, such as those for potato–soil separation, stem and leaf separation, and non-destructive testing, have been adopted to lower the rate of potato damage and improve harvesting quality. In contrast, research on large potato harvesters in China started relatively late but has developed rapidly in recent years. Some companies and research institutions have continuously innovated through the introduction and absorption of foreign technologies, resulting in products like the 1710A combine harvester from Zhongji Meinuo and the 4U-1600 potato combine harvester. While these harvesters meet certain demands of large domestic potato planting bases, there remains a gap in performance, reliability, and intelligence compared to advanced foreign harvesters. Therefore, further efforts in research and development and technological innovation are needed. Statistical analysis of large potato combine harvesters revealed that the majority are from abroad, As shown in Table 2.with several advanced models produced by Grimme, such as the VARITRON series and the VENTOR 4150 series. The potato harvesting machinery produced by the German company GRIMME and the Belgian company AVR can be adapted for the mechanized harvesting of other root vegetables, such as sugar beets and onions, by replacing the front digging and harvesting devices and adjusting the parameters of the separating and conveying devices. This design significantly improves the utilization rate of the equipment. However, domestic potato harvesting machinery still struggles to adapt to the harvesting of potatoes in different soil types.

Table 2. Structure and technical characteristics of large potato harvesters.

Name	Technical Characteristics	Appearance	Place of Production
VARITRON 470 Potato Combine Harvester	1. This machine is a 4-row self-propelled potato harvester; 2. Overall length: 12,810–15,600 mm; 3. Full load weight: 26,000 kg; 4. Collection box capacity: 7000 kg; 5. Discharge height: 4350 mm; 6. Translation: The 7 ton continuous operation hopper utilizes a circulating conveyor belt to ensure that the effective volume is always fully utilized. 7. The harvesting process features multiple automatic functions [77].		Germany
VENTOR 4150 Potato Combine Harvester	1. This machine is a 4-row self-propelled potato harvester; 2. Overall machine length: 15,000 mm; 3. Height: 4000 mm; 4. Full load weight: 30,000 kg; 5. Collection box capacity:15,000 kg; 6. Discharge height: 2630–4640 mm 7. Equipped with 530 horsepower and an innovative folding feature [77].		Germany
Double-L 859 Potato Combine Harvester	1. This machine is equipped with a recirculation device, which allows temporarily recirculating harvested potato chunks when changing working tools or loading trucks, thereby improving working efficiency during operations; 2. Overall machine weight: 7260 kg; 3. Power: 200 horsepower [85].		America

Table 2. *Cont.*

Name	Technical Characteristics	Appearance	Place of Production
AVRSpirit6200 Potato Combine Harvester	1. This machine is a multi-row potato harvester; 2. Collection box capacity: 6000 kg; 3. Power: 90 horsepower 4. Ensure the efficiency and continuity of the machine's operations [77].		Belgium
Dewulf r3060-3 Potato Combine Harvester	1. This machine is a self-propelled potato harvester; 2. It is equipped with a potato vine-killing device at the front, and the soil and potato separating device at the rear can maximize the separation of potatoes from the soil [83].		Belgium
Ropa keiler classic 2 Potato Combine Harvester	1. This machine is a 2-row trailed potato harvester; 2. Bunker: 7.5 t; 3. It has a fully hydraulic drive for optimal adjustment for cleaning; 4. It operates independently of the PTO speed and thus maintains all cleaning units in the optimal speed range; 5. It has reduced fuel consumption and tractor engine speed [83].		Germany
Standen T3	1. This machine features a rear-mounted lifting arm trailed structure; 2. Equipped with power above 147 kW and a 12 V DC electrical system; 3. Digging width: 2.5 m; 4. Adaptable ridge spacing: 800–900 mm; 5. Works on 3 ridges simultaneously and can either follow the loading vehicle or lay the potatoes sideways [84].		England

4. Application of DEM in Potato Harvesting Machinery Research

DEM, as a new discrete numerical analysis method, has been widely applied in the field of agricultural engineering [86]. Its main applications include the simulation of soil and mechanical interactions [87], simulation of seed and fertilizer distribution [88], design and optimization of harvesting machinery [89], study of material conveyance and separation [90], research on the wear of agricultural machinery [91], agricultural waste management, and environmental impact assessments [92]. DEM also provides an advanced method for the design, optimization, and performance evaluation of agricultural machinery, contributing to the improvement of efficiency and sustainability in agricultural production [34,93,94]. This section mainly analyzes and summarizes the application of DEM in research on potato harvesting machinery. The application of DEM in the study of potato harvesting machinery includes discrete simulation processes, such as digging, transportation, collection, and separation. By simulating the interaction between the components of the potato harvester and the soil or potato, the structure of the potato harvester can be optimized to improve the harvesting rate, reduce the mechanical damage rate and bruising rate, and provide a solution for the development of mechanized harvesting equipment for tuber crops.

4.1. Simulation and Analysis of the Excavation Process

Currently, potato digging methods are mainly divided into traditional manual and modern mechanical methods [36,95]. Mechanical digging methods use fixed digging shovels, which have high cutting resistance, energy consumption, and severe wear of the digging blade [96]. Another method is biomimetic digging, which involves designing digging shovels based on biomimetic principles; for example, shovels are designed with a contour resembling the claws of a mole cricket, which can reduce horizontal resistance and normal stress, thereby improving the soil fragmentation and soil discharge efficiency [97]. Since the potato digging process involves interactions between potatoes and machinery, as well as the soil system and machinery, research on the mechanization of potato digging will provide an effective way of reducing mechanical damage to potatoes and improving digging efficiency. However, at this stage, the discrete element simulation model for potatoes is a multi-sphere polymer model derived from the Hertz–Mindlin contact model within the discrete element simulation software (EDEM). This model assumes that the potato skin and flesh have the same mechanical properties. This simplistic approach to modeling potato harvesting has hindered the study of potato damage mechanisms during mechanized harvesting by the discrete element method. Gai et al. proposed a new type of potato double-layer flexible bonding model based on the discrete element method. This model accurately simulates the mechanical properties of potatoes and simultaneously mimics the damage characteristics of the skin and flesh. The bonding parameters of the double-layer flexible model for potato samples were refined through shear calibration tests, resulting in the optimal parameter combination. Shear and compression validation tests were conducted on the whole potato stem, and the average relative error between experimental and simulation test results was found to be e = 3.25% [98]. The potato model proposed in this study can more accurately simulate the mechanical properties of potatoes.

Mechanical damage during potato harvesting accounts for 70% of the total damage in the entire potato production system. In order to reduce the rate of mechanical damage to potatoes, Liang et al. [99] used the discrete element method to study the dynamics and kinematics of the soil–potato sorting process. They determined the equivalent Young's modulus of two potato varieties through quasi-static loading experiments. The mechanical parameters of the potatoes were obtained through experimental setups, and a contact force model was developed. The results showed that the newly developed model ($R^2 = 0.91 \pm 0.06$) outperformed the Kuwabara and Kono model (KK model) ($R^2 = 0.49 \pm 0.09$) and the Yigit–Christoforou model (YC model) ($R^2 = 0.28 \pm 0.13$). In this study, the quasi-static and dynamic material properties of two potato varieties were measured using a universal testing system and a self-developed dynamic impact information acquisition test device. The accuracy of different contact models for transient impact processes was checked by calculating Young's modulus and analyzing the impact information of the potato samples. A viscoelastic–plastic contact model designed specifically for potatoes was developed and validated. The analysis revealed significant differences between the measured and fitted values of peak forces for the KK model, indicating a possible overestimation of the dissipation factor during the unloading phase. The less satisfactory fitting performance of the YC model compared to the measured curves may be due to its reliance on purely elastic or purely plastic equations, which may not adequately capture the complex behavior of the potato. On the other hand, the newly developed model fits the measured data very well and is able to better describe the dynamic mechanical behavior of the potato independent of the impact velocity. This study not only provides a feasible experimental method and contact mechanics model for studying the dynamic mechanical properties of potatoes but also lays the foundation for the future development of discrete elemental models for soil–potato sorting systems. Li et al. [96] designed a potato-digging shovel specifically for clay environments based on the bionic principle of pangolin scales. They used the DEM to simulate the resistance encountered by the bionic shovel during the digging process, as shown in Figure 6a. The movement of soil particles on the flat shovel is chaotic, and the number of green arrows is also fewer than on the bionic shovel. Most soil particles have velocities less than 0.788 m/s, while the shovel's speed is 0.84 m/s. However, the velocities of the soil particles moving on the planar shovel are usually less than the velocity of the shovel. This clearly indicates that soil particles are deposited on the shovel, resulting in an overall reduction in velocity and potentially hindering the movement of soil particles. This result explains why the resistance of the flat shovel is greater than that of the bionic shovel. The study demonstrated that the best bionic prototype had a reduction rate of 22.26% during soil box testing and 14.19% during field testing. Ye et al. [100] simulated the distribution and digging trajectory of particles using DEM-MBD coupling. The simulation results found a critical point at a moisture content of 10% in a sandy soil environment, where the shear strength decreased with increasing moisture content. The change pattern in the simulation results was consistent with theoretical calculations, showing that the reduction in digging resistance was proportional to the cohesive strength when particles were fully sheared. The digging curves obtained from the experiment were consistent with the simulation results in terms of key parameter errors. Li et al. [97] designed a longitudinal wave potato digging shovel based on bionic principles and conducted a simulation study using the DEM to compare the digging process of the bionic longitudinal wave shovel with a regular digging shovel, as shown in Figure 6b. The bionic longitudinal wave potato digging shovel caused greater soil disturbance, leading to better soil flow. The authors also theoretically explained the resistance reduction mechanism of the bionic elements. The bionic longitudinal wave potato digging shovel reduced the resistance by 14.45%, the potato burial rate by 1.34%, and the potato breakage rate by 0.736%. In order to improve the harvesting efficiency of potato harvesters, Li et al. [95] based their work on simulations combined with experimental approaches. They improved the design of the separation and conveying mechanical equipment of potato diggers in heavy soils, focusing on the selection of a mesh-type digging shovel. This digging shovel ensures that minimal soil enters the subsequent separation device while maximizing the protection of the potatoes from damage, thereby reducing the burden on the downstream separation device.

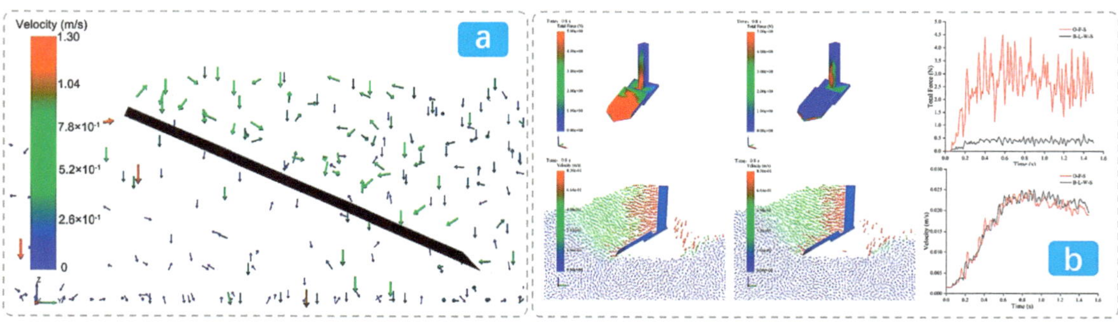

Figure 6. (**a**) Analysis of the digging resistance faced by the bionic shovel [96]. (**b**) Analysis of the digging process of the longitudinal wave potato shovel [97].

In summary, the discrete element method (DEM) can simulate the interaction between soil and potatoes, allowing for the calculation of friction and compression forces between the two. This helps to clarify the potential for these forces to cause damage to the potato skin and enables tracking of the potatoes' movement trajectories in the soil, providing a basis for optimizing digging equipment. In terms of equipment design, DEM can simulate the working conditions of different shapes and structures of digging components, comparing the forces experienced by various shovel designs during excavation and their impact on potato damage, thus facilitating improvements in component shapes. Additionally, it can simulate the effects of operational parameters such as digging speed and soil penetration depth on excavation outcomes, allowing for the adjustment of optimal parameters to enhance digging efficiency and reduce damage to potatoes, ultimately supporting efficient and high-quality potato harvesting operations.

In summary, the potato excavation process is a complex mechanical operation that involves interaction between soil and machinery, potatoes and machinery, and potatoes and soil. The DEM can provide an intuitive understanding of the potato-digging process, including the relationship between soil particles and potatoes. According to the simulation results, the mechanical structure of the potato-digging machine can be optimized, thereby increasing the yield of harvested potatoes.

4.2. Simulation of the Potato–Soil Separation Process

Simulation of the potato–soil separation process is a crucial step in the mechanized harvesting of potatoes [101,102]. Its purpose is to separate potato granules from the soil [103]. This process also has the highest probability of causing mechanical damage and bruising the potato granules [104–106]. Several major issues are currently faced during the mechanized harvesting of potatoes: how to thoroughly separate potatoes, soil, and impurities while controlling the mechanical damage rate and bruising rate of potatoes [107,108]. Existing potato harvesters result in unsatisfactory separation of potatoes and soil, high mechanical damage rate and bruise rate, and low potato harvesting rate [109]. Application of the DEM in agricultural engineering has made it possible to simulate the potato–soil separation process. A large number of simulation studies have been reported in recent years. Li et al. [110] investigated the working principle and separation performance of a potato–soil separation device, focusing on the collision characteristics between potatoes and the screen surface and bars. Key factors such as the linear velocity of the separation screen, the forward speed of the harvester, and the inclination angle of the separation screen were considered. Simulation tests were conducted based on the DEM-MBD coupling method. The optimal parameter combination for the potato–soil separation device was determined to be a linear velocity of 1.25 m/s, a forward speed of 0.83 m/s, and an inclination angle of 25°. Field harvesting experiments showed that the potato loss rate was 1.8%, the damage rate was 1.2%, the impurity rate was 1.9%, the skin breakage rate was 2.1%, and the yield was 0.15–0.21 hectares/h. This research provides valuable theoretical references for simulating potato–soil separation in potato harvesters and optimizing the parameters of these devices.

To solve the problems of poor potato–soil separation and high potato damage rate, Chen et al. [101] designed a rotary vibrating potato separation device and revealed its operating process using DEM-MBD coupling, as shown in Figure 7. The optimal working parameters were determined for the separation process as a vibration point position of 646.5 mm, conveyance speed of the potato–soil separation lifting chain of 1.08 m/s, rotor amplitude of 26.7 mm, and rotor vibration frequency of 5.9 Hz. Field verification experiments were conducted based on these optimal parameters, with potato–soil separation efficiency and potato damage rate of 97.8% and 1.16%, respectively, in the rotor vibration potato separation device. The field test results were consistent with the simulation results, confirming the accuracy of the simulation model. The theoretical DEM provides the references for simulating the rotary vibrating potato–soil separation process and optimizing device parameters. To address the issue of low yield rates of intact potatoes, Du et al. [111] developed a potato–soil separation device combining left and right rotation. They analyzed a DEM model for the potato separation roller, as shown in Figure 8. The force threshold for skin damage to potatoes was demonstrated to be between 190 and 195 N. When the device's incline angle was set at 6°, the separation roller's rotational speed was 100 r/min, the center distance between the separation rollers was 79 mm, the damage rate was 1.25%, the yield rate of intact potatoes was 99.01%, and the skin damage rate was 1.58%. In contrast, with an incline angle of 8°, a separation roller speed was 80 r/min, the center distance was the same at 79 mm, the damage rate increased to 1.43%, the yield rate of intact potatoes dropped to 98.64%, and the skin damage rate was 1.77%. To understand the dynamic relationship between damage and soil separation during potato harvesting using potato harvesters, Li et al. [112] investigated the operating mechanism of the rod-type separator in a small self-propelled potato combine harvester and the performance of a tuber–soil separation. The coupling model was analyzed using DEM-MBD and single-factor simulation experiments. The field tests were validated and revealed an error of 3.81% between the simulation model and field harvesting. However, Yan et al. [113] conducted a simulation study on the process of crushing and separating potato–soil chunks based on the DEM. The study modeled the mechanism of breaking and separating soil clods. Primarily, the three-dimensional model of the separation device was simplified in SolidWorks and imported into the discrete element analysis software (EDEM 2022). The soil breaking and separation simulation is shown in Figure 9a. As the blade rotates above the potato–soil aggregates, the soil is beaten due to collisions. Subsequently, the velocity difference between the blade and the lifting chain drives the potato–soil aggregates to move along the surface of the lifting chain towards the end, continuously breaking and separating the adhered soil particles. Potato–soil breaking and separation were studied in areas with sandy-loam soils.

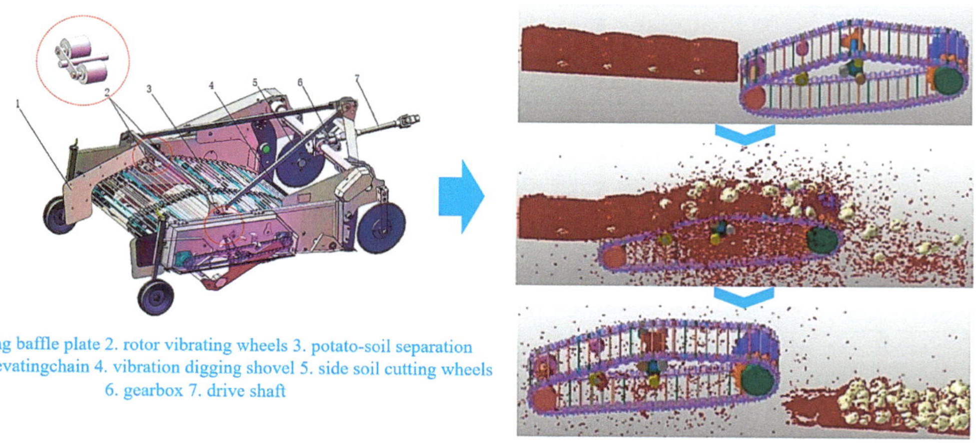

1. collecting baffle plate 2. rotor vibrating wheels 3. potato-soil separation conveying elevatingchain 4. vibration digging shovel 5. side soil cutting wheels 6. gearbox 7. drive shaft

Figure 7. Simulation of the rotor vibration of the potato–soil separation process using a potato separation device [101].

In summary, the potato–soil separation process is random. Particles in different positions are subjected to slight variations in the soil–potato mixture and the conveying chain; this is correlated with the quality of the soil–potato mixture at different harvest speeds. When the angle of the separation belt increases, fine soil particles can be screened out from the gaps more quickly, improving the soil removal effect. The faster the linear velocity of the separation belt, the quicker the soil is screened; however, this results in shorter soil screening times on the separation device, leading to poor soil cleaning efficiency. The energy required to break the soil clods mainly depends on the shape and characteristics of the clods, such as moisture content. The breakage energy generated by collisions is related not only to the mass of the separation belt and the soil body but also to the initial velocity before the collision. The vibration and breakage of the soil clods are determined by a combination of different linear speeds of the separation belt and the total mass of the soil–potato mixture on the separation mechanism. The increase in the linear velocity of the rod affects its vibration and promotes the breaking of soil clods. To ensure that the potatoes suffer as little damage as possible during the first separation process in the combine harvester, the tubers can be protected by the soil. This is because, after the first potato–soil separation, this portion of the soil will pass through a series of mechanisms of secondary soil removal, including the lifting mechanism. This also ensures the soil removal capability of the combine harvester. The simulation of the potato–soil separation process was reported based on the DEM, which was used to analyze the separation mechanism of potatoes. The mechanical structure was effectively optimized for the potato harvester, and the best operating parameters were determined. This led to an improvement in the efficiency of potato–soil separation, a reduction in the rates of mechanical damage and bruising of potatoes during the separation process, and an increase in the number of marketable potatoes.

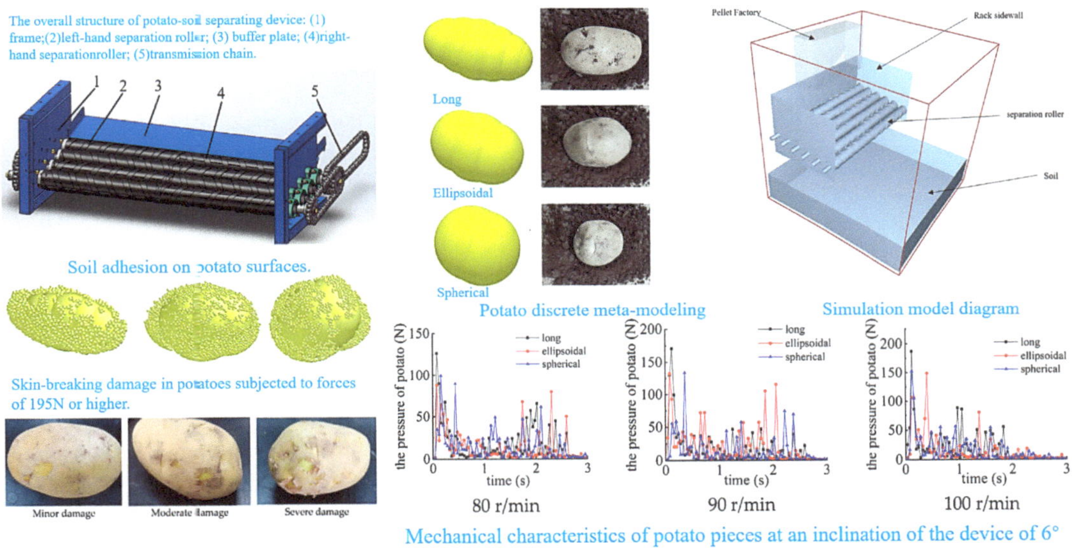

Figure 8. Left–right rotating combined potato separator and discrete element analysis [111].

4.3. Simulation of the Transportation Process

The potato transportation process can be divided into three stages. The first stage involves the conveying of potatoes into the harvester, where the potatoes are transported through the lifting conveyor [114], soil separation device, and other mechanisms to the screening unit. The second stage consists of the screening and conveyance of potatoes to the collection unit. The third stage involves the transportation of potatoes to the storage facility via a transport vehicle [115]. Compared to the digging and soil separation processes, the transportation process inflicts lower mechanical damage and bruising on the potatoes. However, the friction between potatoes results in a higher bruising rate; therefore, the related simulation analysis of this process is an important means of improving transportation efficiency and reducing damage during transportation.

Xie et al. [116] utilized the EDEM software to simulate the potato–soil conveying and separation mechanism and obtained the optimal combination of working parameters. The linear velocity of the conveyor chain was set at 2.3 m/s, with an inclination angle of 18 degrees and an amplitude of 8 mm. Field experiments were conducted to validate the results. The simulation results demonstrated a significant improvement in performance. Lastly, by determining the optimal working parameters and implementing structural innovations, optimization of the potato soil transportation and separation mechanism was achieved. Li et al. [95] used numerical simulation to optimize the potato harvester separation and conveyance machinery. They selected a grille-type digging shovel, which ensures that the amount of soil entering the subsequent separation device is minimized without digging up potatoes and reduces the load on the subsequent separation devices Zhang et al. [117] conducted research on the soil-lifting device of potatoes based on the TRIZ theory. They used the DEM-MBD coupling model to simulate the operation process of the device, obtaining the optimal operating parameters. Sun et al. [118] simulated the transport process of non-spherical particles in a screw mechanism based on DEM, as shown in Figure 9b. The DEM simulation shows that the particle shape has an impact on the behavior of the flow of the particles and the wear of the conveyor. Subsequently, quantitative assessments of mass flow rate and power consumption were obtained to study the influence of particle sphericity under different operating parameters. The particle collision frequency and collision energy consumption were obtained to investigate possible particle breakage between particles and the screw blade. A comparison of particle–particle collisions and particle–wall collisions showed that particles with larger shape indexes are more likely to be damaged in particle–wall collisions. This research broadens the methods of potato transportation. To reduce the rate of damage to potatoes during the transport process, McRae et al. [119] studied a potato conveyor equipped with an automatic discharge height control device. This control device consists of a sensor ball installed at the end of the conveyor and an air switch that activates an electromagnetic valve to operate a hydraulic lift plunger. The conveyor is divided into two sections, with the distal section capable of swinging laterally when it impacts the rear tailgate of the trailer. Geng et al. [115] studied the influence of initial direction on the trajectory of potatoes under different initial velocities using quadratic coefficients. They found that the influence decreases as the initial velocity increases, and there is a relatively stable state when the initial velocity exceeds 2.0 m/s. Torres-Serra et al. [120] utilized DEM simulations to understand the influence of particle-scale effects on flowability. They investigated the impact of particle size, shape, and hygroscopicity on various bulk-handling issues. The study compared the flow propagation of dry monodisperse spherical particles with particle systems containing size dispersion, prolate spheroid particles, and liquid content in capillary states. This research demonstrates that DEM can be an effective tool for characterizing complex particle flows and improving bulk-handling technologies. Due to the discrete and nonlinear characteristics of materials, the macroscopic characterization of materials depends on the changes in the scale of motion within them. Ren [121] established a discrete element model to study the dynamic characteristics of mesoscale-oriented material flow, obtaining the mechanism of action of various operating parameters on the material movement state. By analyzing mesoscale data, it was found that translation speed and rotation speed are the main factors affecting long-distance movement and material velocity, respectively. Shi et al. [122] also conducted a comprehensive survey and analysis of DEMs in powder transportation systems in the field of powder transportation.

4.4. Simulation of the Collection Process

The collection process is usually designed for the convenience of transport vehicles. There are three main technologies. The first one is the potato collection technology, which includes the adaptive height potato collection box technology and the automatic lifting arm conveyor loading potato collection technology. The adaptive height potato collection box technology can automatically adjust the fall height of potatoes, reducing the damage caused by the collision of potato tubers with the box body when they completely enter the separated hopper due to excessive fall height. The second one is the automatic lifting arm conveyor loading potato collection technology. This technology equips the end of the lifting arm with distance sensors such as ultrasonic sensors. Adjusting the hydraulic cylinder changes the tilt angle of the end of the lifting arm and the distance between the lifting arm and the transport vehicle for dropping potatoes. The speed of the lifting arm conveyor can be adjusted according to the quantity of potatoes. This technology has been adopted for potato harvesting models in multiple countries. The third one is human–machine interaction technology, which requires real-time monitoring of the machine's walking status, digging status, conveying and separation status, impurity removal effect, and hopper capacity. This technology integrates mechanical, electrical, hydraulic, and instrumentation technologies and is equipped with a cab terminal human–machine interaction operating system based on PDA/GPS/GPRS/GIS technologies. Dai et al. [123] conducted a simulation study on the solid–liquid two-phase flow during the hydraulic collection process based on DEM-CFD and then confirmed the correctness of the numerical simulation through experiments. Zhang et al. [124] performed a numerical simulation of the sulfurized mineral collection process using DEM-CFD coupling technology. They found that when the roll-up reaches the effective range of the collection hard tube's negative pressure suction, the particles are subjected to the coupling effect of the swirling flow and the negative pressure suction flow and are ultimately transported into the interior of the collection hard tube. Hou et al. [125] established a discrete element model for castor bean three-chamber separation and comminution to analyze the damage to castor beans during mechanical harvesting. The elastic modulus was measured using a TMS pro food physical properties analyzer. The contact parameters were calibrated through the stacking angle. The results showed that the maximum breaking force was 52.65 N. The Box–Behnken response surface design generated the optimal bonding parameters: the normal stiffness per unit area was 9.47×10^7 N/m^3, the shear stiffness per unit area was 4.59×10^7 N/m^3, the normal strength was 7.92×10^3 Pa, and the shear strength was 1.13×10^4 Pa. This study provides support for discrete element simulation parameters to analyze the mechanism of crop damage during mechanized collection processes.

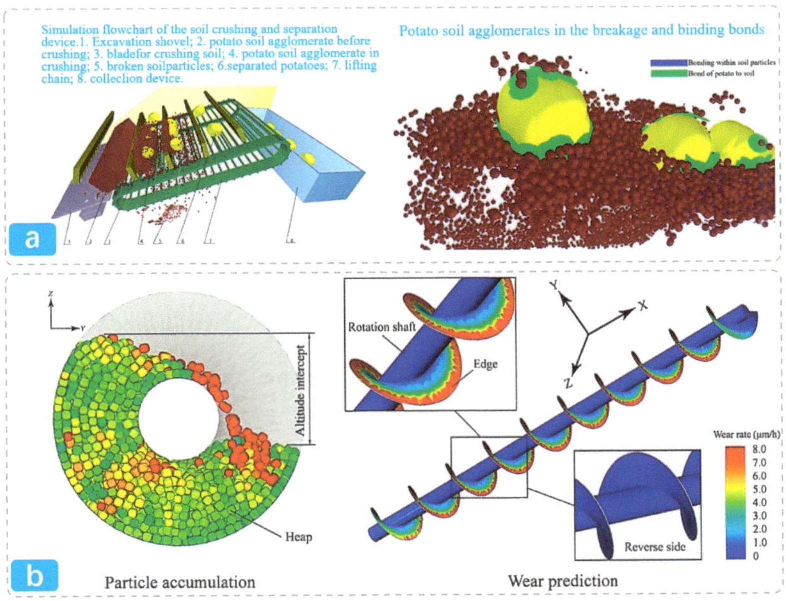

Figure 9. (**a**) Discrete meta-analysis of the potato–soil separation process [113]. (**b**) Discrete meta-analysis of the process of conveying non-spherical particles using the screw conveyor mechanism [118].

In summary, due to the relatively lower damage to and bruising of potatoes during the collection process, researchers have paid less attention to the study of potato collection and focused more on the digging and soil separation processes during potato harvesting. To reduce secondary damage to potatoes during the collection and transport processes, it is recommended that research be conducted on the movement, collision, and friction between potatoes.

5. Summary and Outlook

With the development of computer software and hardware, the DEM has been widely applied in the field of agricultural engineering, particularly in the design of agricultural machinery. However, agricultural machinery design is a knowledge-intensive process. At present, researchers use numerical simulation technology, with numerical analysis and calculation software such as EDEM, FLUENT, and ANSYS, to conduct finite element analysis and discrete element analysis of the structure and operation of agricultural machinery. This approach reduces the design cost and cycle of agricultural machinery, maximizes the design efficiency, and enriches the design concepts. Although the DEM has demonstrated superior design performance in the design of agricultural machinery, the discrete element method still faces some issues and challenges during the actual design process.

- Limited model accuracy: Although the DEM has certain advantages when it comes to simulating interactions between granular materials and objects, the physical properties of materials such as potatoes and soil are very complex in practical applications, making precise modeling challenging. For example, potatoes have irregular shapes and varying sizes, while the structure and mechanical properties of soil can differ significantly across different regions and conditions. Current models often struggle to fully and accurately represent these real-world scenarios, which may lead to discrepancies between simulation results and actual outcomes.
- Difficulties in parameter calibration: The accuracy of discrete element models largely depends on the precision of model parameters, such as the friction coefficient between particles, elastic modulus, and restitution coefficient. However, accurately determining and calibrating these parameters can be challenging, requiring extensive experimental and empirical data. Moreover, the parameter values may vary under different materials and conditions, which increases the uncertainty and error in the model.
- High computational cost: The computational volume of the discrete element method increases dramatically as the complexity of the simulation increases, for example, considering higher numbers of particles, more complex mechanical structures, and motion processes. The requirements for computer hardware and computation time also increase dramatically, limiting its application in large-scale, complex systems and rapid optimization of the design in practical engineering to a certain extent.
- Insufficient treatment of multi-physical field coupling problems: The potato harvest process not only involves mechanical interactions between particles, but it may also involve multi-physical field coupling problems such as soil moisture migration and heat transfer. At present, the discrete element method has some limitations in dealing with these multi-physical field coupling problems, and it is difficult to fully and accurately simulate various physical phenomena and interactions that occur during the actual harvesting process.
- Lack of systematic research: Currently, the application of the discrete element method (DEM) in potato harvesting machinery research is mostly focused on specific components or processes, such as optimization of the digging shovel or separation screen, and systematic research and comprehensive optimization of the entire harvesting machinery system is lacking. However, potato harvesting machinery is a complex

system where various components are interrelated and influence each other. To achieve optimal harvesting results, a holistic and coordinated optimization approach is necessary.

The application of the discrete element method (DEM) in potato harvesting machinery offers significant economic and environmental benefits to the potato industry. By optimizing machine structure and operational parameters through DEM simulations, harvesting operations can become more efficient, reducing work time, labor costs, and fuel consumption. Accurate simulations can help optimize key processes, such as potato–soil separation, minimizing tuber damage and losses, thereby improving harvest quality and increasing farmers' actual income. Utilizing DEM for virtual testing and optimization can reduce the number of actual field trials, shorten the research and development cycle, and lower R&D costs and risks. Precise control of digging depth and force can minimize damage to soil structure and reduce soil erosion and compaction. The improved efficiency of optimized machinery leads to lower energy consumption, which in turn reduces exhaust emissions and other environmental pollutants. In general, DEM has been widely and successfully applied in the field of agricultural engineering, but the research aspect of DEM modeling of soil particles still faces great challenges and has great research prospects, such as modeling soil particles and calibration of particle parameters, simulation of particle loss, and calibration of parameters between soil particles, which still need to be studied in depth.

Author Contributions: Y.Y.: Data curation, Methodology, Writing—original draft; Q.Z.: Formal analysis, Writing—original draft; B.D.: Writing—original draft; J.L.: Funding acquisition, Supervision, Writing—review and editing. All authors have read and agreed to the published version of the manuscript.

Funding: This work was supported by the QinChuangYuan Innovation and Entrepreneurship Talent Project of Shaanxi Province (QCYRCXM-2023-111).

Data Availability Statement: No new data were created or analyzed in this study. Data sharing is not applicable to this article.

Conflicts of Interest: The authors declare no conflicts of interest.

References

1. Han, X.; Yang, R.; Zhang, L.; Wei, Q.; Zhang, Y.; Wang, Y.; Shi, Y. A Review of Potato Salt Tolerance. *Int. J. Mol. Sci.* **2023**, *24*, 10726. [CrossRef] [PubMed]
2. Lindqvist-Kreuze, H.; Bonierbale, M.; Gruneberg, W.J.; Mendes, T.; De Boeck, B.; Campos, H. Potato and sweetpotato breeding at the international potato center: Approaches, outcomes and the way forward. *Theor. Appl. Genet.* **2024**, *137*, 12. [CrossRef] [PubMed]
3. Boivin, M.; Bourdeau, N.; Barnabe, S.; Desgagne-Penix, I. Sprout Suppressive Molecules Effective on Potato (*Solanum tuberosum*) Tubers during Storage: A Review. *Am. J. Potato Res.* **2020**, *97*, 451–463. [CrossRef]
4. Jennings, S.A.; Koehler, A.-K.; Nicklin, K.J.; Deva, C.; Sait, S.M.; Challinor, A.J. Global Potato Yields Increase Under Climate Change with Adaptation and CO_2 Fertilisation. *Front. Sustain. Food Syst.* **2020**, *4*, 519324. [CrossRef]
5. Zhang, H.; Xu, F.; Wu, Y.; Hu, H.; Dai, X. Progress of potato staple food research and industry development in China. *J. Integr. Agric.* **2017**, *16*, 2924–2932. [CrossRef]
6. Dupuis, B.; Nkuriyingoma, P.; Ballmer, T. Economic Impact of Potato Virus Y (PVY) in Europe. *Potato Res.* **2024**, *67*, 55–72. [CrossRef]
7. Raymundo, R.; Asseng, S.; Robertson, R.; Petsakos, A.; Hoogenboom, G.; Quiroz, R.; Hareau, G.; Wolf, J. Climate change impact on global potato production. *Eur. J. Agron.* **2018**, *100*, 87–98. [CrossRef]
8. Ojeda, J.J.; Rezaei, E.E.; Kamali, B.; McPhee, J.J.; Meinke, H.; Siebert, S.; Webb, M.A.; Ara, I.; Mulcahy, F.; Ewert, F. Impact of crop management and environment on the spatio-temporal variance of potato yield at regional scale. *Field Crops Res.* **2021**, *270*, 108213. [CrossRef]
9. Adekanmbi, T.; Wang, X.; Basheer, S.; Liu, S.; Yang, A.; Cheng, H. Climate change impacts on global potato yields: A review. *Environ. Res. Clim.* **2023**, 3, 012001. [CrossRef]

10. Johnson, C.M.; Auat Cheein, F. Machinery for potato harvesting: A state-of-the-art review. *Front. Plant Sci.* **2023**, *14*, 1156734. [CrossRef]
11. Wang, L.; Liu, F.; Wang, Q.; Zhou, J.; Fan, X.; Li, J.; Zhao, X.; Xie, S. Design of a Spring-Finger Potato Picker and an Experimental Study of Its Picking Performance. *Agriculture* **2023**, *13*, 945. [CrossRef]
12. Wei, Z.; Li, H.; Sun, C.; Su, G.; Liu, W.; Li, X. Experiments and Analysis of a Conveying Device for Soil Separation and Clod-Crushing for a Potato Harvester. *Appl. Eng. Agric.* **2019**, *35*, 987–996. [CrossRef]
13. Wang, H.; Zhao, W.; Sun, W.; Liu, X.; Shi, R.; Zhang, H.; Chen, P.; Gao, K. The Design and Experimentation of a Wheeled-Chassis Potato Combine Harvester with Integrated Bagging and Ton Bag-Lifting Systems. *Agriculture* **2024**, *14*, 1461. [CrossRef]
14. Lu, K.; Xie, S.; Gai, X.; Ji, X. Design and Experiment of Toggle Lever-Type Potato Picker. *Agriculture* **2024**, *14*, 826. [CrossRef]
15. Chen, Y.; Wang, Z.; Zhang, H.; Liu, X.; Li, H.; Sun, W.; Li, H. Investigation of the Traveling Performance of the Tracked Chassis of a Potato Combine Harvester in Hilly and Mountainous Areas. *Agriculture* **2024**, *14*, 1625. [CrossRef]
16. Chen, K.J.; Wood, J.D.; Mohammed, I.K.; Echendu, S.; Jones, D.; Northam, K.; Charalambides, M.N. Mechanical Characterisation and modelling of the rolling process of potato-based dough. *J. Food Eng.* **2020**, *278*, 109943. [CrossRef]
17. Lenartowicz, T.; Piepho, H.P.; Przystalski, M. Stability Analysis of Tuber Yield and Starch Yield in Mid-Late and Late Maturing Starch Cultivars of Potato (*Solanum tuberosum*). *Potato Res.* **2019**, *63*, 179–197. [CrossRef]
18. Jaiswal, A.K.; Singh, B.; Mehta, A.; Lal, M. Post-Harvest Losses in Potatoes from Farm to Fork. *Potato Res.* **2023**, *66*, 51–66. [CrossRef]
19. Ping, Z.; Guo, R.; Jin, T.; Zhang, G.; Ning, X. Collision simulation of potato tubers for mechanized harvesting. *J. Food Process Eng.* **2023**, *46*, e14278. [CrossRef]
20. Visse-Mansiaux, M.; Tallant, M.; Brostaux, Y.; Delaplace, P.; Vanderschuren, H.; Dupuis, B. Assessment of pre- and post-harvest anti-sprouting treatments to replace CIPC for potato storage. *Postharvest Biol. Technol.* **2021**, *178*, 111540. [CrossRef]
21. Zhang, Q.; Zhang, Z.; Li, C.; Xu, R.; Yang, D.; Sun, L. Van der Waals materials-based floating gate memory for neuromorphic computing. *Chip* **2023**, *2*, 100059. [CrossRef]
22. Deng, X.; Kang, N.; Zhang, Z. Carbon-Based Cryoelectronics: Graphene and Carbon Nanotube. *Chip* **2023**, *22*, 100064. [CrossRef]
23. Kim, J.; Jeon, B. "Digital design" and three flows of ideas. *J. Asian Archit. Build. Eng.* **2022**, *21*, 1891–1907. [CrossRef]
24. Han, J.S.; Kim, G. Virtual Constructions Design using 3ds Max. *J. Digit. Converg.* **2013**, *11*, 273–278.
25. Li, C.; Luo, Y.; Li, W.; Yang, B.; Sun, C.; Ma, W.; Ma, Z.; Wei, Y.; Li, X.; Yang, J. The On-chip Thermoelectric Cooler: Advances, Applications and Challenges. *Chip* **2024**, *3*, 100096. [CrossRef]
26. Gazquez, J.A.; Castellano, N.N.; Manzano-Agugliaro, F. Intelligent low cost telecontrol system for agricultural vehicles in harmful environments. *J. Clean. Prod.* **2016**, *113*, 204–215. [CrossRef]
27. Guo, Q.; Xia, H. A review of the Discrete Element Method/Modelling (DEM) in agricultural engineering. *J. Agric. Eng.* **2023**, 54. [CrossRef]
28. Xin, C.; Yin, Y.; Song, B.; Fan, Z.; Song, Y.; Pan, F. Machine Learning-Accelerated Discovery of Novel 2D Ferromagnetic Materials with Strong Magnetization. *Chip* **2023**, *2*, 100071. [CrossRef]
29. Yang, K.-S.; He, C.; Fang, J.; Cui, X.; Sun, H.; Yang, Y.; Zuo, C. Advanced RF filters for wireless communications. *Chip* **2023**, *2*, 100058. [CrossRef]
30. Hao, Z.; Zou, K.; Meng, Y.; Yan, J.-Y.; Li, F.; Huo, Y.; Jin, C.-Y.; Liu, F.; Descamps, T.; Iovan, A.; et al. High-performance eight-channel system with fractal superconducting nanowire single-photon detectors. *Chip* **2024**, *3*, 100087. [CrossRef]
31. Siegmann, E.; Enzinger, S.; Toson, P.; Doshi, P.; Khinast, J.; Jajcevic, D. Massively speeding up DEM simulations of continuous processes using a DEM extrapolation. *Powder Technol.* **2021**, *390*, 442–455. [CrossRef]
32. Qi, L.; Chen, Y.; Sadek, M. Simulations of soil flow properties using the discrete element method (DEM). *Comput. Electron. Agric.* **2019**, *157*, 254–260. [CrossRef]
33. Gong, H.; Chen, Y.; Wu, S.; Tang, Z.; Liu, C.; Wang, Z.; Fu, D.; Zhou, Y.; Qi, L. Simulation of canola seedling emergence dynamics under different soil compaction levels using the discrete element method (DEM). *Soil Tillage Res.* **2022**, *223*, 105461. [CrossRef]
34. Zhao, H.; Huang, Y.; Liu, Z.; Liu, W.; Zheng, Z. Applications of Discrete Element Method in the Research of Agricultural Machinery: A Review. *Agriculture* **2021**, *11*, 425. [CrossRef]
35. Du, B.; Zhao, C.; Dong, G.; Bi, J. FEM-DEM coupling analysis for solid granule medium forming new technology. *J. Mater. Process. Technol.* **2017**, *249*, 108–117. [CrossRef]
36. Horabik, J.; Molenda, M. Parameters and contact models for DEM simulations of agricultural granular materials: A review. *Biosyst. Eng.* **2016**, *147*, 206–225. [CrossRef]
37. Huang, L.; He, R.; Yang, Z.; Tan, P.; Chen, W.; Li, X.; Cao, A. Exploring hydraulic fracture behavior in glutenite formation with strong heterogeneity and variable lithology based on DEM simulation. *Eng. Fract. Mech.* **2023**, *278*, 109020. [CrossRef]
38. Wang, P.; Yin, Z.-Y.; Hicher, P.-Y.; Cui, Y.-J. Micro-mechanical analysis of one-dimensional compression of clay with DEM. *Int. J. Numer. Anal. Methods Geomech.* **2023**, *47*, 2706–2724. [CrossRef]

39. Zhong, W.; Yu, A.; Liu, X.; Tong, Z.; Zhang, H. DEM/CFD-DEM Modelling of Non-spherical Particulate Systems: Theoretical Developments and Applications. *Powder Technol.* **2016**, *302*, 108–152. [CrossRef]
40. Cundall, P.A. The Measurement and Analysis of Accelerations in Rock Slopes. Ph.D. Thesis, Imperial College, London, UK, 1971.
41. Cundall, P.A. A computer model for simulating progressive, large-scale movements in blocky rock systems. *Proc. Int. Symp. Rock Mech.* **1971**, *8*, 129–136.
42. Martinez Morillo, G.C ; Bandeira, A.A. Discrete Element Method applied to the simulation of the stress state in granular materials. *Soil Res.* **2019**, *57*, 85–100. [CrossRef]
43. Meng, J.; Huang, J.; Li, H.; Laue, J.; Li, K. A static discrete element method with discontinuous deformation analysis. *Int. J. Numer. Methods Eng.* **2019**, *120*, 918–935. [CrossRef]
44. Fleissner, F.; Gaugele, T.; Eberhard, P. Applications of the discrete element method in mechanical engineering. *Multibody Syst. Dyn.* **2007**, *18*, 81–94. [CrossRef]
45. Tsunazawa, Y.; Soma, N.; Iijima, M.; Tatami, J.; Mori, T.; Sakai, M. Validation study on a coarse-grained DEM-CFD simulation in a bead mill. *Powder Technol.* **2024**, *440*, 119743. [CrossRef]
46. Napolitano, E.; Di Renzo, A.; Di Maio, F.P. Coarse-grain DEM-CFD modelling of dense particle flow in gas-solid cyclone. *Sep. Purif. Technol.* **2022**, *287*, 120591. [CrossRef]
47. Mudarisov, S.; Farkhutdinov, I.; Khamaletdinov, R.; Khasanov, E.; Mukhametdinov, A. Evaluation of the significance of the contact model particle parameters in the modelling of wet soils by the discrete element method. *Soil Tillage Res.* **2022**, *215*, 105228. [CrossRef]
48. Li, J.; Qiao, T.; Ji, S. General polygon mesh discrete element method for arbitrarily shaped particles and complex structures based on an energy-conserving contact model. *Acta Mech. Sin.* **2023**, *39*, 722245. [CrossRef]
49. Xue, B.; Que, Y.; Pei, J.; Ma, X.; Wang, D.; Yuan, Y.; Zhang, H. A state-of-the-art review of discrete element method for asphalt mixtures: Model generation methods, contact constitutive models and application directions. *Constr. Build. Mater.* **2024**, *414*, 134842. [CrossRef]
50. Tykhoniuk, R.; Tomas, J.; Luding, S.; Kappl, M.; Heim, L.; Butt, H.-J. Ultrafine cohesive powders: From interparticle contacts to continuum behaviour. *Chem. Eng. Sci.* **2007**, *62*, 2843–2864. [CrossRef]
51. He, H.; Zheng, J.; Chen, Y.; Ning, Y. Physics engine based simulation of shear behavior of granular soils using hard and soft contact models. *J. Comput. Sci.* **2021**, *56*, 101504. [CrossRef]
52. Wong, C.P.Y.; Coop, M.R. The contact mechanics of a UK railway ballast. *Géotechnique* **2023**, *74*, 1700–1712. [CrossRef]
53. Jamshidi, H.; Ahmadian, H. A modified rough interface model considering shear and normal elastic deformation couplings. *Int. J. Solids Struct.* **2020**, *203*, 57–72. [CrossRef]
54. Gao, W.; Feng, Y.T.; Wang, C. A coupled isogeometric/multi-sphere discrete element approach for the contact interaction between irregular particles and structures. *Powder Technol.* **2023**, *430*, 118971. [CrossRef]
55. Chen, G.; Schott, D.L.; Lodewijks, G. Sensitivity analysis of DEM prediction for sliding wear by single iron ore particle. *Eng. Comput.* **2017**, *34*, 2031–2053. [CrossRef]
56. Li, X.; Du, Y.; Liu, L.; Mao, E.; Yang, F.; Wu, J.; Wang, L. Research on the constitutive model of low-damage corn threshing based on DEM. *Comput. Electron. Agric.* **2022**, *194*, 106722. [CrossRef]
57. Ghodki, B.M.; Kumar, K.; Goswami, T.K. Modeling breakage and motion of black pepper seeds in cryogenic mill. *Adv. Powder Technol.* **2018**, *29*, 1055–1071. [CrossRef]
58. Shi, Y.; Jiang, Y.; Wang, X.; Thuy, N.T.D.; Yu, H. A mechanical model of single wheat straw with failure characteristics based on discrete element method. *Biosyst. Eng.* **2023**, *230*, 1–15. [CrossRef]
59. Aleshin, V.; Van Den Abeele, K. Preisach analysis of the Hertz-Mindlin system. *J. Mech. Phys. Solids* **2009**, *57*, 657–672. [CrossRef]
60. Ahmad, F.; Qiu, B.; Ding, Q.; Ding, W.; Khan, Z.M.; Shoaib, M.; Chandio, F.A.; Rahim, A.; Khaliq, A. Discrete element method simulation of disc type furrow openers in paddy soil. *Int. J. Agric. Biol. Eng.* **2020**, *13*, 103–110. [CrossRef]
61. Wang, Z.; Li, J.; Yu, T.; Cheng, Q.; Li, F.; Li, Z. Sampling dynamic analysis and discrete element simulation on a deep layer regolith drill for extraterrestrial celestial bodies. *Acta Astronaut.* **2024**, *225*, 141–157. [CrossRef]
62. Shi, J.; Shan, Z.; Yang, H. Research on the macro- and meso-mechanical properties of frozen sand mold based on Hertz-Mindlin with Bonding model. *Particuology* **2024**, *88*, 176–191. [CrossRef]
63. Liu, X.; Gui, N.; Yang, X.; Tu, J.; Jiang, S. A DEM-embedded finite element method for simulation of the transient heat conduction process in the pebble bed. *Powder Technol.* **2021**, *377*, 607–620. [CrossRef]
64. Gui, N.; Yan, J.; Xu, W.; Ge, L.; Wu, D.; Ji, Z.; Gao, J.; Jiang, S.; Yang, X. DEM simulation and analysis of particle mixing and heat conduction in a rotating drum. *Chem. Eng. Sci.* **2013**, *97*, 225–234. [CrossRef]
65. Oschmann, T.; Kruggel-Emden, H. A novel method for the calculation of particle heat conduction and resolved 3D wall heat transfer for the CFD/DEM approach. *Powder Technol.* **2018**, *338*, 289–303. [CrossRef]
66. Wu, Z.; Zhou, Y.; Fan, L. A fracture aperture dependent thermal-cohesive coupled model for modelling thermal conduction in fractured rock mass. *Comput. Geotech.* **2019**, *114*, 103108. [CrossRef]

67. Mukherjee, R.; Sansare, S.; Nagarajan, V.; Chaudhuri, B. Discrete Element Modeling (DEM) based investigation of tribocharging in the pharmaceutical powders during hopper discharge. *Int. J. Pharm.* **2021**, *596*, 120284. [CrossRef]
68. Zhang, T.; Lu, Y. A method to deal with constant wall flux boundary condition in a fluidized bed by CFD-DEM. *Chem. Eng. J.* **2021**, *406*, 126880. [CrossRef]
69. Jiang, S.; Wan, H.; Cao, G.; Tan, Y.; Liu, J.; Yang, S.; Xiao, X.; Tong, Z.; Yu, Q.B. Optimization of the Stirring Blade Structure of the Pumping Unit Based on the Improvement of Concrete Suction Efficiency. *Adv. Civ. Eng.* **2022**, *2022*, 1255348. [CrossRef]
70. Ouyang, Y.; Yang, Q.; Chen, X. Bonded-Particle Model with Nonlinear Elastic Tensile Stiffness for Rock-Like Materials. *Appl. Sci.* **2017**, *7*, 686. [CrossRef]
71. Caserta, A.J.; Navarro, H.A.; Cabezas-Gómez, L. Damping coefficient and contact duration relations for continuous nonlinear spring-dashpot contact model in DEM. *Powder Technol.* **2016**, *302*, 462–479. [CrossRef]
72. Wojtkowski, M.B.; Pecen, J.; Horabik, J.; Molenda, M. Rapeseed impact against a flat surface: Physical testing and DEM simulation with two contact models. *Powder Technol.* **2010**, *198*, 61–68. [CrossRef]
73. Burns, S.J.; Hanley, K.J. Establishing stable time-steps for DEM simulations of non-collinear planar collisions with linear contact laws. *Int. J. Numer. Methods Eng.* **2017**, *110*, 186–200. [CrossRef]
74. Wang, C.; Liu, G.; Zhai, Z.; Guo, X.Y.; Wu, Y. CFD-DEM study on the interaction between triboelectric charging and fluidization of particles in gas-solid fluidized beds. *Powder Technol.* **2023**, *419*, 118340. [CrossRef]
75. Pei, C.; Wu, C.-Y.; England, D.; Byard, S.J.; Berchtold, H.; Adams, M.J. Numerical analysis of contact electrification using DEM-CFD. *Powder Technol.* **2013**, *248*, 34–43. [CrossRef]
76. Rasera, J.N.; Cruise, R.D.; Cilliers, J.J.; Lamamy, J.; Hadler, K. Modelling the tribocharging process in 2D and 3D. *Powder Technol.* **2022**, *407*, 117607. [CrossRef]
77. Yang, X. Design and experimental study of potato harvester in hilly and mountainous areas. *INMATEH Agric. Eng.* **2021**, *64*, 151–158. [CrossRef]
78. Mino, Z. 1700/1710 potato combine harvester. *Mod. Agric. Machinery* **2012**, *40*. Available online: https://www.nongjitong.com/product/maen_1710b_potato_harvester.html (accessed on 23 January 2025).
79. No.8 2014 Model Guide. *Farm Machinery* **2014**, 63–70. Available online: https://kns.cnki.net/kcms2/article/abstract?v=OJyTKzW6Fepy5vUkJVPEWtlQecCtHJ-H7mS1OxQu6L29hKanbBbDnWeTOMYGjn8i_WFd56B7n-0aooK4QdYZ3ADXZXBmzTDXZYrmFMcEOZPUfsKmK8ieR6eYbnH1QNnFHkP899WFWOz_Tg_2ITXH_y-zYJ4IiwQqi0LQ-0pZKi7_uL2SJ9_EN_wNPqOVLeuc&uniplatform=NZKPT&language=CHS (accessed on 23 January 2025).
80. Hongzhu 4U-170 potato harvester. *Farmers Get Rich* **2015**, *27*. Available online: https://www.nongjitong.com/product/qdhongzhu_4u-170_potato_harvester.html (accessed on 23 January 2025).
81. Wei, H.; Zhang, J.; Yang, X.; Huang, X.; Dai, L.; Sun, G.; Liu, X. Improved design and experiment of 4UFD-1400 potato combine harvester. *Trans. Chin. Soc. Agric. Eng.* **2014**, *30*, 12–17.
82. Zhao, M. Design of Crawler Self-Propelled Potato Harvester. Master's Thesis, Ningxia University, Ningxia, China, 2023. [CrossRef]
83. Wang, H. Design and experiment of mountain self-propelled potato combine harvester. Master's Thesis, Kunming University of Science and Technology, Kunming, China, 2021. [CrossRef]
84. Wang, H.; Zhao, W.; Sun, W.; Zhang, H.; Liu, X.; Li, H. Research progress on potato mechanized harvesting technology and equipment. *Trans. Chin. Soc. Agric. Eng.* **2023**, *39*, 1–22.
85. Shi, M.; Wei, H.; Hu, Z.; Liu, X.; Yang, X. Investigation on the current situation of potato harvesting machinery products at home and abroad. *Farm Mach.* **2019**, *40*, 206–210.
86. Zhao, Z.; Wu, M.; Jiang, X. A Review of Contact Models' Properties for Discrete Element Simulation in Agricultural Engineering. *Agriculture* **2024**, *14*, 238. [CrossRef]
87. Kim, Y.-S.; Siddique, M.A.A.; Kim, W.-S.; Kim, Y.-J.; Lee, S.-D.; Lee, D.-K.; Hwang, S.-J.; Nam, J.-S.; Park, S.-U.; Lim, R.-G. DEM simulation for draft force prediction of moldboard plow according to the tillage depth in cohesive soil. *Comput. Electron. Agric.* **2021**, *189*, 106368. [CrossRef]
88. Pasha, M.; Hare, C.; Ghadiri, M.; Gunadi, A.; Piccione, P.M. Inter-particle coating variability in a rotary batch seed coater. *Chem. Eng. Res. Des.* **2017**, *120*, 92–101.
89. Ren, D.; Yu, H.; Zhang, R.; Li, J.; Zhao, Y.; Liu, F.; Zhang, J.; Wang, W. Research and Experiments of Hazelnut Harvesting Machine Based on CFD-DEM Analysis. *Agriculture* **2022**, *12*, 2115. [CrossRef]
90. Capdeville, P.J.; Kuang, S.; Yu, A. Unrevealing energy dissipation during iron ore transfer through chutes with different designs. *Powder Technol.* **2024**, *435*, 119446. [CrossRef]
91. Tong, J.; Mohammad, M.; Zhang, J.; Ma, Y.; Rong, B.; Chen, D.; Menon, C. DEM Numerical Simulation of Abrasive Wear Characteristics of a Bioinspired Ridged Surface. *J. Bionic Eng.* **2010**, *7*, 175–181. [CrossRef]
92. Zidek, M.; Zegzulka, J.; Jezerská, L.; Rozbroj, J.; Gelnar, D.; Nečas, J. Simulation model of loading bin bottom by bulk material. *Chem. Eng. Res. Des.* **2020**, *154*, 151–161. [CrossRef]

93. Kesner, A.L.; Chotěborský, R.; Linda, M.; Hromasová, M.; Katinas, E.; Sutanto, H. Stress distribution on a soil tillage machine frame segment with a chisel shank simulated using discrete element and finite element methods and validate by experiment. *Biosyst. Eng.* **2021**, *209*, 125–138. [CrossRef]
94. Aikins, K.A.; Ucgul, M.; Barr, J.B.; Awuah, E.; Antille, D.L.; Jensen, T.A.; Desbiolles, J.M.A. Review of Discrete Element Method Simulations of Soil Tillage and Furrow Opening. *Agriculture* **2023**, *13*, 541. [CrossRef]
95. Li, H.; Gao, F. Improvement design of separation and conveying machinery and equipment of potato excavator in heavy soil. *Phys. Chem. Earth Parts A/B/C.* **2023**, *130*, 103363. [CrossRef]
96. Li, J.; Jiang, X.; Ma, Y.; Tong, J.; Hu, B. Bionic Design of a Potato Digging Shovel with Drag Reduction Based on the Discrete Element Method (DEM) in Clay Soil. *Appl. Sci.* **2020**, *10*, 7096. [CrossRef]
97. Li, J.; Li, X.; Hu, B.; Gu, T.; Wang, Z.; Wang, H. Analysis of the resistance reduction mechanism of potato bionic digging shovels in clay and heavy soil conditions. *Comput. Electron. Agric.* **2023**, *214*, 108315. [CrossRef]
98. Gai, X.; Xie, S.; Deng, W.; Lu, K.; Ji, X. Construction and parameter calibration of potato double-layer flexible bonding model based on discrete-element method. *J. Food Process Eng.* **2024**, *47*, e14602. [CrossRef]
99. Liang, Z.; Huang, Y.; Li, D.; Wada, M.E. Parameter determination of a viscoelastic–plastic contact model for potatoes during transient collisions. *Biosyst. Eng.* **2023**, *234*, 156–171. [CrossRef]
100. Ye, F.; Lu, T.; Xu, C. Digging characteristics of grab based on DEM-MBD simulation and experiment. *Comput. Part. Mech.* **2024**. [CrossRef]
101. Chen, M.; Liu, X.; Hu, P.; Zhai, X.; Han, Z.; Shi, Y.; Zhu, W.; Wang, D.; He, X.; Shang, S. Study on rotor vibration potato-soil separation device for potato harvester using DEM-MBD coupling simulation. *Comput. Electron. Agric.* **2024**, *218*, 108638. [CrossRef]
102. Xie, S.; Zhang, Y.; Li, J.; Liu, F. Analysis of Breaking and Separating Characteristics of Potato-Soil Aggregates Based on the New Type of Swing Separation Sieve. *Agronomy* **2024**, *14*, 1272. [CrossRef]
103. McRae, D.C. Improving the Quality of the Potato Crop. *Outlook Agric.* **1990**, *19*, 237–242. [CrossRef]
104. Xie, S.; Wang, C.; Deng, W. Experimental study on collision acceleration and damage characteristics of potato. *J. Food Process Eng.* **2020**, *43*, e13457. [CrossRef]
105. Shang, Z.; Wang, F.; Xie, S.; Deng, W.; Li, J.; Guo, Y.; Lu, K.; Gai, X.; Ji, X. Construction and analysis of the forced vibration model of separating screen under a dropped potato. *J. Food Process Eng.* **2023**, *46*, e14346. [CrossRef]
106. Deng, W.; Wang, C.; Xie, S. Collision simulation of potato on rod separator. *Int. J. Food Eng.* **2020**, *17*, 435–444. [CrossRef]
107. Dorokhov, A.; Didmanidze, O.; Aksenov, A.; Sibirev, A.; Sazonov, N.; Mosyakov, M.; Godyaeva, M. The Results of Studies on the Assessment of the Destruction of Soil Clods during Combine Harvesting of Potatoes. *Agriculture* **2022**, *12*, 2024. [CrossRef]
108. Milne, F.N.J. The Canon of Potato Science: 34. Potato Harvesting. *Potato Res.* **2007**, *50*, 347–349. [CrossRef]
109. Dorokhov, A.; Ponomarev, A.; Zernov, V.; Petukhov, S.; Aksenov, A.; Sibirev, A.; Sazonov, N.; Godyaeva, M. The Results of Laboratory Studies of the Device for Evaluation of Suitability of Potato Tubers for Mechanized Harvesting. *Appl. Sci.* **2022**, *12*, 2171. [CrossRef]
110. Li, Y.; Fan, J.; Hu, Z.; Luo, W.; Yang, H.; Shi, L.; Wu, F. Calibration of Discrete Element Model Parameters of Soil around Tubers during Potato Harvesting Period. *Agriculture* **2022**, *12*, 1475. [CrossRef]
111. Du, X.; Liu, J.; Zhao, Y.; Zhang, C.; Zhang, X.; Wang, Y. Design and Test of Discrete Element-Based Separation Roller Potato-Soil Separation Device. *Agriculture* **2024**, *14*, 1053. [CrossRef]
112. Li, Y.; Hu, Z.; Gu, F.; Wang, B.; Fan, J.; Yang, H.; Wu, F. DEM-MBD Coupling Simulation and Analysis of the Working Process of Soil and Tuber Separation of a Potato Combine Harvester. *Agronomy* **2022**, *12*, 1734. [CrossRef]
113. Yan, D.; Deng, W.; Xie, S.; Liu, C.; Ren, Z.; Zhao, H.; Cai, Y.; Zhao, Z. Discrete Element-Based Simulation Analysis and Research of Potato Soil Agglomerate Fragmentation and Separation. *Appl. Sci.* **2023**, *13*, 8416. [CrossRef]
114. Zhao, H.; Deng, W.; Xie, S.; Zhao, Z. Performance Optimization and Experimental Study of Small-Scale Potato-Grading Device. *Agriculture* **2024**, *14*, 822. [CrossRef]
115. Geng, J.; Wang, S.; Gao, Z.; Liu, Z.; Rao, X. Influence of initial orientations on potato conveyor trajectories by machine vision. *Comput. Electron. Agric.* **2019**, *163*, 104838. [CrossRef]
116. Xie, H.K.; Gao, G.; Tian, B.; Li, B.; Zhang, S.; Huang, J. Optimization of Potato Soil Transportation Separation Mechanism Based on Discrete Element Method and TRIZ Theory. *J. Phys. Conf. Ser.* **2019**, *1267*, 012071. [CrossRef]
117. Zhang, H.; Li, H.; Sun, W.; Li, H.; Liu, X.; Sun, G.; Lu, Y.; Chen, Y.; Xing, W. Optimization of Potato Planter Soil Lifting Device Based on TRIZ Theory. *Agriculture* **2024**, *14*, 1695. [CrossRef]
118. Sun, H.; Ma, H.; Zhao, Y. DEM investigation on conveying of non-spherical particles in a screw conveyor. *Particuology* **2022**, *65*, 17–31. [CrossRef]
119. McRae, D.C. A potato harvester delivery conveyor with automatic discharge height control. *Potato Res.* **1974**, *17*, 138–151. [CrossRef]

120. Torres-Serra, J.; Rodriguez-Ferran, A.; Romero, E. Study of grain-scale effects in bulk handling using discrete element simulations *Powder Technol.* **2021**, *382*, 284–299. [CrossRef]
121. Ren, H.; Meng, W.; Sun, X.; Zhao, Z.; Zhao, X. Discrete element analysis on dynamic characteristics of directional material flow driven by horizontal trough-free screw conveyor. *Powder Technol.* **2023**, *418*, 118276. [CrossRef]
122. Shi, Q.; Sakai, M. Recent progress on the discrete element method simulations for powder transport systems: A review. *Adv. Powder Technol.* **2022**, *33*, 103664. [CrossRef]
123. Dai, H.; Li, Y. Research on the Collection Characteristics of a Hydraulic Collector for Seafloor Massive Sulfides. *J. Mar. Sci. Eng.* **2024**, *12*, 1534. [CrossRef]
124. Zhang, B.; Lu, H.; Yang, J.; Sun, P.; Deng, L. Numerical and experimental study on ore-collecting characteristics of deep-sea seafloor massive sulfide. *Ocean Eng.* **2024**, *310*, 118729. [CrossRef]
125. Hou, J.; Ren, Z.; Liu, D.; Ma, Z.Y.; Liu, X.; Wang, W. Parameter Calibration of a Discrete Element Model for Three-Compartment Separation and Fragmentation of Castor Capsules During Harvesting. *J. Biosyst. Eng.* **2023**, *48*, 198–214. [CrossRef]

agriculture

MDPI

Article

An Ultrasonic Ridge-Tracking Method Based on Limiter Sliding Window Filter and Fuzzy Pure Pursuit Control for Ridge Transplanter

Wei Liu [1], Jinhao Zhou [1], Yutong Liu [1], Tengfei Zhang [1], Meng Yan [1], Ji Chen [2], Chunjian Zhou [2], Jianping Hu [1] and Xinxin Chen [1,*]

1 School of Agricultural Engineering, Jiangsu University, Zhenjiang 212013, China; mario_liu@ujs.edu.cn (W.L.); 2112316042@stmail.ujs.edu.cn (J.Z.); 2222016041@stmail.ujs.edu.cn (Y.L.); 2112016001@stmail.ujs.edu.cn (T.Z.); 2212016059@stmail.ujs.edu.cn (M.Y.); hujp@ujs.edu.cn (J.H.)

2 Shanghai Agricultural Machinery Research Institute, Shanghai 201106, China; shnjscj@163.com (J.C.); shnjszcj@163.com (C.Z.)

* Correspondence: xxchen@ujs.edu.cn; Tel.: +86-195-9996-2668

Abstract: There are various types of fruits and vegetables that need to be planted on ridges. In order to allow for seedlings with a certain row space and seedling space, the ridge transplanter should be able to track along the ridge. Therefore, an ultrasonic ridge-tracking method and system were developed to let the ridge transplanter track the ridge accurately. The ultrasonic ridge-tracking method mainly contains a limiter sliding window filtering algorithm and a fuzzy look-ahead distance decision model. The limiter sliding window filtering algorithm was proposed to filter the abnormal measuring results to avoid disoperation of the steering mechanism. Moreover, the fuzzy look-ahead distance decision model was proposed to determine the optimal look-ahead distance in order to obtain a desirable tracking performance. Additionally, a comparison experiment of the proposed ultrasonic ridge-tracking method and the universal pure pursuit method was conducted. The experimental results show that the greatest mean absolute errors of the lateral deviations of the ultrasonic ridge-tracking method and universal pure pursuit were 10.56 mm and 13.11 mm. The greatest maximum absolute errors of the lateral deviations of the ultrasonic ridge-tracking method and universal pure pursuit were 18.87 mm and 23.23 mm. In addition, the greatest root mean square error of the lateral deviation of the ultrasonic ridge-tracking method and the universal pure pursuit method were 13.52 mm and 15.66 mm. According to the ridge-tracking performance of the proposed ultrasonic ridge-tracking method, it can be used in practical transplanting conditions. Moreover, in other fields, robots or intelligent machinery can also apply the proposed ultrasonic ridge-tracking method to track objects similar to ridges.

Keywords: ultrasonic measuring; path tracking; ridge; transplanter; filter; pure pursuit

Citation: Liu, W.; Zhou, J.; Liu, Y.; Zhang, T.; Yan, M.; Chen, J.; Zhou, C.; Hu, J.; Chen, X. An Ultrasonic Ridge-Tracking Method Based on Limiter Sliding Window Filter and Fuzzy Pure Pursuit Control for Ridge Transplanter. *Agriculture* **2024**, *14*, 1713. https://doi.org/10.3390/agriculture14101713

Academic Editor: Mustafa Ucgul

Received: 30 August 2024
Revised: 25 September 2024
Accepted: 27 September 2024
Published: 29 September 2024

1. Introduction

A variety of fruits and vegetables, such as the strawberry, tomato, cucumber, and so on, need to be planted on ridges. The fruit and vegetable seedlings are usually transplanted onto ridges with a certain row spacing and line spacing by ridge transplanters. Thus, the ridge transplanter needs to track along the ridge, so as to keep accurate and stable row spaces and avoid the seedlings planting out of the ridge [1].

However, because in most cases the ridgers are driven manually, the curvatures of ridges may not be straight. Hence, the ridge-tracking performance of the ridge transplanter is crucial to the transplanting quality. Moreover, due to the wheels of the transplanter being in the furrow, the yaw of the transplanter was caused by the forces of the uneven ground on the wheels. Hence, according to the above two reasons, the heading angle of the transplanter should be controlled in real time [2]. Up to now, the steering of ridge transplanter has been mainly controlled by the operator [3]. Therefore, in order to reduce the labor costs and improve the transplanting quality, the autonomous ridge-tracking method of the ridge transplanter needs to be developed.

Essentially, ridge tracking is a special type of path tracking with position constraints between the agricultural machinery and the ridge. There are three path-tracking methods commonly used in agricultural machinery, namely the kinematic method [4], dynamic method [5], and model-free method [6]. The kinematic method is easy to implement and is thus widely used [7]. The kinematic methods mainly consist of the pure pursuit tracking (PPT) method, the linear quadratic regulator (LQR) method, and the model predictive control (MPC) method. The PPT method selects a series of points in the reference path as look-ahead points, and then the agricultural machinery tracks the preset path by approaching the look-ahead points in order [8,9].

The LQR method and MPC method conduct enormous iterative computations to predict the future behaviors of the agricultural machinery. The optimal control inputs are selected according to the computation results [10]. A linear time-varying MPC algorithm was developed for path tracking based on the autonomous front-wheel steering system [11], and a conventional MPC algorithm was also improved based on the recursive least squares method with an adaptive forgetting factor [12]. Due to massive calculations, the LQR and MPC methods are usually time-consuming and model-dependent [13], so they may not be ideal for real-time ridge-tracking situations. If the steering of the ridge transplanter cannot be controlled in time, then the ridge will be damaged by the ridge transplanter.

The kinematic methods are easily affected by disturbances and thus lack robustness [7]. Therefore, some researchers developed path-tracking algorithms based on the dynamic models of agricultural machinery [14]. Among the dynamic methods, the sliding mode control (SMC) algorithm stands out as the most classical one. Raghavendra et al. proposed an intelligent SMC method to improve the path-tracking performance of the autonomous vehicle under the condition of parameter uncertainties. This method employed fractional order sliding manifolds, optimized sliding coefficients, and used fuzzy adaptive switching gain to enhance the tracking performance [15]. Although the SMC method can effectively eliminate environmental noise and has strong robustness, when system states slide along the slide mode surface, the machinery system may be chattered, which could affect the tracking performance of the agricultural machinery [16].

Because some parameters, such as the friction coefficient between the tire and ground, the vehicle load distribution, and so on, are difficult to measure, building a precise kinematic or dynamic model of agricultural machinery is still a challenge [17]. Therefore, more and more studies are focused on model-free path-tracking methods to overcome the time-varying and nonlinear characteristics of the agricultural machinery.

The PID control algorithm is the most widely applied among the model-free path-tracking methods. Zhang et al. applied a comprehensive performance index to optimize the PID controller for the agricultural wheeled vehicle whose kinematic model was the Ackerman model. The results showed that the overshoot and regulation time of the optimized control system were 10.03% and 3.95 s [18]. The PID control algorithm is easy to implement in practice. However, the typical PID controller has certain coefficients, which are not adaptable to the agricultural environment with time-varying parameters and external disturbances [19].

So as to overcome the above limitations, some nonlinear path-tracking algorithms that combine the PID and coefficient-determining algorithms were proposed. Fuzzy inference can determine the optimal parameters according to variable environmental conditions [20]. The steering angle of agricultural machinery was regulated in real time by a parameter-variable PID-controlling algorithm whose proportional, integral, and differential coefficients were determined by fuzzy rules [21]. The experimental results showed that the maximal lateral error and heading error of the developed path-tracking system were 4.39 cm and 2.13°.

Although some studies have been conducted to let the agricultural machinery track a reference path on a flat field ground [22], few studies have considered how to navigate machinery along a ridge with a variable curvature. Some positioning sensors, like GPS, cannot detect the position of the ridge, while machine vision or LiDAR would be affected by hard light. In order to overcome the mentioned sensing problems, the developed ridge-tracking method in this research used ultrasonic sensors to monitor the relative position between the ridge and the transplanter.

Hence, the goal of this study was to develop an ultrasonic ridge-tracking (URT) method for ridge transplanters to track along the ridge accurately. The proposed URT method mainly consists of two sections. One is a limiter sliding window filtering algorithm and the other is a fuzzy look-ahead distance decision method. In detail, the limiter sliding window (LSW) filtering algorithm was used to reduce the influence of the abnormal measuring results caused by ridge damage. Using the developed fuzzy look-ahead distance decision model, the optimal look-ahead distance can be determined and adjusted in real time based on the current lateral deviation and traveling velocity. Additionally, the field experiments were conducted to compare the performance of the proposed URT method and the universal pure pursuit (UPP) method.

2. Materials and Methods

2.1. Overall Research and Development Approach

In this research, the development of the ridge-tracking system used the reverse engineering approach [23]. The procedures of the development approach are displayed in Figure 1. First, the functions of the ridge-tracking system were analyzed, and then the kinematic model of the transplanter was developed. Moreover, in order to improve the tracking accuracy, the key sub-methods, i.e., the LSW filtering algorithm and fuzzy look-ahead distance decision model, were proposed. Finally, the ridge-tracking performance of the developed ultrasonic ridge-tracking method was validated in a greenhouse. The developed ultrasonic ridge-tracking method used the above sequence.

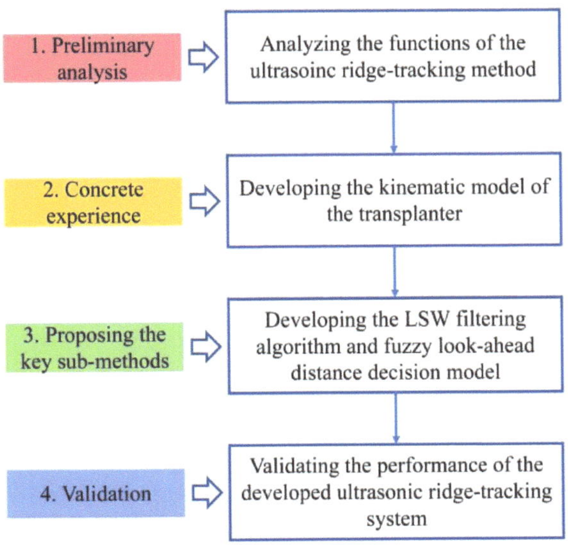

Figure 1. Flowchart of the research and development sequence of the ultrasonic ridge-tracking method.

2.2. Structure and Working Principle of the Transplanter Chassis

The proposed ultrasonic ridge-tracking method was applied to the developed transplanter chassis. The structure of the developed transplanter chassis is illustrated in Figure 2. The chassis contained two sections. One was the steering mechanism and the other was the driving mechanism. The steering mechanism consisted of a steering motor, a steering reducer, a rack and gear mechanism, a slide guide rail, two sliders, and two steering rods.

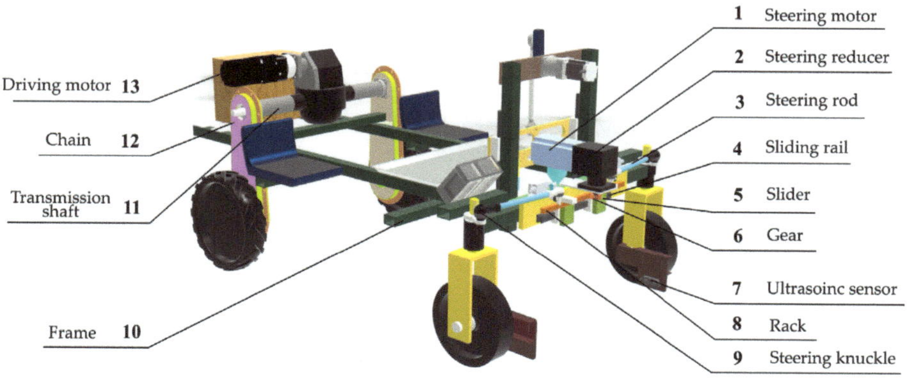

Figure 2. Structure of the developed transplanter chassis.

The working principle of the steering mechanism can be explained as follows. The output shaft of the steering motor connected to the input shaft of the reducer, and the output shaft of the reducer was fixed to a gear that was used to drive the rack for transverse motions. Two sliders mounted on the sliding rail were used to support the rack. Moreover, the steering rod connected to the slider and the steering knuckle of front wheel. Therefore, according to the theory of the aforementioned steering mechanism, the steering angle of the front wheel was controlled by the rotating angle of the steering motor.

As for the driving mechanism, its working principle can be explained as follows. The driving mechanism was composed of a driving motor, a reducer, two transmission shafts, and two chain systems. The driving motor connected to the reducer to expand the driving torque. The reducer had two symmetric output shafts, which connected to the left and right transmission shaft, respectively. The transmission shafts connected to the reducer and the sprocket wheel, which used chains to drive the rear wheels. The structural parameters of the chassis of the ridge transplanter are shown in Table 1.

Table 1. Structural parameters of the chassis of the ridge transplanter.

Structural Parameters	Value
Length (mm)	1700
Width (mm)	1300
Distance between wheels (mm)	800
Wheel base (mm)	1230
Maximum steering angle (°)	30
Suitable height of ridge (mm)	250–360

2.3. Developed Ultrasonic Ridge-Tracking System

The developed ultrasonic ridge-tracking system contained two ultrasonic sensors (DYP-A02YY-V2.0, Shenzhen Dianyingpu Technology Ltd., Shenzhen, China), a programmable logical controller PLC (XD3-32T-C, Wuxi XINJIE company limited by shares, Wuxi, China), an embedded signal acquisition system (ESAS) based on a micro control unit, MCU (STM32F103ZET6, STMicroelectronics, Switzerland), a steering stepper motor (60BEP115LC, Times Brilliant Electrical LLC, China), a steering motor driver (HBS86HAC, Times Brilliant Electrical LLC, China), a brushless direct current motor (86BL130-430, Times Brilliant Electrical LLC, China), and a brushless direct current motor driver (ZM6618, Times Brilliant Electrical LLC, China). The detailed specifications of the ultrasonic ridge-tracking system components are listed in Table 2 and the schematic of the developed ultrasonic ridge-tracking system is shown in Figure 3.

Table 2. Specifications of the ultrasonic ridge-tracking system components.

Component	Manufacturer	Model	Function
Ultrasonic sensor	Shenzhen Dianyingpu Technology Ltd., Shenzhen, China	DYP-A02YY-V2.0	Monitoring the distance from ultrasonic sensor to the sidewall
PLC	Wuxi Xinjie Company Limited by share, Wuxi, China	XD3-32T-C	Controlling the rotational speed and rotational angle of the motors
MCU	STMicroelectronics, Geneva, Switzerland	STM32F103ZET6	Processing the measuring results and developed ultrasonic ridge-tracking algorithm
ESAS	Guangzhou Xingyi Electronic Co., Ltd., Guangzhou, China	ELITE	Integrating the peripheral circuit of the MCU
Steering stepper motor	Times Brilliant Electrical LLC, Beijing, China	60BEP115LC	Controlling the steering angle of the front wheels
Steering motor driver	Times Brilliant Electrical LLC, Beijing, China	HBS86HAC	Driving the stepper motor
Brushless direct current motor	Times Brilliant Electrical LLC, Beijing, China	86BL130-430	Powering the driving mechanism of the ridge transplanter
Brushless direct current motor driver	Times Brilliant Electrical LLC, Beijing, China	ZM6618	Driving the brushless direct current motor

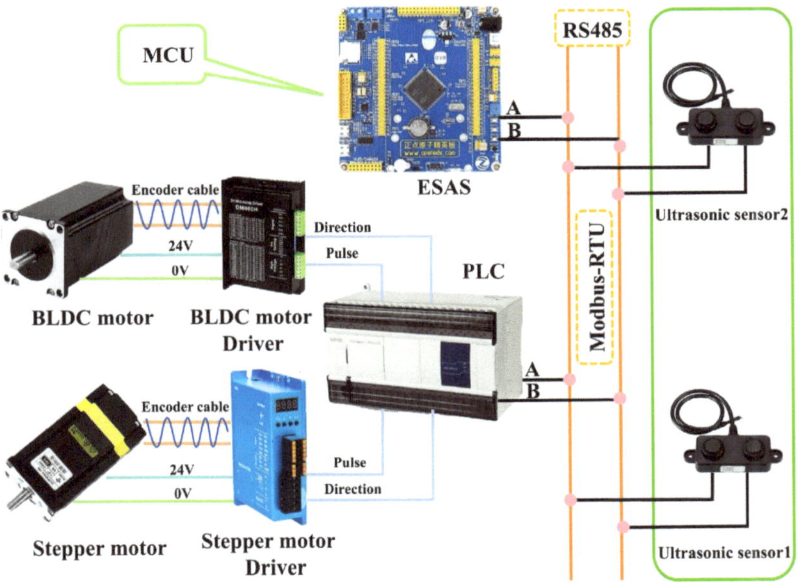

Figure 3. Components of the developed ultrasonic ridge-tracking system.

The ultrasonic sensors were used to measure the distances from the ridge sidewalls to the transplanter. The ESAS was applied to process the ultrasonic measurement information and operate the control algorithm. Then, the ESAS sent the rotational angle of the steering motor and the rotational speed of the driving motor into the PLC. The PLC controlled the rotational angle of the steering motor as well as the rotational speed of the driving motor by changing the frequency of the pulses. Moreover, the rotational directions of the steering and driving motors were regulated by the different electronic level output by the PLC. The stepper motor was used to drive the steering mechanism and the brushless direct current motor was applied to drive the driving mechanism.

The hardware connection and the protocol from the ultrasonic sensors to the ESAS were the RS485 and the Modbus-RTU. The sample frequency of the ultrasonic sensor was 14 Hz. Moreover, the physical and protocol that connected the ESAS to the PLC were RS485 and Modbus-RTU protocol.

2.4. Principle of the Proposed Ultrasonic Ridge-Tracking Method

The principle of the proposed ultrasonic ridge-tracking method can be explained as follows. The ultrasonic sensors were used to measure the distance from the ultrasonic sensor to the ridge sidewall, and then the raw data were input into the limiter sliding window filtering algorithm to filter abnormal measuring results. After that, the center deviation was computed according to the distances from the ultrasonic sensors to the ridge sidewalls. In the next step, the optimal front sight point was selected from the optional ridge centers according to the current lateral deviation and traveling velocity. The look-ahead distance was equal to the norm of the vector from the center of the rear axle of the transplanter to the selected front sight point. According to the computed look-ahead distance, the steering angle model outputted the theoretical steering angle of the front wheels. After all of the above controlling processes were finished, the control process in the sampling period *t* was accomplished. The block diagram of the proposed URT method is demonstrated in Figure 4. The key innovations (i.e., the limiter sliding window filter and the fuzzy look-ahead distance decision model) are marked in red in Figure 4.

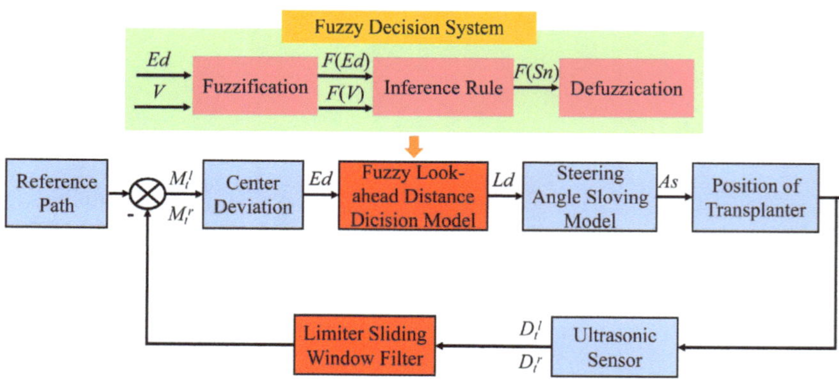

Figure 4. Block diagram of the proposed ultrasonic ridge-tracking method.

In Figure 4, $D_t{}^l$ and $D_t{}^r$ are the distances to the left and right ridge sidewalls at the sampling period *t*; $M_t{}^l$ and $M_t{}^r$ are the filtered measured distances to the left and right ridge sidewalls at sampling period *t*; *Ed* is the lateral deviation between the ridge center and transplanter center; *V* means the traveling velocity of the transplanter; *F*(*Ed*), *F*(*V*), and *F(Sn)* denote the fuzzy membership of the lateral deviation, traveling velocity, and serial number of the optional ridge center, respectively; *Ld* denotes the optimal look-ahead distance; *As* means the theoretical steering angle of the front wheel.

2.5. Limiter Sliding Window (LSW) Filtering Algorithm

In order to acquire the deviation between the transplanter and the ridge, the distances from the two ridge sidewalls to the transplanter were measured by left and right ultrasonic sensors, respectively.

The transverse distance between the two ultrasonic sensors was 804 mm. In this research, the ridges for planting strawberry were selected as the application object. The width (i.e., the transverse distance between two vertical sidewalls) and height of the ridges was 600 mm and 250 mm. If the transplanter center was misaligned with the ridge center, then the transverse distance from one sidewall of ridge to the ultrasonic sensor was more than 102 mm, and the transverse distance from the other side of ridge to the sensor was less than 102 mm. Regarding the height of the ultrasonic sensors, it should be lower than the height of ridge; thus, in this research, the installation height was set as 50 mm. The installation positions of the ultrasonic sensors are shown in Figure 5.

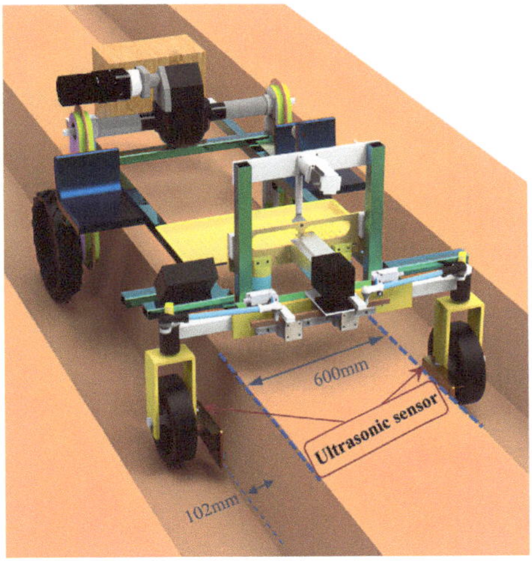

Figure 5. Installation positions of the ultrasonic sensors.

As the transplanter travels, the distance needs to be measured in real time from the ultrasonic sensor to the surface of the sidewall. The direction of the transplanter was determined by the left and right distances monitored by the ultrasonic sensors. If a sidewall of the ridge was damaged, the ultrasonic wave would not reflected on the surface but on the irregular pit of the sidewall, which may lead to malfunctions of the steering mechanism. For example, if the left sidewall was damaged, then the left ultrasonic sensor would obtain a large distance compared to the distance to the sidewall. Accordingly, the transplanter would judge that it was far away to the left sidewall, and then the steering mechanism would move to the right direction, as seen in Figure 6. After that, the center of the transplanter may offset in the right direction and the transplanter would travel in a wrong direction. In practice, the transplanter should monitor the distance to the sidewall rather than the local pits. Therefore, in order to reduce the influence of ridge damage, the limiter sliding window (LSW) filter was developed to filter the measured distances to the pits.

The limiter algorithm can effectively overcome the interference caused by accidental factors, and the sliding window algorithm can use several previous measuring results to smooth the final result. If the limiter algorithm and the sliding window algorithm can be fused, then the measuring distances at damaged pits can be removed and the several previous measuring distances can be used to obtain a precise measuring result. Therefore, the limiter sliding window (LSW) filtering algorithm was proposed in this study.

The processes of the LSW filtering algorithm will be explained as follows. The first step was a limiter process that was used to judge the range of the current measuring results. If the difference between the current measuring value D_t and the last measuring value D_{t-1} was less than the limiter threshold ΔT, then the current value D_t would be selected as the effect value. Otherwise, the last measuring value D_{t-1} was chosen as the effect value.

(a)

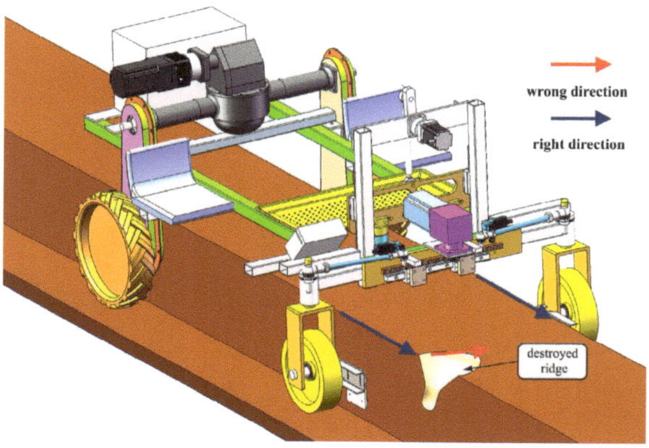

(b)

Figure 6. Direction change affected by the damaged ridge. (**a**) Actual damaged ridges; (**b**) malfunction caused by the damaged ridge.

The formula of the limiter process is listed below:

$$EV_t = \begin{cases} D_t & |D_t - D_{t-1}| \leq \Delta T \\ D_{t-1} & |D_t - D_{t-1}| > \Delta T \end{cases} \tag{1}$$

where EV_t implies the effect value of the limiter process at sampling period t, mm;

D_t denotes the measuring result at sampling period t, mm;

D_{t-1} denotes the measuring result at sampling period $t-1$, mm;

ΔT is the limiter threshold, in this study, ΔT was selected as 100 mm.

The second step was the sliding window process. The sliding window had 7 effect values from EV_{t-1} to EV_{t-7}. When the newest effect value EV_t was added to the sliding window, the oldest one EV_{t-7} would be removed from the sliding window.

The third step was the sorting process. The effect values in the sliding window were sorted by the bubble method. The fourth step was the removing process, in which the maximum and minimum in the sliding window were removed. The fifth step was solving for the average processing value, followed by computing the mean of the remaining 5 effect values as the consequence of the LSW filtering algorithm.

The equation of the LSW filtering algorithm can be illustrated as follows.

$$M_t = \frac{1}{TN-2} \times \left(\sum_{k=(t-6)}^{t} EV_t - Max - Min \right) \tag{2}$$

where TN indicates the total number of effect values in the sliding window. In this research, TN is equal to 7:

k represents the serial number of effect values in the sliding window;
EV_t implies the effect value of the limiter process at sampling period t, mm;
Max means the maximum of the elements in the sliding window, mm;
Min means the minimum of the elements in the sliding window, mm;
M_t is the filtered data in the sampling period t;
t implies the serial number of the sampling periods.

The schematic of the LSW filtering algorithm is illustrated in Figure 7.

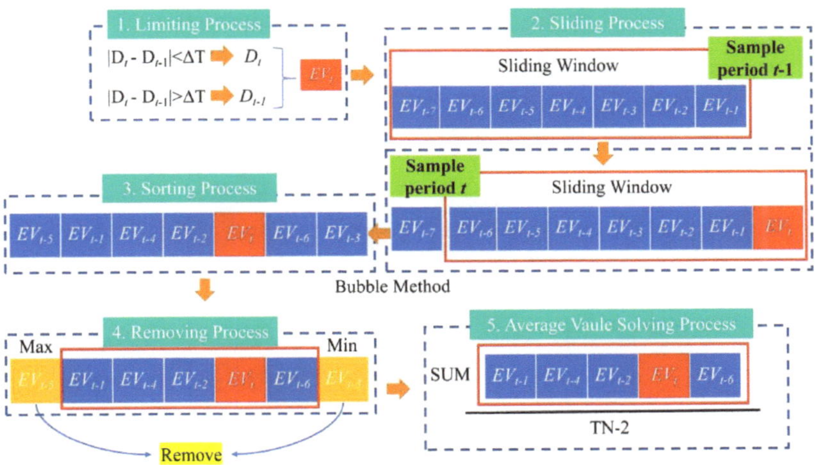

Figure 7. Schematic of the LSW filtering algorithm. The red block means the effect value at the current sample period t; the blue blocks denote the effect values in the previous sampling periods from $t-1$ to $t-7$; the yellow blocks are the max and min of the effect value sequence.

2.6. Kinematic Model of the Ridge Transplanter

In this research, the kinematic model of the ridge transplanter was the relationship between the position of front sight point and the theoretical steering angle of the front wheel. The norm of the vector from the center of the rear axle of the transplanter to the front sight point was defined as the look-ahead distance. In this study, the center in the width direction of the ridge, which was abbreviated as the ridge center, was viewed as the optional front sight point. Hence, it is important to obtain the accurate coordinates of the ridge centers. The measured ridge center at the sampling period t in the coordinate of the transplanter was $P_t(x_t, y_t)$, whose coordinates can be computed as follows.

$$D_l = S - M_t^l \tag{3}$$

$$D_r = S - M_t^r \tag{4}$$

where S is the theoretical distance between the ultrasonic sensor and the middle line of the transplanter, mm. In this study, S is equal to 402 mm.

M_t^l is the filtered result between the left ultrasonic sensor and the left sidewall of ridge, mm;

$M_t{}^r$ is the filtered result between the right ultrasonic sensor and the right sidewall of ridge sidewall, mm;

D_l is the distance between the left sidewall and the ridge center, mm;

D_r is the distance between the right sidewall and the ridge center, mm.

The x-coordinate of the ridge center in the coordinate system of transplanter can be computed by Equation (5).

$$x_t = D_r - D_l = M_t^r - M_t^l \tag{5}$$

where x_t is the x-coordinate of the ridge center.

D_l is the distance between the left sidewall and the ridge center, mm;

D_r is the distance between the right sidewall and the ridge center, mm.

$M_t{}^l$ is the filter result between the left ultrasonic sensor and the left sidewall of the ridge, mm;

$M_t{}^r$ is the filter result between the right ultrasonic sensor and the right sidewall of the ridge, mm;

The y-coordinate of the ridge center was the distance between the center of the rear axle (i.e., the origin of the coordinate system of the transplanter) and the center of the two ultrasonic sensors. The exact value of the y-coordinate at a sampling period $t(y_t)$ was 1253 mm. The coordinate of the ridge center $P_t(x_t, y_t)$ is demonstrated in Figure 8.

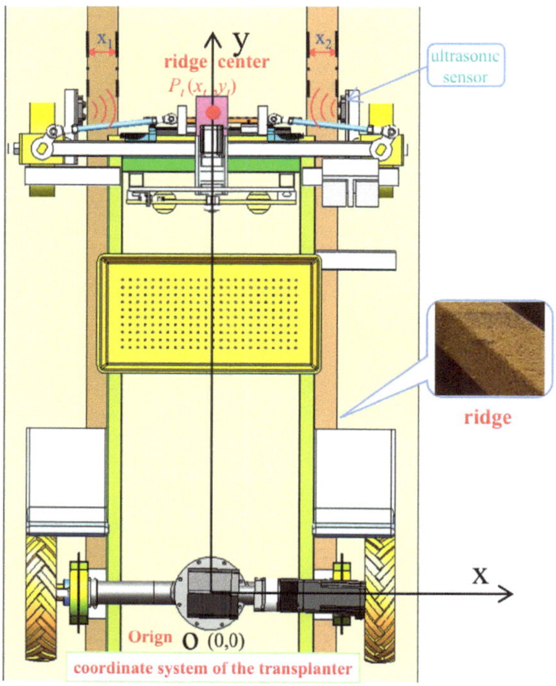

Figure 8. Coordinate of the ridge center in the coordinate system of transplanter.

As the transplanter traveled, the coordinates of ridge centers were updated. The coordinate of the nth ridge center $P_t{}^n(x_t{}^n, y_t{}^n)$ at sampling period t transformed into the $P^n{}_{t+1}$ $(x^n{}_{t+1}, y^n{}_{t+1})$ at sampling period $t + 1$, which is demonstrated in Equation (6).

$$\begin{bmatrix} x_{t+1}^n \\ y_{t+1}^n \\ 1 \end{bmatrix} = \begin{bmatrix} cos\phi & sin\phi & v \cdot t \cdot sin\phi \\ -sin\phi & cos\phi & v \cdot t \cdot cos\phi \\ 0 & 0 & 1 \end{bmatrix} \cdot \begin{bmatrix} x_t^n \\ y_t^n \\ 1 \end{bmatrix} \tag{6}$$

where v is the traveling velocity of the transplanter, mm/s.

φ is the difference of the heading angle of transplanter between the sampling period $t+1$ and the sampling period t, rad;

n indicates the serial number of the ridge centers.

The schematic of the coordinate transform of the ridge centers is illustrated in Figure 9.

$$n = \frac{y_t}{V} \tag{7}$$

where n denotes the total number of the optional ridge centers.

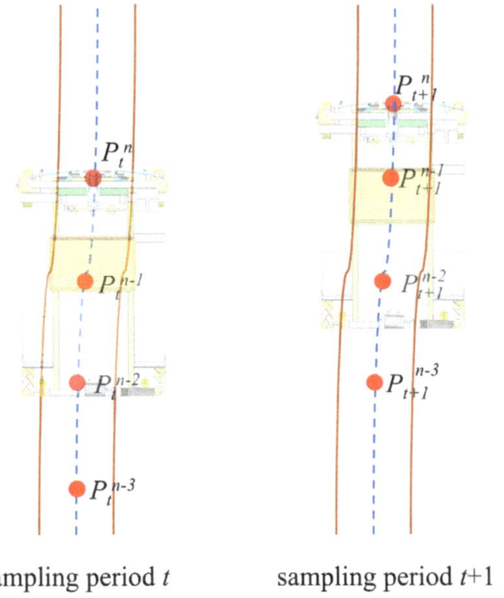

Figure 9. Coordinate transform of ridge centers.

y_t means the distance between the center of the rear axle to the center of the two ultrasonic sensors, mm;

V indicates the traveling velocity of the transplanter, mm/s.

If a new ridge center was added to the optional queue, then the oldest one was discarded. According to Equation (6), when the transplanter at sampling period t moved to a new position at sampling period $t+1$, the coordinates of the ridge centers in the optional queue (from $P_t{}^1$ to $P_t{}^n$) were updated to the new coordinates (from $P^1{}_{t+1}$ to $P^n{}_{t+1}$).

Moreover, the kinematic model of the transplanter can be simplified as a two-wheel model, as seen in Figure 10. In addition, the relationship between the look-ahead distance, turning radius, and angle deviation of the two-wheel model is demonstrated in Equation (8).

$$\frac{l_d}{sin(2\alpha)} = \frac{R}{sin(\frac{\pi}{2} - \alpha)} \tag{8}$$

where l_d is the look-ahead distance, mm.

R represents the turning radius, mm;

α denotes the heading deviation, rad.

Equation (8) can be simplified as the following Equation (9)

$$\frac{1}{R} = \frac{2 \times sin\alpha}{l_d} \tag{9}$$

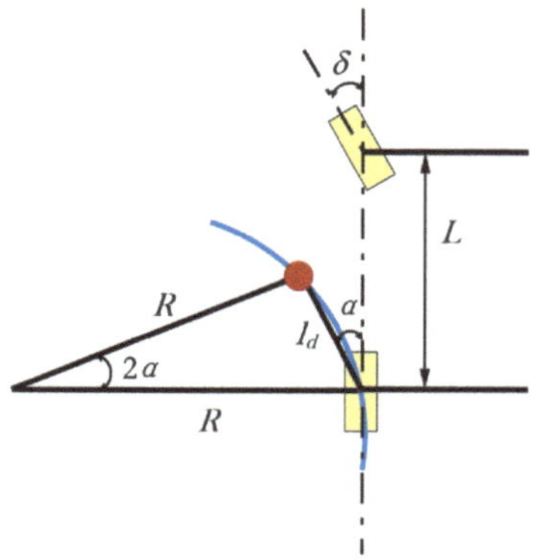

Figure 10. Two-wheel model of the ridge transplanter. *L* indicates the wheel base of the transplanter; α is the heading deviation; δ represents the steering angle of the front wheel; *R* means the turning radius, and L_d denotes the look-ahead distance.

So as to ensure that the center of the rear axle can overlap with the selected front sight points, the front wheel steering angle must be controlled. The steering angle of the front wheel can be computed by Equation (10).

$$\delta = arctan\frac{L}{R} \tag{10}$$

where δ represents the steering angle of front wheel, rad.

L indicates the wheel base of the transplanter, mm;

R represents the turning radius, mm.

Substituting Equation (9) into Equation (10), the steering angle of the front wheel can be deduced as Equation (11) [24].

$$\delta = \arctan\left(\frac{2 \times L \times sin\alpha}{l_d}\right) \tag{11}$$

According to Equation (11), the front wheel steering angle was determined by the wheel base, look-ahead distance, and heading deviation. Because the sampling period was short and the traveling velocity was slow, the heading deviation between two continuous sampling periods was viewed as the same, so that the influence of the heading deviation can be neglected. The heading deviation from the center of the rear axle to the front sight point can be expressed by Equation (12).

$$\alpha = arcsin\left(\frac{M_t^r - M_t^l}{l_d}\right) \tag{12}$$

where ${M_t}^l$ is the filtered result between the left ultrasonic sensor to the left sidewall of ridge, mm.

${M_t}^r$ is the filtered result between the right ultrasonic sensor to the right sidewall of ridge, mm;

l_d is the look-ahead distance, mm.

Substituting Equation (12) into Equation (11), the steering angle model can be calculated by Equation (13).

$$\delta = \arctan\left[\frac{2 \times L \times \left(M_t^r - M_t^l\right)}{{l_d}^2}\right] \tag{13}$$

Equation (13) shows that the front wheel steering angle was determined by the wheel base, left ridge distance, right ridge distance, and look-ahead distance. Both of the left ridge distance and right ridge distance can be measured by the ultrasonic sensors. According to the aforementioned steering angle model, the look-ahead distance was the only variable to be determined.

2.7. Fuzzy Look-Ahead Distance Decision Model

In this research, the optimal front sight point was selected from the optional ridge centers measured in the latest n sampling periods. The total number of ridge centers in the optional queue can be computed by Equation (8). When the look-ahead distance was short, the lateral adjustment ability was enhanced and the transplanter quickly approached the front sight point with a large curvature. However, a short look-ahead distance led to the trajectory-tracking oscillation when approaching the front sight point. When the look-ahead distance was long, the transplanter slowly approached the reference path with a small curvature. Under such a condition, the tracking route did not oscillate, but the adjustment time was longer. Therefore, according to the above analyses, the look-ahead distance needs to be regulated according to different lateral deviation conditions.

As for the relationship between the traveling velocity and look-ahead distance, if the traveling velocity of the transplanter was large, then the look-ahead distance should become large in order to improve the stability in the tracking process. In contrast, if the velocity of the transplanter was low, then the look-ahead distance should be small to improve the precision of tracking.

According to the analyses above, the lateral deviation and the traveling velocity were used as the factors for deciding the look-ahead distance. Based on different lateral deviations and velocities, a fuzzy look-ahead distance decision model was proposed to determine the appropriate look-ahead distance. The inputs of the fuzzy look-ahead decision method were the lateral deviation and traveling velocity. Additionally, the output of the fuzzy look-ahead decision model was the serial number of the front sight points. In the next step, the look-ahead distance can be computed by the norm of the vector from the center of the rear axle (the original of the coordinate system of the transplanter) to the selected front sight point, which is listed in Equation (14).

$$l_d = \sqrt{\left(x_t^k\right)^2 + \left(y_t^k\right)^2} \quad (14)$$

where l_d denotes the look-ahead distance, mm.

$x_t{}^k$ implies the x-coordinate of the kth ridge center at the sampling period t, mm;

$y_t{}^k$ implies the y-coordinate of the kth ridge center at the sampling period t, mm;

k means the serial number of selected front sight point;

t means the current sample period.

The distance between the two ultrasonic sensors was 804 mm and the width of the standard ridge was 600 mm. If one side of the transplanter contacted the ridge sidewall, then the maximum lateral deviation was 204 mm. According to above analyses, the basic domain of the lateral deviation was set as {0 mm, 210 mm}. As for the traveling velocity, the universal traveling velocity range of the ridge transplanter was from 37.5 mm/s to 62.5 mm/s, so the range {30 mm/s, 70 mm/s} was set as the basic domain of the traveling velocity.

The longitudinal separation from the center of ultrasonic sensors to the center of the rear axle was 1250 mm. So as to leave transplanter enough distance to regulate, the smallest serial number of the optional front sight points was chosen as 10th. If the transplanter traveled at velocity of 37.5 mm/s, then the max serial number of optional front sight points was 33rd. Therefore, the basic domain of the serial number of the optional front sight points was {10, 30}. The membership functions of the lateral deviation, traveling velocity, and serial number of optional front sight points were triangular, as shown in Figure 11.

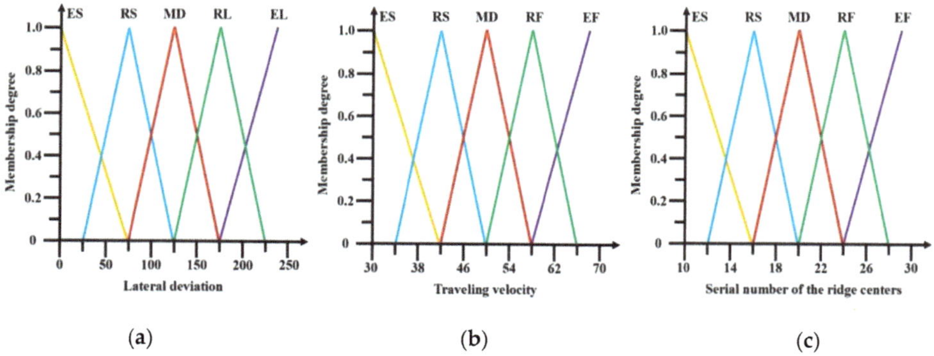

Figure 11. Membership function distributions of the lateral deviation, traveling speed, and serial number of front sight points. (**a**) The membership function of the lateral deviation; (**b**) the membership function of the traveling speed; (**c**) the membership function of the serial number of front sight points.

The fuzzy sets of lateral deviations were classified as {ES (extremely small), RS (relatively small), MD (medium), RL (relatively large), and EL (extremely large)}. The fuzzy sets of the travel velocity were classified as {ES (extremely slow), RS (relatively slow), MD (medium), RF (relatively fast), and EF (extremely fast)}. The fuzzy sets of the serial number of the optional front sight points were classified as {ES (extremely small), RS (relatively small), MD (medium), RL (relatively large), and EL (extremely large)}. The inference rules used to obtaining the optimal serial number of the front sight points are listed in Table 3.

Table 3. Fuzzy inference rules for the optimal serial number of the front sight points.

Serial Number of Front Sight Points		Traveling Velocity				
		ES	RS	MD	RF	EF
Lateral Deviation	ES	RL	RL	EL	EL	EL
	RS	MD	MD	RL	EL	EL
	ZO	RS	RS	MD	RL	RL
	RL	ES	RS	MD	MD	MD
	EL	ES	ES	ES	RS	RS

The centroid method was used to defuzzification of the serial number of the front sight points. If the consequence of the defuzzification was a decimal, the rounding-off method was applied to round up the decimal. After obtaining the serial number of the front sight points, the look-ahead distance and the front wheel steering angle were calculated by Equations (13) and (14).

2.8. Field Experiments

2.8.1. Validation Experiment of the LSW Filtering Algorithm

The premise of accurate ridge tracking was to obtain the precise lateral deviation between the transplanter and ridge. Thus, a full factorial experiment was conducted to compare the performance of the proposed LSW filtering algorithm and the commonly used direct measuring method.

The theoretical measuring distance and traveling velocity of the ridge transplanter were selected as the experimental factors. Under each factorial combination, the experiment was repeated three times. The top width, bottom width, and height of the ridge were 400 mm, 600 mm, and 300 mm, respectively. The theoretical measuring distance between the ultrasonic sensor and the ridge sidewall was set manually at the start point and then the transplanter traveled forward automatically. The picture of the validation experiment of the LSW filtering algorithm is displayed in Figure 12.

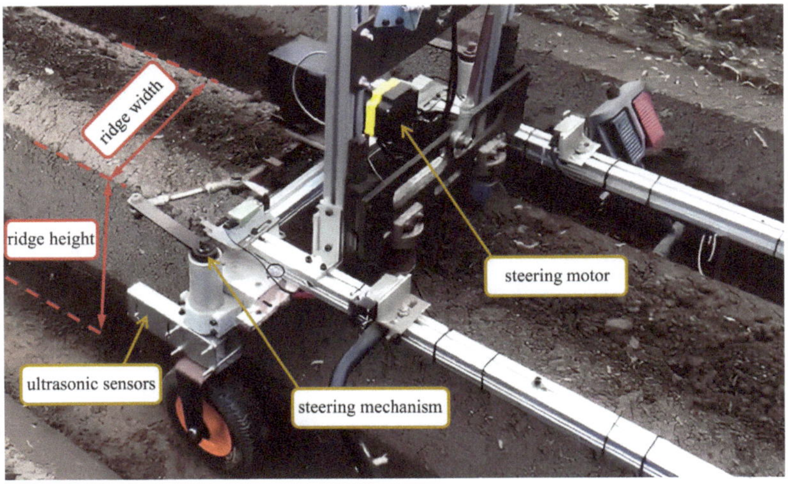

Figure 12. Validation experiment of the LSW filtering algorithm.

According to the actual velocity range of the ridge transplanter, the experimental levels of traveling velocity were 37.5 mm/s, 50 mm/s, and 62.5 mm/s. Moreover, the factorial levels of the theoretical measuring distance were 30 mm, 90 mm, 150 mm, and 210 mm. The indicator of the validating experiments was the mean absolute percentage error (MAPE), which can be calculated according to Equation (15).

$$\mathrm{MAPE} = \frac{1}{N} \cdot \sum_{i=1}^{N} \left[\left| \frac{\mathrm{TD} - MD_i}{\mathrm{TD}} \right| \times 100\% \right] \tag{15}$$

where MAPE represents mean absolute percentage error, %.

TD means theoretical measuring distance, mm;

MD_i indicates the measuring distance, mm;

i is the serial number of repetitions;

N is the repetition of each level combination; in this study, N is equal to 3.

2.8.2. Comparison Experiment of the Proposed Ultrasonic Ridge-Tracking (URT) Method and the Universal Pure Pursuit (UPP) Method

In order to test the performance and robustness of the proposed URT method and the universal pure pursuit (UPP) method, field experiments were conducted. The experimental greenhouse is named DaXiaoHua family farm (119.34° E, 31.98° N) and is located in Jurong city, Jiangsu province. The top width, bottom width, height, and length of the experimental ridge was 400 mm, 600 mm, 250 mm and 50 m, respectively.

The traveling velocity of the transplanter was regarded as the experimental factor, and the experimental levels of the forward velocity were 37.5 mm/s, 50 mm/s and 62.5 mm/s. At each experimental level, the transplanter traveled 20 m.

Moreover, the lateral deviation was used as the experimental index to judge the performance of the proposed URT method and conventional UPP method. The URT method used the variable look-ahead distances and the conventional UPP method applied the current measured ridge center as the front sight point. The procedures of the experiment of the proposed ridge-tracking system are demonstrated below.

Before the test, red liquid dye was added to a container that had a hole in its cap. Then, the container was fixed to the center of the rear axle of the transplanter. Additionally, the traveling velocity on human machine interface was set. When the transplanter traveled along the ridge, a red line of transplanter trace was formed on the top surface of the experimental ridge. The red trace line was the projection of the transplanter center on the ridge top.

After each experiment, ten ridge centers were measured with an interval of 100 mm at the start points of 0 m, 5 m, 10 m, and 15 m. The schematic of the whole experimental route is shown in Figure 13. The centers of the ridge top surface were marked with green dye. The deviation between the green markers and the red track line were measured by a vernier caliper with a precision of 0.01 mm. Then, the lateral deviation between the trace of the transplanter and the ridge centers was obtained. In order to validate the deviation correction capability of the two tracking methods, the center of the transplanter was set to deviate 50 mm from that of the ridge at the starting point. The field experiment of the proposed URT method and conventional UPP method is illustrated in Figure 14.

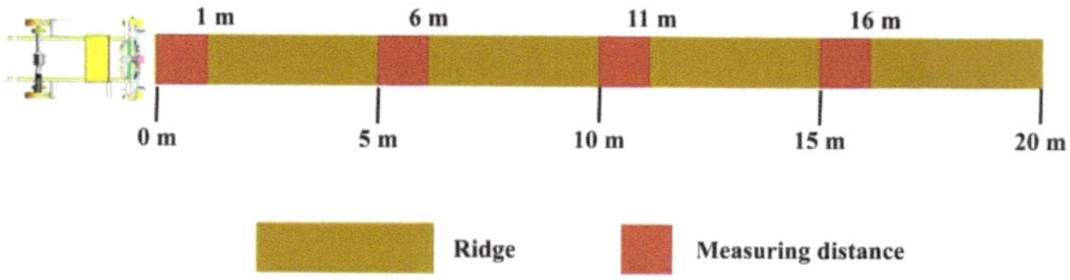

Figure 13. The measuring positions along the ridge.

Figure 14. Validation experiment of the URT method.

Furthermore, the indicators of the validation experiment were mean absolute lateral deviation (MALD), maximum absolute lateral deviation (MXALD), and root mean square lateral deviation (RMSLD). The formulas of the MALD and RMSLD are listed in Equations (16) and (17).

$$\text{MALD} = \frac{1}{N} \cdot \sum_{i=1}^{N} |E_i| \tag{16}$$

$$\text{RMSLD} = \sqrt{\frac{1}{N} \cdot \sum_{i=1}^{N} (E_i)^2} \tag{17}$$

where MALD is the mean absolute lateral deviation, mm.

RMSLD denotes the root mean square lateral deviation, mm;

N means the total number of lateral deviations;

E_i is the lateral deviation between the trance of transplanter and the ridge center, mm; i implies the serial number of the lateral deviations.

3. Results

3.1. Performance Comparison between the LSW Filtering Method and the Direct Measuring Method

The final results of the LSW filtering method and the direct measuring method are displayed in Table 4 and the average values of the final results are displayed in Figure 15. Because the raw outputs of the ultrasonic sensor were integrals, the final results of the direct measuring method were also integrals. Moreover, the final results of the LSW filtering method were decimals because they were the average value of the data in the sliding window. The ranges of the absolute measuring errors of the LSW filtering method and the direct measuring method were 0.6 mm~5.4 mm and 1 mm~7 mm.

Table 4. The final results of the LSW filtering method and direct measuring method.

Theoretical Distance (mm)	Velocity (mm/s)	Final Results (mm)		Theoretical Distance (mm)	Velocity (mm/s)	Final Results (mm)	
		LSW Filtering Method	Direct Measuring Method			LSW Filtering Method	Direct Measuring Method
30	37.5	31.4	31	90	37.5	87.8	86
		31.6	32			92.4	93
		30.8	31			88.2	87
	50	30.6	32		50	93.2	95
		31.4	30			92.4	94
		31.2	31			87.6	88
	62.5	31.2	32		62.5	87.4	87
		31	31			93.4	94
		32.2	32			93.6	95
150	37.5	148.6	148	210	37.5	206.8	205
		152.8	153			207.4	214
		153.8	155			212.6	213
	50	153.4	145		50	213.6	214
		146.4	151			215.4	215
		148.6	147			208.2	206
	62.5	152.8	153		62.5	206.2	203
		154.6	147			216.4	214
		148.8	154			208.2	206

In most cases, the final results of the LSW filtering method were closer to the theoretical measuring distance. However, under the theoretical measuring distance of 30 mm and the traveling velocity of the 50 mm/s, the final result of the direct measuring method was equal to 30 mm, leading to a more accurate average and a lower MAPE value. The reason might be that the dead zone of the ultrasonic sensor was 0 mm~30 mm, and if the measuring distance was less than 30 mm, then the outputs of the ultrasonic sensor were saturated and the output was 30 mm. When conducting the experiment, the distance between the ultrasonic sensor and the sidewall of the ridge might be less than 30 mm, so the output of the ultrasonic was 30 mm.

Under each combination of the levels of factors, the MAPE of the LSW filtering method and direct measuring method are demonstrated in Figure 16. The range of the MAPE of the LSW filtering method was from 1.33% to 4.89%, and that of the direct measuring method was from 1.90% to 5.56%.

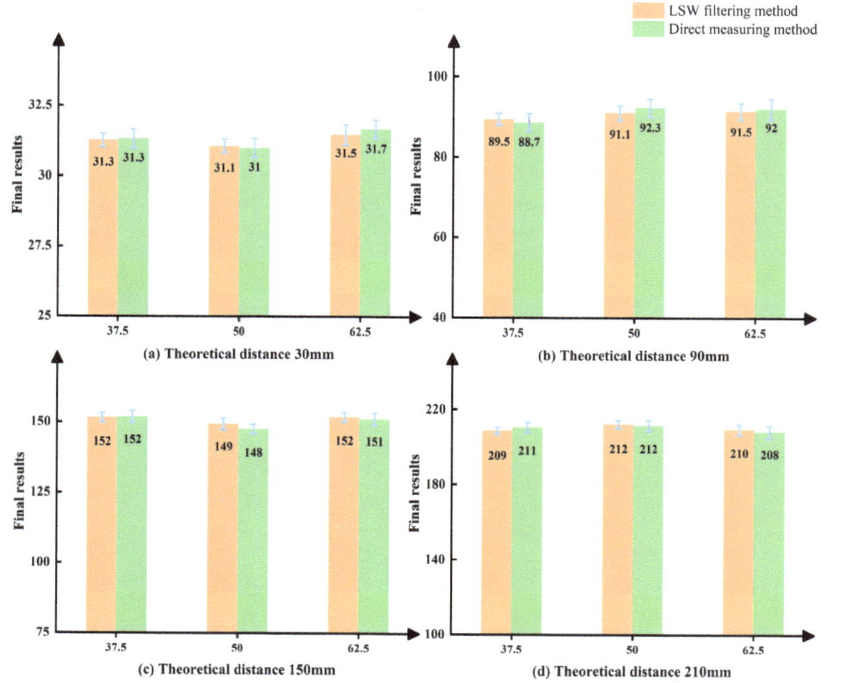

Figure 15. Average value of the final results of the LSW filtering method and the direct measuring method.

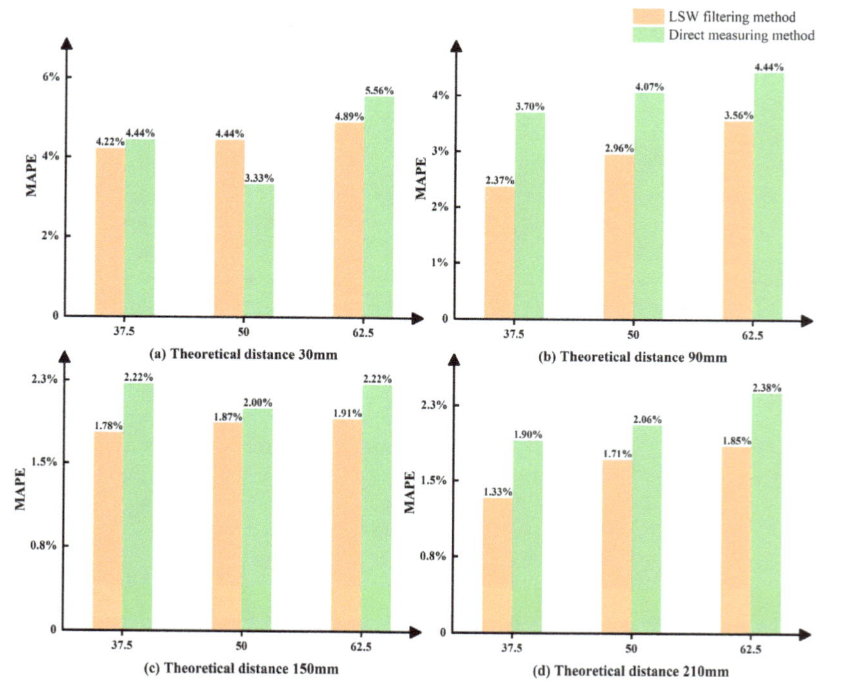

Figure 16. MAPE of the LSW filtering method and the direct measuring method.

As seen in Figure 16, when the theoretical measuring distance was the same, the MAPE of both measuring methods also rose as the traveling velocity increased, but the differences were not obvious. The velocity of the ultrasonic wave was greater than the traveling velocity of the transplanter, so the traveling velocity of the transplanter was not a significant factor in the accuracy of the final results.

As the theoretical measuring distance increased, the trend of the MAPE of both methods decreased at the same traveling velocity. As the theoretical measuring distance increased, the MAPE of both methods decreased, although the absolute errors of the final results increased. Because the measuring errors of the ultrasonic sensor were within a certain range, the relative error rate was lower as the theoretical measuring distance increased.

Under each combination of the levels of factors, the MAPE of the LSW filtering method was less than that of the direct measuring method. The reason is analyzed below. The raw outputs of the ultrasonic sensor had random measuring noise. The direct measuring method used the raw outputs as the measuring results, and hence the final results also had the random noise. Regarding to the LSW filtering method, the largest and the smallest measuring data in the sliding window were removed and the average value of the remaining data were reviewed as the final measuring result. According to the above analysis, the positive random measuring noise and negative ones were counteracted in the average solving process, which led to the final results with less random noise. That was the reason why the measuring performance of the proposed LSW filtering method was better than that of the direct measuring method.

3.2. *Tracking Performance of the Proposed URT Method and the UPP Method*

After each experiment, the traveling trajectory (red line) of transplanter was acquired and the ridge centers (green points) were measured, which is shown in Figure 17.

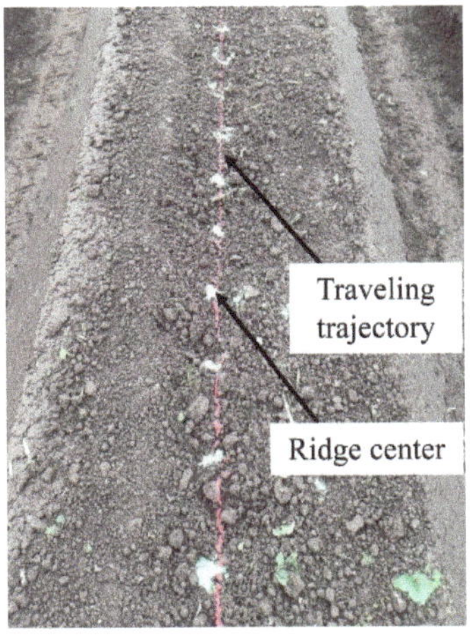

Figure 17. The marks of the traveling trajectory and ridge centers.

When the transplanter traveled at different velocities, the trend of the lateral deviations of the proposed URT method and conventional UPP method are shown in Figure 18. Positive lateral deviation meant the transplanter skewed to the right direction and negative lateral deviation meant the chassis skewed to the left direction. When the traveling distance exceeded 5 m, the transplanter was viewed as traveling in the steady state. At the traveling velocity of 37.5 mm/s, 50 mm/s, and 62.5 mm/s, the steady-state lateral deviations of the URT method were in the range of 1.25 mm~12.44 mm, 1.66 mm~14.74 mm, and 2.35 mm~18.87 mm, respectively; the steady-state lateral deviations of the UPP method ranged from 2.74 mm to 18.55 mm, 3.29 mm to 19.54 mm, and 3.24 mm to 23.23 mm, respectively.

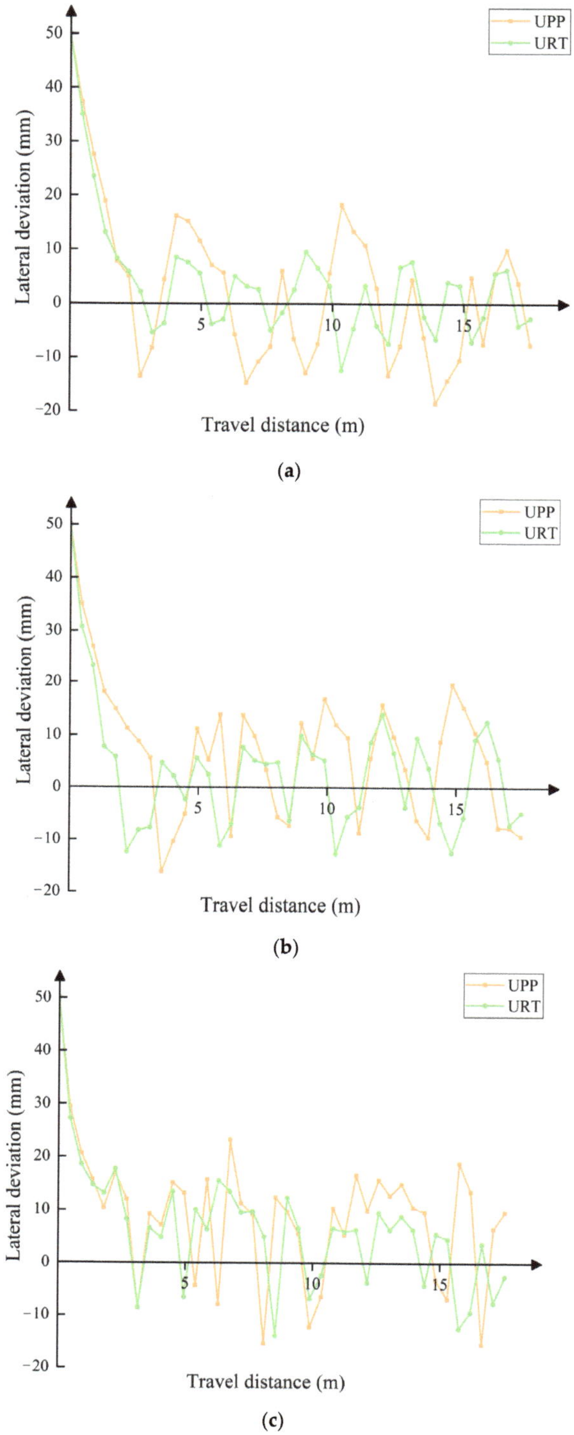

Figure 18. The lateral deviations controlled by the URT method and UPP method. (**a**) Lateral deviations at traveling velocity of 37.5 mm/s. (**b**) Lateral deviations at traveling velocity of 50 mm/s. (**c**) Lateral deviations at traveling velocity of 62.5 mm/s.

The MALD, MXALD, and RMSLD of the proposed URT method and conventional UPP method are shown in Table 5. When traveling at the same velocity, the MALD, MXALD, and RMSLD were less than their corresponding indicators in the UPP method, respectively. No matter the URT method or the UPP method, the MALD value at the traveling velocity of 62.5 mm/s was the highest, and that at the traveling velocity of 37.5 mm/s was the lowest. The MXALD values and RMSLD values also conformed to the same trend. The experimental results show that as the traveling velocity rose, the tracking performance of both of the URT method and the UPP method declined.

Table 5. The ridge-tracking indicators of the URT method and the UPP method.

Traveling Velocity (mm/s)	MALD (mm)		MXALD (mm)		RMSLD (mm)	
	URT	UPP	URT	UPP	URT	UPP
37.5	7.39	11.67	12.44	18.55	11.85	15.05
50	8.96	11.35	14.74	19.54	12.56	14.61
62.5	10.56	13.11	18.87	23.23	13.52	15.66

The reason why the faster the velocity, the worse the ridge-tracking performance will be analyzed. In the same sampling period, if the velocity was faster, then the transplanter traveled a longer distance compared to that at a low velocity. According to the tangent triangle function, the same heading error at a high velocity resulted in a significant lateral deviation because the hypotenuse was greater. Moreover, under the high-velocity condition, the traveling direction of a transplanter was more likely to be affected by the vibrations from the ground. The aforementioned two reasons may mean that the lateral deviations at a high velocity were greater than those at a low velocity.

4. Discussion

4.1. Tracking Performance Comparison

The curvature of the ridge is approximately a straight line, and thus some trajectory-tracking methods for straight line were compared to the research outcomes of Yang et al. [24], Hu et al. [25], and Zhang et al. [26] in Table 6, all of whom focused on path-tracking control for agriculture machinery. The proposed URT method exhibited significant advantages in both MALD and MXLD. Furthermore, it cannot be ignored that the previous results traveled along straight paths on a plat field, which made it much less difficult to follow the ridge. According to the above comparisons, the developed ultrasonic ridge-tracking method in this research had a desirable tracking performance in the ridge condition.

Table 6. Comparison between the developed URT method to previous route-tracking methods of agricultural machinery.

Tracking Method	Mean Absolute Lateral Deviation (mm)	Maximum Absolute Lateral Deviation (mm)
Developed URT method	8.42	18.87
Yang et al. [24]	18	46
Hu et al. [25]	35.4	127.2
Zhang et al. [26]	88	32

4.2. Strengths of the Proposed Ultrasonic Ridge-Tracking Method

In order to allow for seedlings with a certain row space and seedling space, the ridge transplanter should be able to track along the ridge. Some sensors, like GPS, cannot detect the position of the ridge, while machine vision or LiDAR would be affected by hard light in agricultural conditions. In order to overcome the mentioned sensing problems, the developed ridge-tracking method in this research used ultrasonic sensors to monitoring the relative position between the ridge and the transplanter. The robustness of the ultrasonic sensor was the great compared to machine vision and LiDAR. The key innovations of the proposed URT method were the LSW filtering algorithm and the optimal look-ahead distance decision model. The strengths of the LSW filtering algorithm and the optimal look-ahead distance decision model are listed below.

1. The ridge damage phenomenon results in erroneous judgement of the steering angle. However, the proposed LSW filtering algorithm can filter the rapidly changing data, which probably represent the distances from the ultrasonic sensor to the ridge pits. Additionally, the proposed LSW method can improve the accuracy by fusing the measurement results from previous sampling periods. Hence, the tracking performance of the proposed URT method would not be obviously affected by ridge damage.
2. The conventional UPP method uses a certain look-ahead distance to determine the fore sight points, which may lead to a long regulating time under sizeable lateral deviation conditions or oscillations in the small lateral deviation conditions. The proposed URT method regulates the optimal look-ahead distance based on the actual lateral deviations and velocities, so the transplanter can keep a desirable ridge-tracking performance in real time.

4.3. Implication

The proposed URT method can not only be applied to the ridge transplanter, but also to other ridge-operation machinery, such as the harvester, seeder, and so forth. In other fields, the proposed tracking method can also be used in situations where robots or machinery move along a solid object below the chassis. For example, in logistics warehouses, transport robots can use the proposed tracking method to track the invariable route, which uses solid cuboid as the reference path.

4.4. Weaknesses and Future Research

Although the proposed URT method had ideal tracking performance when traveling along ridges, it is still a challenging operation to automate in ridge conditions. Ssymmetrical ultrasonic sensors can only monitor the distances to the ridge currently being tracked; that is to say, only the relative position between the operating ridge to the transplanter can be acquired. The relative position from the transplanter to the other ridges cannot be obtained. When traveling at the end of the ridges, the transplanter needs to be stopped and moved to an adjacent unoperated ridge manually.

The automatic level of the proposed URT method could be improved if the following suggestions are accepted in the future.

When a transplanter completes the operation on the current ridge, it needs to move off the current ridge and move on to an adjacent one. In future work, the URT system could fuse other sensing information to detect adjacent ridges. Based on the above system, the simultaneous location and mapping (SLAM) method of the ridge transplanter could be developed in the future.

Using the proposed URT method, the coordinates of the ridge centers obtained in previous sampling periods need to be transformed into the current coordinate system at the current sampling period by the coordination transform method. Thus, the position of the transplanter in the current sampling period relative to the previous sampling period is crucial. If the Kalman filter is used for estimating the positions and postures of the transplanter, then the coordinate transform of the ridge centers might be more accurate, which could result in a better tracking performance.

5. Conclusions

The main successes in this research were the development of an ultrasonic ridge-tracking method and its corresponding system for the ridge transplanter. The LSW filtering method was proposed to improve the accuracy of the ultrasonic measuring results and remove the abnormal values. Moreover, the fuzzy look-ahead distance decision model was developed to determine the optimal look-ahead distance in real time. The validation experiment of the proposed LSW filtering method was conducted. Additionally, the tracking performances of the proposed URT method and the conventional UPP method were compared under field conditions.

With the same combination of factorial levels, the performance of the LSW filtering method was better than that of the direct measuring method. The MAPE values of the LSW filtering method and the direct measuring method were less than 4.89% and 5.56%. All of the MAPE values of the proposed LSW filtering method were less than 5%, and therefore the LSW filtering method was recommended to obtain the distances from the sidewalls to the ultrasonic sensors.

At different traveling velocities, the tracking performance of the URT method and that of the UPP method were compared. The greatest MALD values of the lateral deviations of the URT method and the UPP method were 10.56 mm and 13.11 mm. The greatest MXALD values of the lateral deviations of the URT method and the UPP method were 18.87 mm and 23.23 mm. Moreover, the greatest RMSLD values of the lateral deviations of the URT and the UPP were 13.52 mm and 15.66 mm. The experimental results showed that under the same combination of factorial levels, the tracking performance of the proposed URT method was better than that of the conventional UPP method.

According to the ridge-tracking performance of the proposed URT method, it can be used in practical agricultural ridge conditions, for example, transplanting, seeding, harvesting, and so on. Moreover, in other fields, robots or intelligent machinery can also apply the proposed URT method to track objects similar to ridges.

Author Contributions: Conceptualization, W.L. and J.Z.; methodology, W.L., M.Y. and Y.L.; software, W.L. and J.C.; validation, T.Z. and J.H.; formal analysis, W.L., J.H. and X.C.; investigation, M.Y. and T.Z.; resources, J.H.; data curation, W.L. and J.Z.; writing—original draft preparation, W.L.; writing—review and editing, W.L., C.Z. and J.H. All authors have read and agreed to the published version of the manuscript.

Funding: This work was supported by the Shanghai Agricultural Science and Technology Innovation Project "Research and Development of Key Intelligent Technologies for Fully Automated Lettuce Transplanting Equipment (2023-02-08-00-12-F04592)", the Priority Academic Program Development of Jiangsu Higher Education Institutions (no. PAPD-2023-87) and National Natural Science Foundation of China (Grant number 31901455).

Institutional Review Board Statement: Not applicable.

Data Availability Statement: Data are contained within the article.

Conflicts of Interest: The authors declare no conflict of interest.

References

1. Li, J.; Shang, Z.; Li, R ; Cui, B. Adaptive Sliding Mode Path Tracking Control of Unmanned Rice Transplanter. *Agriculture* **2022**, *12*, 1225. [CrossRef]
2. Fu, D.; Chen, Z.; Yao, Z.; Liang, Z.; Cai, Y.; Liu, C.; Tang, Z.; Lin, C.; Feng, X.; Qi, L. Vision-based trajectory generation and tracking algorithm for maneuvering of a paddy field robot. *Comput. Electron. Agric.* **2024**, *226*, 109368. [CrossRef]
3. Zhang, T.; Zhou, J.; Liu, W.; Yue, R.; Yao, M.; Shi, J.; Hu, J. Seedling-YOLO: High-Efficiency Target Detection Algorithm for Field Broccoli Seedling Transplanting Quality Based on YOLOv7-Tiny. *Agronomy* **2024**, *14*, 931. [CrossRef]
4. Tang, L.; Yan, F.; Zou, B.; Wang, K.; Lv, C. An Improved Kinematic Model Predictive Control for High-Speed Path Tracking of Autonomous Vehicles. *IEEE Access* **2020**, *8*, 51400–51413. [CrossRef]
5. Zhong, S.; Peng, D.; Huang, B.; Ma, L. Path-Tracking Ability of the ASV on Different Adhesion Coefficient Roads Based on Slide Mode Control. *Electronics* **2024**, *13*, 105. [CrossRef]
6. Cheng, J.; Zhang, B.; Zhang, C.; Zhang, Y.; Shen, G. A model-free adaptive predictive path-tracking controller with PID terms for tractors. *Biosyst. Eng* **2024**, *242*, 38–49. [CrossRef]
7. Shi, Y.; Chen, X.; Xi, X.; Shan, X.; Jin, Y.; Zhang, R. Research progress on the path tracking control methods for agricultural machinery navigation. *Trans. Chin. Soc. Agric. Eng. (Trans. CSAE)* **2023**, *39*, 1–14.
8. Ahn, J.; Shin, S.; Kim, M.; Park, J. Accurate Path Tracking by Adjusting Look-Ahead Point in Pure Pursuit Method. *Int. J. Automot. Technol.* **2021**, *22*, 119–129. [CrossRef]
9. Xia, Q.; Chen, L.; Xu, X.; Cai, Y.; Jiang, H.; Pan, G. Expected yaw rate–based trajectory tracking control with vision delay for intelligent vehicle. *Sci. Prog.* **2020**, *103*, 1–23. [CrossRef]

10. Xu, S.; Peng, H.; Tang, Y. Preview Path Tracking Control With Delay Compensation for Autonomous Vehicles. *IEEE Trans. Intell. Transp. Syst.* **2021**, *22*, 2979–2989. [CrossRef]
11. Xie, J.; Xu, X.; Wang, F.; Tang, Z.; Chen, L. Coordinated control based path following of distributed drive autonomous electric vehicles with yaw-moment control. *Control Eng. Pract.* **2021**, *106*, 104659. [CrossRef]
12. Wang, Z.; Chen, L.; Cai, Y.; Sun, X.; Wang, H. Adaptive coordinated control strategy for autonomous vehicles based on four-wheel steering. *Proc. Inst. Mech. Eng. Part D J. Automob. Eng.* **2023**, *2023*, 09544070231213755. [CrossRef]
13. Zhang, Y.; Liu, K.; Gao, F.; Zhao, F. Research on Path Planning and Path Tracking Control of Autonomous Vehicles Based on Improved APF and SMC. *Sensors* **2023**, *23*, 7918. [CrossRef]
14. Huang, W.; Ji, X.; Wang, A.; Wang, Y.; Wei, X. Straight-Line Path Tracking Control of Agricultural Tractor-Trailer Based on Fuzzy Sliding Mode Control. *Appl. Sci.* **2023**, *13*, 872. [CrossRef]
15. Shet, R.M.; Lakhekar, G.V.; Iyer, N.C. Intelligent fractional-order sliding mode control based maneuvering of an autonomous vehicle. *J. Ambient Intell. Humaniz. Comput.* **2024**, *15*, 2807–2826. [CrossRef]
16. Wu, L.; Liu, J.; Vazquez, S.; Mazumder, S.K. Sliding Mode Control in Power Converters and Drives: A Review. *IEEE/CAA J. Autom. Sin.* **2022**, *9*, 392–406. [CrossRef]
17. Kebbati, Y.; Ait-Oufroukh, N.; Ichalal, D.; Vigneron, V. Lateral control for autonomous wheeled vehicles: A technical review. *Asian J. Control* **2023**, *25*, 2539–2563. [CrossRef]
18. Zhang, M.; Lin, X.; Ding, Y.; Yin, W.; Qian, Y. Design of path following controllers based on performance index for agricultural vehicle. *Trans. Chin. Soc. Agric. Eng. (Trans. CSAE)* **2012**, *28*, 40–46.
19. Somefun, O.; Akingbade, K.; Dahunsi, F. The dilemma of PID tuning. *Annu. Rev. Control* **2021**, *52*, 65–74. [CrossRef]
20. Liu, W.; Wang, R.; Xie, C.; Ye, Q. Investigation on adaptive preview distance path tracking control with directional error compensation. *Proc. Inst. Mech. Eng. Part I J. Syst. Control Eng.* **2020**, *234*, 484–500. [CrossRef]
21. An, G.; Zhong, Z.; Yang, S.; Yang, L.; Jin, C.; Du, J.; Yin, X. EASS: An automatic steering system for agricultural wheeled vehicles using fuzzy control. *Comput. Electron. Agric.* **2024**, *217*, 108544. [CrossRef]
22. Lee, K.; Choi, H.; Kim, J. Development of Path Generation and Algorithm for Autonomous Combine Harvester Using Dual GPS Antenna. *Sensors* **2023**, *23*, 4944. [CrossRef] [PubMed]
23. Fargnoli, M.; Vita, L.B.; Gattamelata, D.; Laurendi, V.; Tronci, M. A Reverse Engineering Approach to Enhance Machinery Design for Safety. In Proceedings of the DS 70: Proceedings of DESIGN 2012, the 12th International Design Conference, Dubrovnik, Croatia, 21–24 May 2012; Volume 627, pp. 627–636.
24. Yang, Y.; Li, Y.; Wen, X.; Zhang, G.; Ma, Q.; Cheng, S.; Qi, J.; Xu, L.; Chen, L. An optimal goal point determination algorithm for automatic navigation of agricultural machinery: Improving the tracking accuracy of the Pure Pursuit algorithm. *Comput. Electron. Agric.* **2022**, *194*, 106760. [CrossRef]
25. Hu, C.; Ru, Y.; Li, X.; Fang, S.; Zhou, H.; Yan, X.; Liu, M.; Xie, R. Path tracking control for brake-steering tracked vehicles based on an improved pure pursuit algorithm. *Biosyst. Eng.* **2024**, *242*, 1–15. [CrossRef]
26. Zhang, C.; Dong, W.; Xiong, Z.; Hu, Z.; Wang, D.; Ding, Y. Design and Experiment of Fuzzy Adaptive Pure Pursuit Control of Crawler-type Rape Seeder. *Trans. Chin. Soc. Agric. Eng. (Trans. CSAE)* **2021**, *52*, 105–114.

agriculture

Article

Optimal Design and Tests of a Pulsating Roll-Cleaning Device for Tiger Nuts

Zhijun Lv, Wanzhang Wang, Dekun Yang, Xun He, Zhe Qu, Wenzhuo Jing, Yong Mei, Huinan Huang *, Baohao Su and Yangfan Zhuang

College of Mechanical and Electrical Engineering, Henan Agricultural University, Zhengzhou 450002, China; lvzhijun@henau.edu.cn (Z.L.); wangwz@henau.edu.cn (W.W.); YDKyangdekun@stu.henau.edu.cn (D.Y.); hexun@henau.edu.cn (X.H.); quzhe071171@henau.edu.cn (Z.Q.); jignwenzhuo@stu.henau.edu.cn (W.J.); meiyong@stu.henau.edu.cn (Y.M.); su18749566795@163.com (B.S.); zhuangyangfan@163.com (Y.Z.)
* Correspondence: huanghuinan@henau.edu.cn

Abstract: Aiming at the problems of low efficiency and poor cleaning quality in the cleaning process of harvested tiger nuts (*Cyperus esculentus*) with lots of soil and fibrous roots on the surface to meet the demand for mechanized harvesting of tiger nuts in the Yellow River, Huaihe River, and Haihe River regions in China, a pulsating roll-cleaning device was devised. The analytical software for meshing the transmission of pitch curves of elliptical gears with arbitrary orders was developed based on Visual Basic 6.0, thus obtaining a pair of elliptical gears that meet the steady transmission requirement. The meshing of this pair of elliptical gears enables pulsating variable-speed rotation of a brush roller, which reduces the breakage and accumulation of tiger nuts. Circular screen holes are distributed uniformly on the upper and lower cylindrical walls. Tiger nuts are cleaned under friction with the brush roller and the circular screen holes, while residues (including soil) fall in the settlement and accumulation tank through the screen holes, which realizes the rapid discharge and collection of residues, including soil, and reduction of dust. The separation process of soil from tiger nuts in the cleaning device was studied using RecurDyn 2023-EDEM 2022 simulation software. Taking the rate of rotation of the brush roller, order of elliptical gears, and eccentricity of the elliptical gears as the test factors while taking the cleaning efficiency, breakage rate, and impurity rate as the evaluation indices, a test bench for the cleaning performance of tiger nuts was established. Design-Expert V8.0.6 was applied to carry out a three-factor three-level Box–Behnken design (BBD) using response surface methodology (RSM), thus ascertaining the optimal parameter combination of the cleaning device. When the rate of rotation of the brush roller, order of elliptical gears, and eccentricity of the elliptical gears are, separately, 280 rpm, 2, and 0.122, the cleaning efficiency, breakage rate, and impurity rate of tiger nuts are 1.83 t·h^{-1}, 0.15%, and 0.16%, respectively. This satisfies the cleaning requirement of tiger nuts and provides a basis for the design of cleaning machines for tiger nuts.

Keywords: tiger nuts; cleaning device; elliptical gears; optimal design; simulation analysis

Citation: Lv, Z.; Wang, W.; Yang, D.; He, X.; Qu, Z.; Jing, W.; Mei, Y.; Huang, H.; Su, B.; Zhuang, Y. Optimal Design and Tests of a Pulsating Roll-Cleaning Device for Tiger Nuts. *Agriculture* **2024**, *14*, 1673. https://doi.org/10.3390/agriculture14101673

Academic Editor: Mustafa Ucgul

Received: 30 August 2024
Revised: 19 September 2024
Accepted: 20 September 2024
Published: 25 September 2024

1. Introduction

Agricultural production underpins the national economy and social development. Edible oil, as a necessity to daily living, is an important component of agricultural production [1]. In recent years, the demand for edible oil in China has increased year by year with the economy development. In the context of economic globalization, various countries have all enlarged their import and export trades of edible oil. The demand for edible oil in China, as a large importer of edible oil, has always exceeded supply due to the influence of factors including farmland, climate, and technology [2]. Therefore, China has no advantage over other countries in terms of the import and export trades of edible oil, giving rise to an urgent need to improve the import and export trades of edible oil in the country. Tiger nuts, as a new oil-bearing crop, do not compete with other oil-bearing crops such as soybeans, peanuts, and maize for cultivation land and are characterized by a high oil content and rich abundance of nutrients therein. Therefore, promoting the cultivation of tiger nuts is key to relieving the shortage of stock of vegetable oil in China [3–7].

Tiger nuts can be easily planted and managed and deliver high-volume oil production. Nevertheless, harvested tiger nuts are mixed with multiple impurities, including grass, gravel, and other soil. Although some cleaning and screening machines are available for removing grass and gravel, soil adhering to tiger nut surfaces must be manually removed (rather than by vibrating screen), owing to its strong bond with the tiger nut skin. This significantly decreases the production efficiency and economic benefit of the harvest.

There is little research into the impurity removal machinery of tiger nuts in foreign countries; the cleaning machinery used in the processing of tiger nuts commonly used in other countries is a type of soil removal device reliant on water washing [8,9]. After the removal of simple impurities, the harvested tiger nuts are delivered to a water washer, in which shells and bad tiger nuts floating in the water can be removed while tiger nuts coated with soil are cleaned by water flows [10]; because such a cleaning method consumes a huge amount of water and needs a perfect sewage disposal system, it is unsuitable for small- and medium-sized farmers [11]. Chinese scholars have also studied tiger nuts: Qi Jiangtao et al. [12] designed mechanical harvesting and conveying devices for tiger nuts and explored the vibration and soil removal rate in the conveying process through use of the extended distinct element method (EDEM). By doing so, they attained a group of optimal structural parameters, which screen out some soil while ensuring the conveying efficiency of tiger nuts. Zhao et al. [13] developed a harvesting and impurity removal system of tiger nuts that separates and removes different impurities mixed with tiger nuts through multi-stage screening. The tiger nut harvesting and screening mechanism devised by Zhang et al. [14] firstly threshed and then separated tiger nuts and was composed of a threshing system and a screening system. The angle of inclination of the screens and vibration amplitude on the screening efficiency were revealed by simulating the screen motion using ADAMS, which influences the rate of rotation of the crank. Finally, a group of optimal operation parameters of the screening mechanism were determined. The tiger nut harvester designed by Liu et al. [15] integrated functions including threshing, separation, and cleaning to good effect. In summary, the common washing cleaning method abroad has problems such as not being conducive to subsequent storage and easily causing tiger nuts to become moldy, while the dry cleaning method, which is less reported at present, has problems such as the uneven surface of tiger nuts and the bond between soil and nut surfaces, which makes it difficult to easily remove soil on its surface or the high breakage rate of nuts. Therefore, the optimization design of a tiger nuts dry cleaning machine needs to be carried out urgently.

The goal of the work is to design a pulsating roll-cleaning device aims to ensure efficient cleaning, reduce the breakage rate, improve the quality of harvested tiger nuts, and investigate the effect of the main parameters on the separation process of soil from tiger nuts in the cleaning device. Using RecurDyn 2023-EDEM 2022 coupling simulations and physical experiments, achieving the objectives to investigate the effect of the main parameters, such as the rate of rotation of the brush roller, order of the elliptical gears, and eccentricity of the elliptical gears, on the operational quality of the cleaning device; to evaluate the operational quality by introducing the cleaning efficiency, breakage rate, and impurity rate of tiger nuts; to determine a group of optimized parameters for the design of the cleaning device, the research findings in this work can provide a favorable material basis for the subsequent drying and other processes.

2. Overall Design and Working Principle of the Pulsating Roll-Cleaning Device

2.1. Overall Structure

To avoid damage to tiger nut tubers while removing soil adhering to tiger nut surfaces and meet the design requirement for the dry state of tiger nuts, a pulsating roll-cleaning device of tiger nuts was composed of a cleaning chamber, a discharge tube, a rack, an adjustable counterweight, an impurity removal blower, a settlement and collection tank, and a variable-speed transmission system. The overall structure is shown in Figure 1, and key technical parameters are shown in Table 1. Therein, the cleaning chamber consisted of a spindle, an upper cylinder, a lower cylinder, circular screen holes, a brush roller, and a feeding hopper. The variable-speed transmission system was composed of a YE3-160M-6 motor (made in Lu'an Yisheng Motor Co., Ltd, Lu'an, China), a pulley drive unit, and elliptical gears. The spindle was fixed with the brush roller.

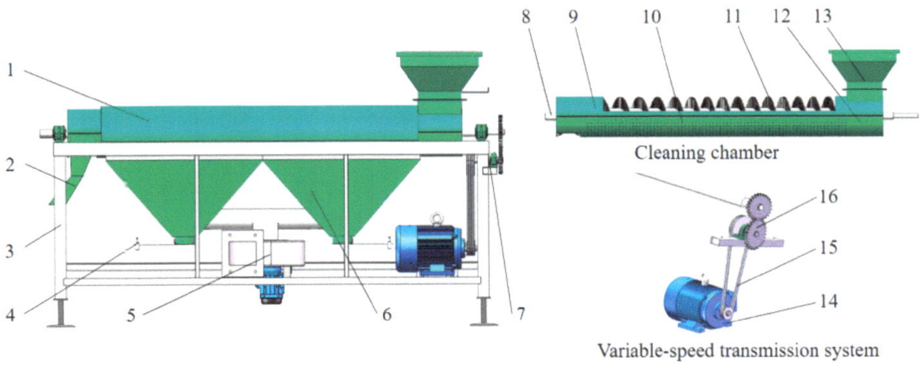

Figure 1. Structural diagram of the pulsating roll-cleaning device of tiger nuts. (1) Cleaning chamber; (2) discharge tube; (3) rack; (4) adjustable counterweight; (5) impurity removal blower; (6) settlement and collection tank; (7) drive system; (8) spindle; (9) upper cylinder; (10) circular screen holes; (11) brush roller; (12) lower cylinder; (13) feeding hopper; (14) motor; (15) pulley drive unit; (16) drive unit of high-order elliptical gears.

Table 1. Key technical parameters of the cleaning device.

Key Technical Parameter	Value	Unit
Length	3000	mm
Width	600	mm
Height	1970	mm
Diameter of brush roller	320	mm
Motor power	7.5	kW
Power of impurity removal blower	1.5	kW

2.2. Working Principle

When the pulsating roll-cleaning device of tiger nuts works, the motor drives a pair of elliptical gears to realize the pulsating variable-speed rotation of the brush roller. Then, tiger nuts enter the cleaning chamber from the feed inlet after adjusting the feed quantity with a push–pull baffle. Soil on the tiger nut surface is removed by friction with the brush roller and the circular screen holes during migration of tiger nuts because of the rotation of the brush roller. Subsequently, removed impurities, including soil and rootstocks, fall into the settlement and collection tank under action of the impurity removal blower. Therein, heavy impurities are discharged from the lower impurity outlet under action of the adjustable counterweight, while their lighter counterparts are expelled by the blower and collected. Being pushed by the brush roller, tiger nuts migrate to, and are discharged from, the discharge hole. In this way, clean tiger nuts are obtained.

3. Design of Key Structures

3.1. Design of the Variable-Speed Transmission System

In existing cleaning devices, lots of tiger nuts are damaged during the high-speed and constant-speed rotation of brush rollers directly driven by a belt drive unit used in the drive system. To solve the problem, the active elliptical gear is driven by a belt drive, and the active elliptical gear is meshed with the driven elliptical gear. The driven elliptical gear rotates coaxially with the brush rollers, so as to realize the pulsating variable-speed rotation of the brush rollers.

To ensure the stability of high-speed transmission, the gear transmission was analyzed by taking the meshing of first-order and second-order elliptical gears as an example, so as to determine the basic parameters of the gears [16,17]. In the equations below, $n = 2$ and $m = 1$.

The schematic diagram for the transmission of the second-order and first-order elliptical gears is displayed in Figure 2. The center of gyration of the first-order elliptical gear is located at one focus, while that of the high-order elliptical gear is located at the center of gravity of the pitch curve. Based on the equations below, the polar equation of the parameters of the high-order elliptical gear was deduced according to the basic parameters of the first-order elliptical gear, namely, the semi-major axis a_1 and the eccentricity e_1:

$$R_2(\Phi_2) = \frac{a_2(1 - e_2^2)}{1 + e_2 \cos nm\Phi_2} \tag{1}$$

$$R_1(\Phi_1) = A - R_2(\Phi_2) \tag{2}$$

$$A = a_1\left[1 + \sqrt{1 + (n^2 - 1)(1 - e_1^2)}\right] \tag{3}$$

$$a_2 = a_1\sqrt{1 + (n^2 - 1)(1 - e_1^2)} \tag{4}$$

$$e_2 = \frac{e_1}{\sqrt{1 + (n^2 - 1)(1 - e_1^2)}} \tag{5}$$

where e_1 and e_2 are, separately, eccentricities of the first-order and high-order elliptical gears; n is the average gear ratio; A is the center-to-center spacing; a_1 and a_2 are semi-major axes of the first-order and high-order elliptical gears (mm), respectively; m is the deformation coefficient.

When the first-order elliptical gear rotates by an angle Φ_1, the following high-order elliptical gear rotates by Φ_2. The arch lengths of the pitch curves of the two gears are identical, so the MN arc length is relative to the MP arc length, then

$$i_{21}(\Phi_2) = \frac{R_1}{R_2} = \frac{A - R_2}{R_2} \tag{6}$$

$$1 + i_{21} + i_{21}'' \geq 0 \tag{7}$$

To ensure the steady transmission of the cleaning device, the pitch curve of the high-order elliptical gear should not be concave, so it needs to calculate the maximum eccentricity e_1 to provide reference for the parametric design of the high-order elliptical gear.

$$e_{1\text{KP}} = \frac{1}{\sqrt{n^2m^4 - 2m^2 + 1}} \tag{8}$$

$$e_1 \leq e_{1\text{KP}} \tag{9}$$

where e_{1KP} is the maximum eccentricity that avoids the high-order elliptical gear having a concave pitch curve.

According to Equations (8) and (9), when the average gear ratio is 2 and the deformation coefficient is 1, the eccentricity of the first-order elliptical gear is shown as follows. Similarly, the eccentricities corresponding to different m and n can be deduced, which ensures the reasonability of the parametric design of key structures of the high-order elliptical gear.

When $n = 2$ and $m = 1$, $$e_1 \leq e_{1KP} = \frac{1}{\sqrt{3}}$$

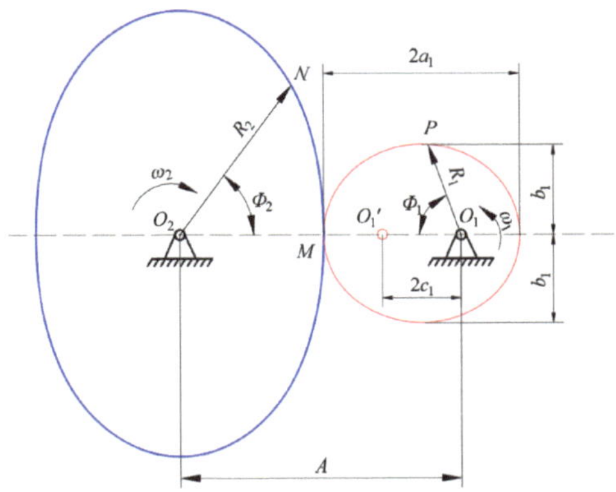

Figure 2. Transmission via second-order and first-order elliptical gears. Legend: ω_1 and ω_2 separately represent the angular velocities of the first-order and second-order elliptical gears; c_1 is the semi-focal length of the first-order elliptical gear; M is the current meshing point; N and P are meshing points after rotation through angle Φ.

On the basis of analyzing the above elliptical gear transmission theory, the analysis software for the meshing transmission of the pitch curves of the elliptical gears with arbitrary orders was developed (Computer Software Copyright Registration Number: 2024SR0352583). The initial interface of the software is shown in Figure 3.

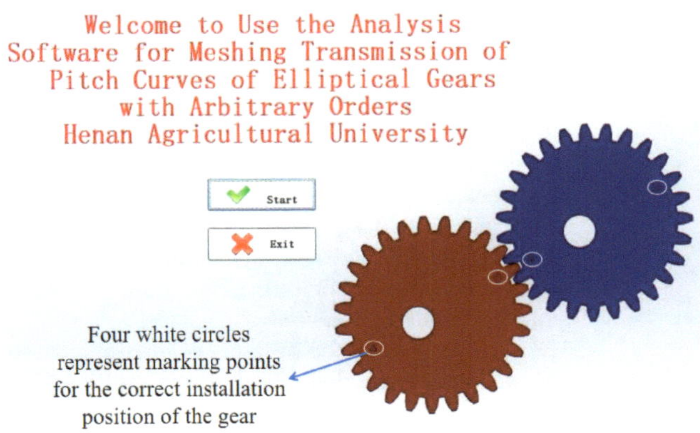

Figure 3. Initial interface of the software.

The work interface of the software includes seven parts, namely, ① menu bar, ② animation display area, ③ functional control area, ④ parameter adjustment area, ⑤ area to display changes in instantaneous gear ratio curves, ⑥ important data display area, and ⑦ area to display the progress toward the optimization goal, as shown in Figure 4.

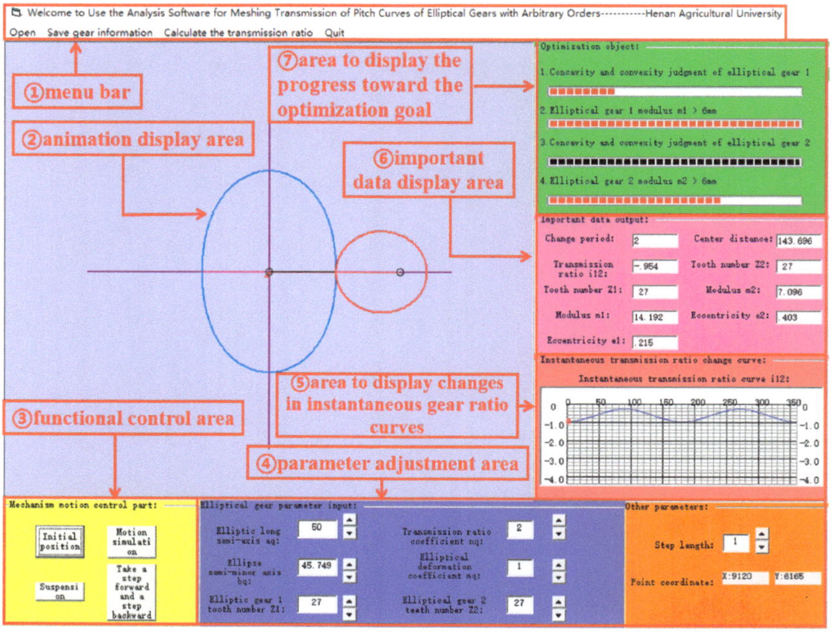

Figure 4. Design interface of the software.

1. Buttons on the "menu bar" can be used to invoke and save data and to calculate the gear ratio of the two gears, which is displayed in the "area to display changes in the instantaneous gear ratio curves".
2. The "animation display area" intuitively shows pitch curves of the designed elliptical gears. The diagrams also change in real time when adjusting the parameters of the elliptical gears, thus fully exhibiting the human–computer interaction function.
3. The "function control area" can control the start and pause of gear meshing and also achieve adjustment of the movement steps.
4. The "parameter adjustment area" can adjust the parameters, including the length of the semi-major axis, length of the semi-minor axis, number of teeth, gear ratio, and deformation coefficient of the elliptical gears. Through adjusting these parameters, elliptical gears with different orders, gear ratios, and eccentricities can be obtained (Figure 5).
5. The "area to display changes in the instantaneous gear ratio curves" can, in real time, show changes to the gear ratio, lending convenience to the computational process.
6. The "important data display area" can accurately display key parameters, including the gear ratio, number of teeth, module, and center-to-center spacing directly as digits, which facilitates a comparative analysis.
7. The "area to display the progress of the optimization goal" can intuitively exhibit whether the designed elliptical gears meet the design requirement or not. The parameters are regarded as satisfying the design requirement when all progress bars are red. As long as one progress bar is black, the parameters do not meet the design requirements and should be adjusted until they do.

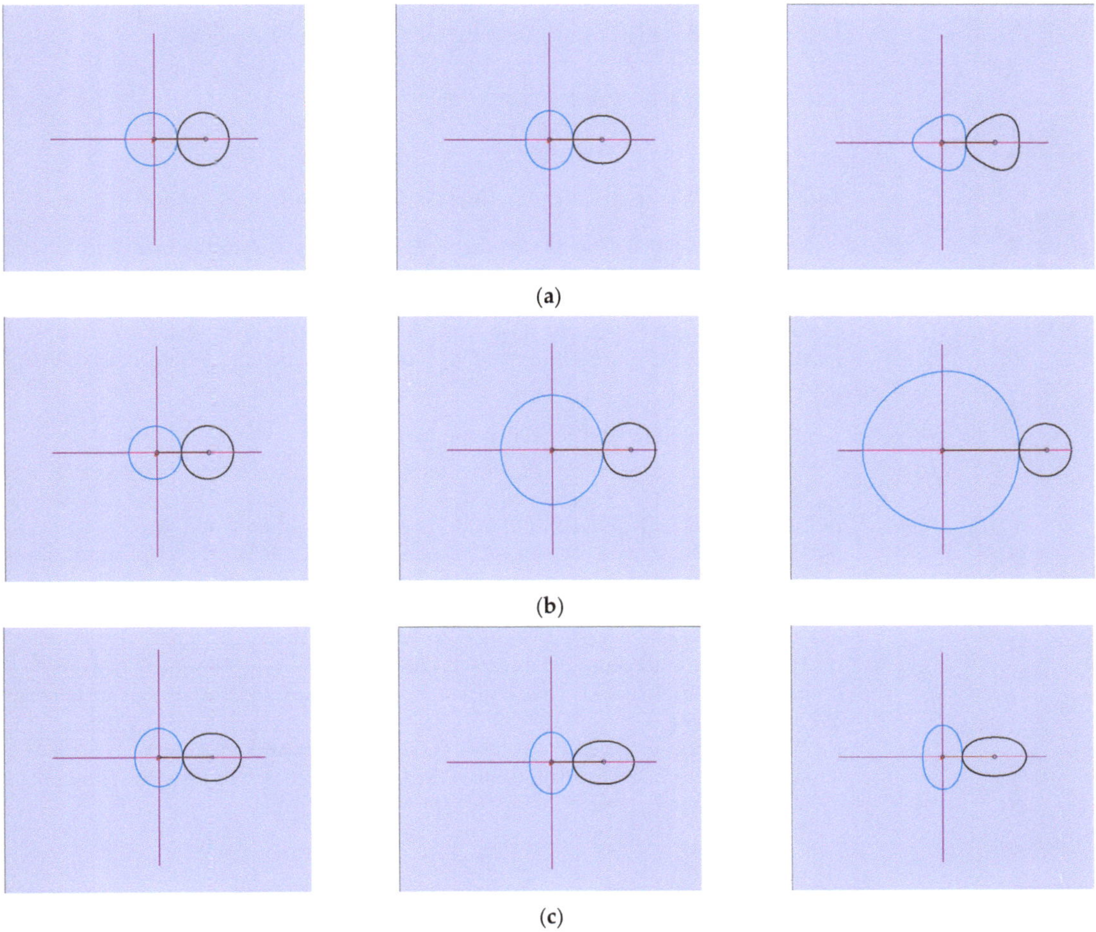

Figure 5. Elliptical gears with different parameters. (a) Elliptical gears with different orders. (b) Elliptical gears with different gear ratios. (c) Elliptical gears with different eccentricities.

SOLIDWORKS 2021 was adopted to establish 3D models of the elliptical gears, as shown in Figure 6.

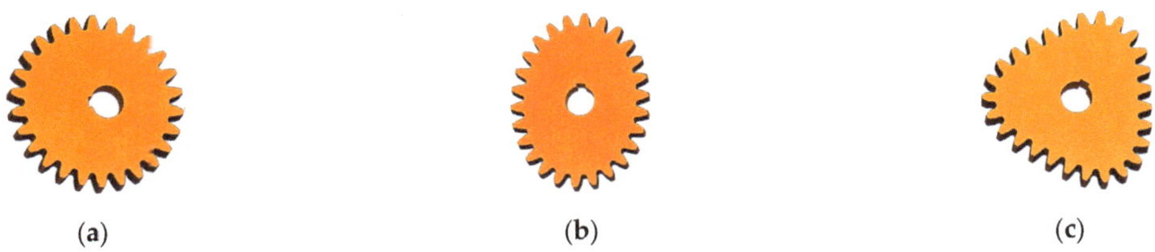

Figure 6. Three-dimensional models of elliptical gears. (a) First-order elliptical gears; (b) second-order elliptical gears; (c) third-order elliptical gears.

3.2. Design of Circular Screen Holes

The cleaning chamber is composed of two semicircular cylinders to facilitate disassembly. Circular screen holes (6-mm diameters) are uniformly distributed on the cylindrical walls, which increase the number of collisions between tiger nuts and cylindrical walls and facilitate the discharge of soil. The structure is illustrated in Figure 7.

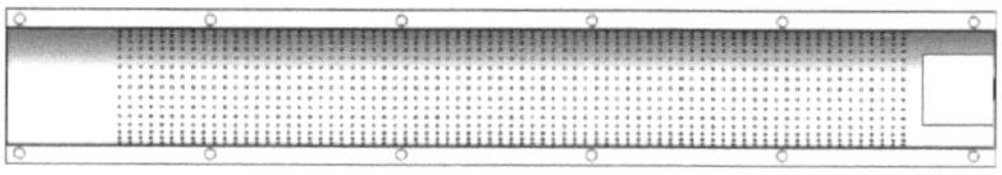

Figure 7. Schematic diagram of the circular screen holes on the lower cylindrical wall.

To ensure that the brush can push tiger nuts stuck onto the cylindrical wall, the clearance between the brush roller and drum screen was designed to be 5 mm, according to the typical dimensions of tiger nuts and the deflection variation of the brush roller shaft. Additionally, a sealed structure was formed by the perforated part of the drum screen and the settlement and collection tank to avoid environmental pollution due to excessive dust emission.

3.3. Design of the Settlement and Collection Tank

In order to improve the cleaning efficiency and reduce the impurity rate of tiger nuts, a settlement and collection tank was designed according to the falling characteristics of the particles, as displayed in Figure 8. The settlement and collection tank envelopes part of the circular screen holes to ensure that the whole device is in an intermittent, sealed state. Airflow can be formed under action of the impurity removal blower, so that particles including dust and fibrous roots fall rapidly and are then collected. Therein, heavy impurities are expelled via the impurity outlet under the action of the adjustable counterweight, while light detritus is discharged by the blower and collected. The dust pollution can be alleviated while efficiently discharging impurities in the settlement and collection process.

To guarantee the overall tightness and compact arrangement of the machinery, the length L_C is set to 2.24 m while the width can be calculated according to the principle of the constant air volume using the following formula [18]:

$$B_{\mathrm{min}} = \frac{Q_F \eta}{L_1 v_F} \tag{10}$$

where B_{min} is the minimum width of the settlement and collection tank (m); Q_F is the airflow (2.87 $\mathrm{m^3 \cdot s^{-1}}$); η is the sealing efficiency factor of the dedusting system, which is set to 0.93; L_C denotes the length of the settlement and collection tank (m); v_F is the falling speed of particles with a diameter of 1 mm ($\mathrm{m \cdot s^{-1}}$), which is taken as 5.47.

In this way, the minimum width is calculated to be 0.218 m. To adapt to the size of the cleaning device, B is set to 0.3 m.

According to the critical formula for the falling of particles, the design is such that

$$\frac{B_{\mathrm{min}}}{v_H} = \frac{h_{\mathrm{min}}}{v_S} \tag{11}$$

where h_{min} is the minimum height of the settlement and collection tank (m); v_H is the horizontal component of the airflow velocity ($\mathrm{m \cdot s^{-1}}$); v_S is the vertical component of the airflow velocity ($\mathrm{m \cdot s^{-1}}$).

When designing the settlement and collection tank, it is necessary to ensure that the dust particles take the same time to arrive at either the side edge or the bottom. Generally, the height of the tank should be larger than or equal to the width, so H is 0.62 m.

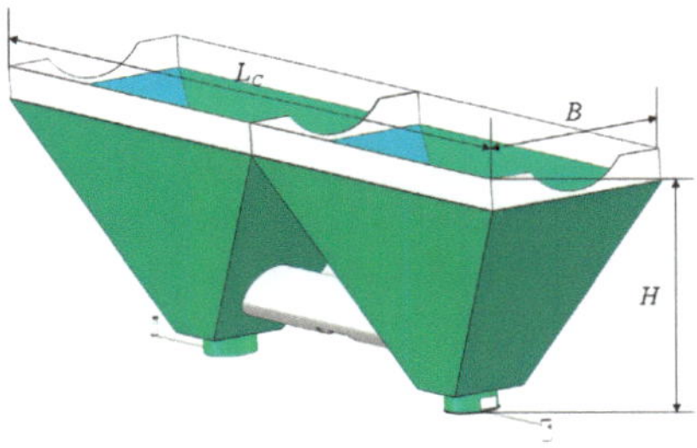

Figure 8. Three-dimensional model of the settlement and collection tank.

4. Simulation of the Working Process of the Cleaning Device Based on RecurDyn 2023-EDEM 2022 Coupling

4.1. Establishment of the Distinct Element Model of Tiger Nut Aggregates

To truthfully reproduce the state of the tiger nuts, the 3D dimensions of five groups of 100 tiger nuts were measured by the 111N-101 digital vernier caliper (accuracy of 0.01 mm, made in Foshan Lenghui Quantity Cutting Tools Co., Ltd., Foshan, China) in basic tests; that is, the average length *L*, width *W*, and thickness *T* are 12.46 mm, 11.53 mm, and 10.15 mm, respectively, as shown in Figure 9.

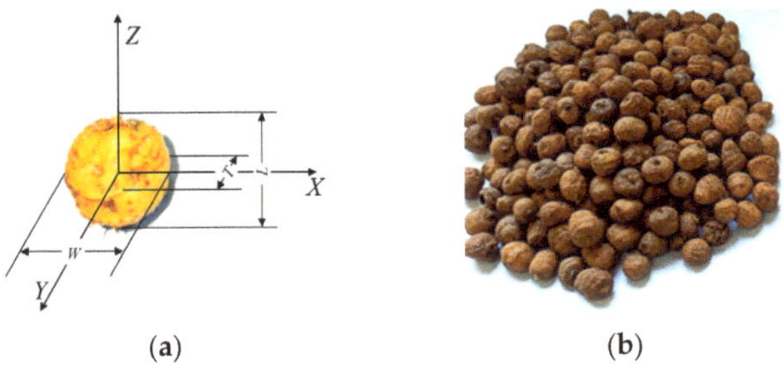

Figure 9. Tiger nuts. (**a**) Dimensions of a tiger nut; (**b**) appearance of tiger nuts.

SOLIDWORKS 2021 was used to establish a model representing a typical tiger nut. The distinct element model of tiger nuts was generated using the auto-filling function in EDEM 2022. The soil particles were appropriately enlarged (to have a diameter of 3 mm), while ensuring the binding force between soil and tiger nuts was unchanged. In this way, the distinct element model of the tiger nut–soil aggregate was developed, as displayed in Figure 10.

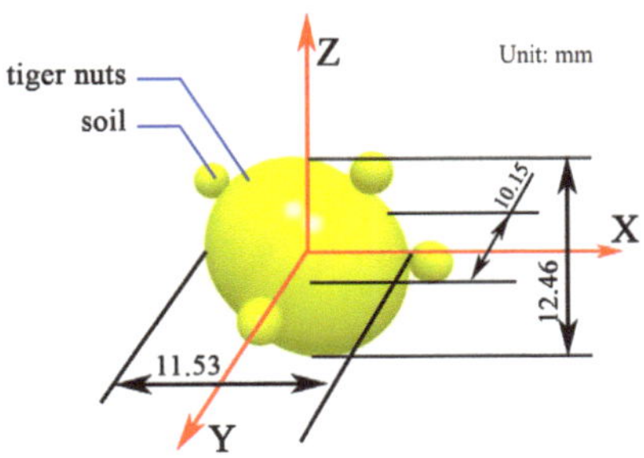

Figure 10. Distinct element model of the tiger nut–soil aggregate.

Soil adheres to harvested tiger nuts [19–22], so the Hertz–Mindlin (with bonding) model was selected for soil particles and tiger nuts to simulate the adhesion between them.

The Hertz–Mindlin (no-slip) model was selected for the relationship between soil particles and tiger nuts. It is the most commonly used model and can simulate the elastic contact between particles [23–27].

The contact radius needs to be set when setting the particle parameters in EDEM 2022. The contact radius is related to the water content of particles and is generally slightly larger than the particle diameter. It is calculated using the following formula:

$$\begin{cases} Q_2 = \dfrac{m_4}{m_3 + m_4} = \dfrac{\rho_3 V_2}{\rho_2 V_1 + \rho_3 V_2} \\ V_1 = \dfrac{4}{3}\pi {R_3}^3 \\ V_2 = \dfrac{4}{3}\pi {R_4}^3 - \dfrac{4}{3}\pi {R_3}^3 \end{cases} \tag{12}$$

where Q_2 is the water content of particles (%); m_3 and m_4 are masses of particles and water (kg); V_1 and V_2 are volumes of particles and water (m^3); ρ_2 and ρ_3 represent the densities of particles and water (kg·m^{-3}); R_3 and R_4 are the particle radius and contact radius (mm), respectively.

The establishment of the EDEM 2022 simulation system mainly involves the setting of physical and contact parameters of tiger nut particles, soil particles, and machine parts [28–34]. Two materials, termed G and MS, are set in the Equipment Material stage to represent steel and brush, respectively. The physical and contact parameters are listed in Table 2.

Table 2. Setting of the EDEM simulation parameters.

Simulation Parameter	Value
Tiger nut density	1102 kg·m^{-3}
Soil density	1780 kg·m^{-3}
Material density of the brush	1760 kg·m^{-3}
Steel density	7801 kg·m^{-3}
Poisson's ratio of tiger nuts	0.360
Poisson's ratio of the soil	0.260
Poisson's ratio of the brush	0.400
Poisson's ratio of steel	0.290
Shear modulus of tiger nuts	3.09×10^7 Pa
Shear modulus of the soil	2.77×10^6 Pa
Shear modulus of the brush	1.02×10^8 Pa
Shear modulus of steel	8.02×10^{10} Pa
Coefficient of restitution for collisions between tiger nuts	0.437
Coefficient of rolling friction between tiger nuts	0.268
Static friction coefficient between tiger nuts	0.780
Coefficient of restitution for collisions between soil particles	0.140
Coefficient of rolling friction between soil particles	0.270
Static friction coefficient between soil particles	0.560
Coefficient of restitution for collisions between tiger nuts and soil particles	0.260
Coefficient of rolling friction between tiger nuts and soil particles	0.320
Static friction coefficient between tiger nuts and soil particles	0.500
Coefficient of restitution for collisions between tiger nuts and steel plates	0.624
Coefficient of rolling friction between tiger nuts and steel plates	0.195
Static friction coefficient between tiger nuts and steel plates	0.420
Coefficient of restitution for collisions between tiger nuts and the brush	0.360
Coefficient of rolling friction between tiger nuts and the brush	0.180
Static friction coefficient between tiger nuts and the brush	0.410
Coefficient of restitution for collisions between soil and steel plates	0.150
Coefficient of rolling friction between soil and steel plates	0.360
Static friction coefficient between soil and steel plates	0.500
Coefficient of restitution for collisions between soil and the brush	0.010
Coefficient of rolling friction between soil and the brush	0.120
Static friction coefficient between soil and the brush	0.190

A particle factory is created to generate 1500 particles per second (and each particle is composed of six bonds). Particles are generated from the beginning (0 s) to the end of simulations, with no limit to the total number.

To accelerate the simulation speed, key parts, including the feeding hopper, drum screen, brush roller, and elliptical gears, were retained, and (.x_t) files were selected when exporting the model files using SOLIDWORKS 2021. The simplified model was exported to RecurDyn V9R4, in which the motion constraints and contacts were set, as shown in Figure 11.

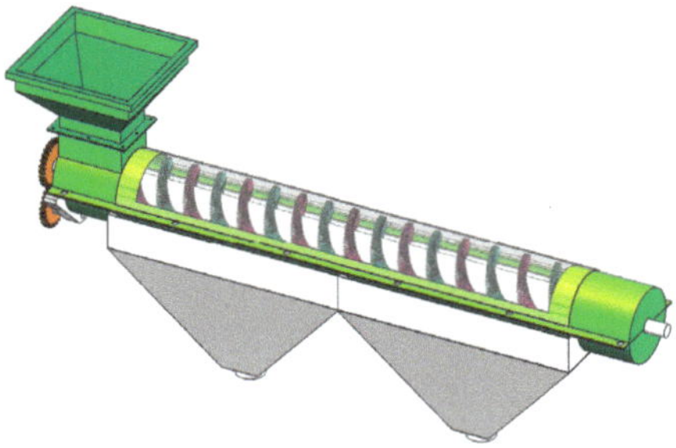

Figure 11. Geometric model of the cleaning device in the simulation.

4.2. Single-Factor Simulation Tests

4.2.1. Test Schemes and Results

According to the aforementioned trials, the rate of rotation of the brush roller, deformation coefficient of the elliptical gears, and eccentricity of the elliptical gears were selected as the test factors to conduct separate single-factor tests.

1. Influence of the rate of rotation of the brush roller on the cleaning effect

EDEM 2022 was adopted to simulate the operation of the cleaning device when the brush roller was rotated at 240, 260, 280, 300, and 320 rpm, and the simulation process is shown in Figure 12.

Figure 12. Number of bonds in each part in the cleaning process. (where the red part among tiger nuts in figure represent the bonds between tiger nuts, the arrows indicates the migration direction from inlet to outlet during the cleaning process of tiger nuts, three big circles are the enlarged display of three black oval circles).

The red part between the tiger nuts and soil particles in the figure is bonds, which may break and fail when forces thereon such as collision, friction, and extrusion exceed a certain limit, allowing simulation of the separation process of soil from tiger nuts. Figure 12 shows that there are many bonds between tiger nuts and soil particles at the feed inlet, and also, a large quantity of soil particles is removed. The front settlement and collection tank holds lots of soil particles; with the movement of tiger nuts towards the discharge hole, the number of bonds reduces, and soil particles are collected in the rear reduce. Basically, no red bonds are observed at the discharge hole, and tiger nuts are found not to be mixed with soil particles. This finding proves that soil particles are effectively removed in the cleaning process and soil particles fall into the settlement and collection tank via circular screen holes, thus accomplishing the separation of soil from tiger nuts.

Changes in the number of bonds with time were plotted using Origin 2022, as illustrated in Figure 13. A total of 9000 bonds are generated per second during simulations, and the cleaning requirement is regarded as being met when the real-time number of bonds less than 6000. The cleaning device is shown to reach the steady state more than 3.5 s, and the number of bonds fluctuates within a certain range. By exporting the data and calculating the means, the average number of bonds after the cleaning device reaches steady-state operation as the optimal rate of rotation is obtained, which can be used to reflect the influence of the rate of rotation on the quality of cleaning.

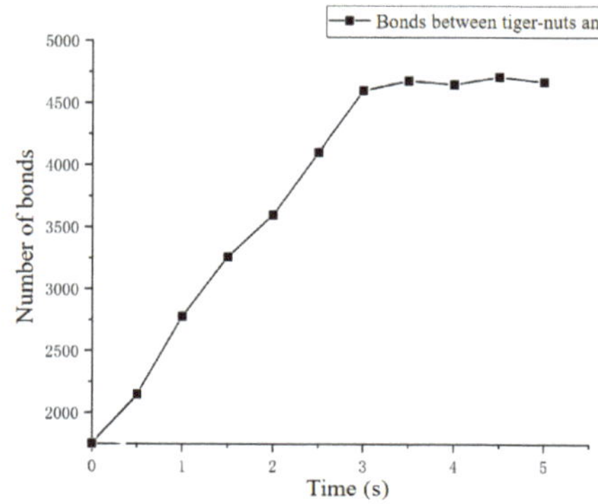

Figure 13. Changes in the number of bonds in the simulation system with time.

2. Influence of the order of elliptical gears on the cleansing effect

The speed of elliptical gears with different orders varies with time in each period, which enables the brush roller to have accelerated and decelerated motions at different frequencies. The tiger nuts in the chamber move to the discharge hole in a tidal manner, which can avoid their excessive wear at a sustained high speed. The four levels selected for the order of elliptical gears include the first, second, third, and fourth orders. The four orders were used to conduct virtual simulation tests driven by different gears, and at the same time, comparative tests driven by circular gears were conducted, as shown in Figure 14.

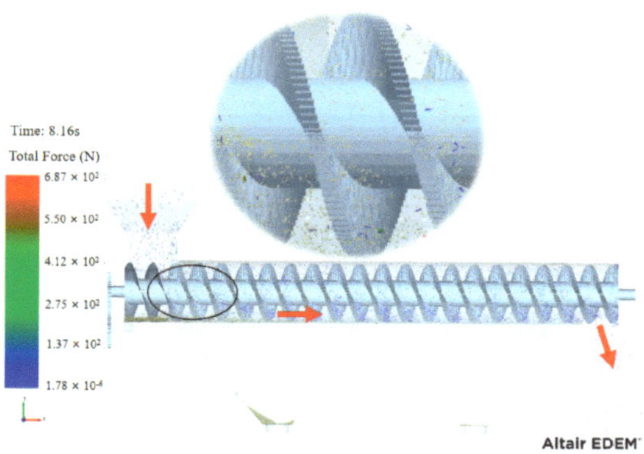

Figure 14. Direction of movement of, and stress on, tiger nuts. (where the arrows indicates the migration direction from inlet to outlet during the cleaning process of tiger nuts, big circles is the enlarged display of black oval circles).

Figure 14 demonstrates that, as the tiger nuts move to the discharge hole, tiger nuts have more contact between each other, and the force between them also increases. This is because, with the separation of soil from tiger nut surfaces, the number of tiger nuts with direct surface contact increases. Origin 2022 was adopted to plot the change in the number of collisions of tiger nuts with time, as shown in Figure 15.

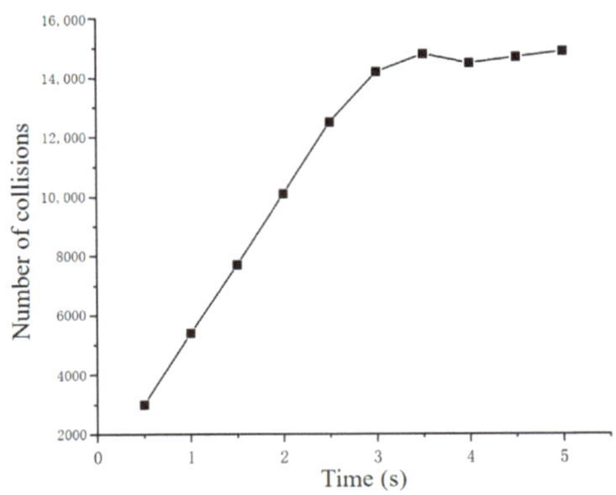

Figure 15. Statistical results pertaining to the number of collisions between tiger nuts.

Through multiple tests, the optimal number of collisions was determined to be between 7000 and 16,000. The larger the number of collisions, the slower the motion of the tiger nuts, where it becomes difficult for soil to fall from the nuts; the lower the number of collisions, the faster the motion of the tiger nuts, at which the excessive contact force resulting from collisions with the drum screen may damage the tiger nut surface and increase the rate of breakage of tiger nuts. Appropriate structural and operation parameters were obtained via statistical analysis of the number of collisions.

3. Influence of the eccentricity of elliptical gears on the cleansing effect

The eccentricity of elliptical gears affects the gear ratio of elliptical gears during operation, which enables pulsating variable-speed rotation of the brush roller and changes the motion state of, and stress on, the tiger nuts. This delivers a large force to the brush roller in the acceleration process, which makes it easier for soil to fall from the nuts. The five levels selected for the eccentricity of the elliptical gears were 0.075, 0.1, 0.125, 0.15, and 0.175. EDEM 2022 was used to simulate the operation of the brush roller driven by elliptical gears with different eccentricities, as illustrated in Figure 16.

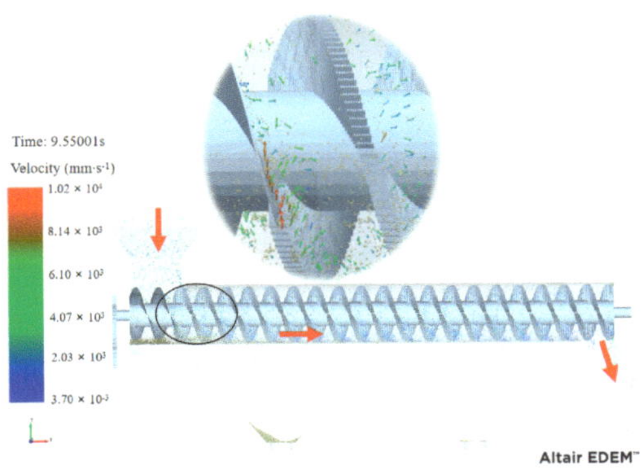

Figure 16. Direction of the motion and speed of tiger nuts. (where the arrows indicates the migration direction from inlet to outlet during the cleaning process of tiger nuts, big circles is the enlarged display of black oval circles).

Origin 2022 was used to plot the change in the stress on the tiger nuts with time in the cleansing process (Figure 17). A good cleaning effect was achieved when the force on the tiger nuts is between 80 and 140 N in the cleansing process.

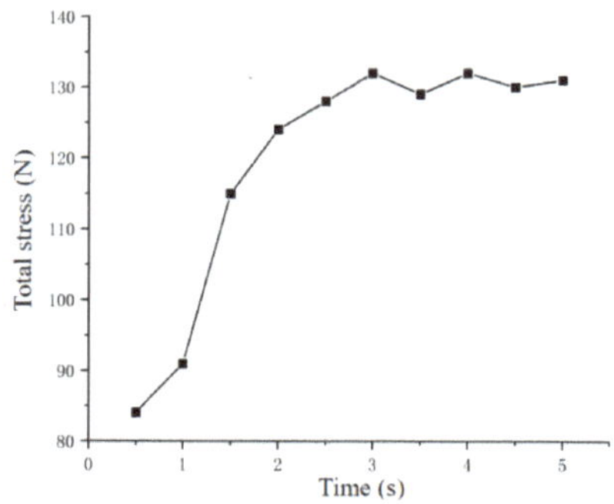

Figure 17. Statistical results pertaining to the total stress on the tiger nuts during cleansing.

4.2.2. Analysis of the Simulation Test Results

Simulation analysis reveals that the cleaning effect is good—that is, there are a small number of bonds—when the rate of rotation of the brush roller accelerates or changes significantly. With the elevation of the order of the elliptical gears, the stress on the tiger nuts decreases and the number of collisions between tiger nuts increases. The cleansing effect of all brush rollers with added elliptical gears is superior to that without elliptical gears. The numbers of bonds are, separately, 4100 and 3770 when using circular gears and fourth-order elliptical gears, under which, the forces on tiger nuts are 90.2 N and 93.7 N, respectively. Comparison shows that the two do not differ obviously; that is, the fourth-order gears do not improve the quality of cleaning to any significant extent. Figure 18 shows the influences of the order of elliptical gears on the number of collisions, number of bonds, and stress on tiger nuts.

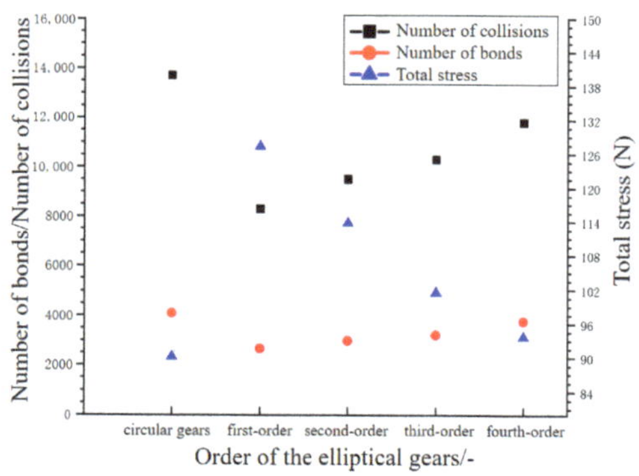

Figure 18. Influence of the order of elliptical gears on the cleansing effect.

With the acceleration of the rate of rotation of the brush roller, the stress on tiger nuts increases while the number of collisions of tiger nuts decreases. The cleaning quality is improved significantly after the rate of rotation of the brush roller reaches 260 rpm. Figure 19 displays the influences of the rate of rotation of the brush roller on the number of collisions, number of bonds, and stress on tiger nuts.

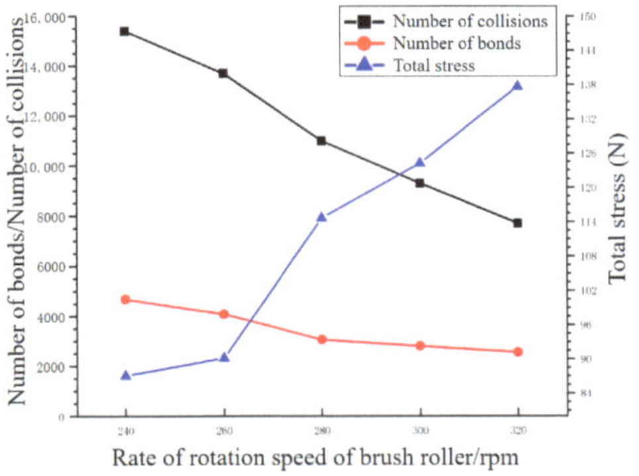

Figure 19. Influence of the rate of rotation of the brush roller on the cleaning effect.

With the increasing eccentricity of the elliptical gears, the stress on the tiger nuts increases while the number of collisions between tiger nuts is reduced. All output parameters increase abruptly when the eccentricity is 0.175, whereupon the number of collisions, number of bonds, and force on the tiger nuts are, separately, 7560, 2580, and 139.3 N. Despite the small number of bonds and a good cleaning effect in the simulation tests, tiger nuts are subjected to significant stress, which may result in a high rate of breakage of the tiger nuts. In the simulation process, tiger nuts are found to bounce out of the feed inlet in the cleaning process using the brush roller driven by elliptical gears with an eccentricity of 0.175. This is a result of the overly rapid motion of tiger nuts, which causes a loss of some tiger nuts in the cleansing process. Figure 20 shows the influences of the eccentricity of the elliptical gears on the number of collisions, number of bonds, and force on tiger nuts.

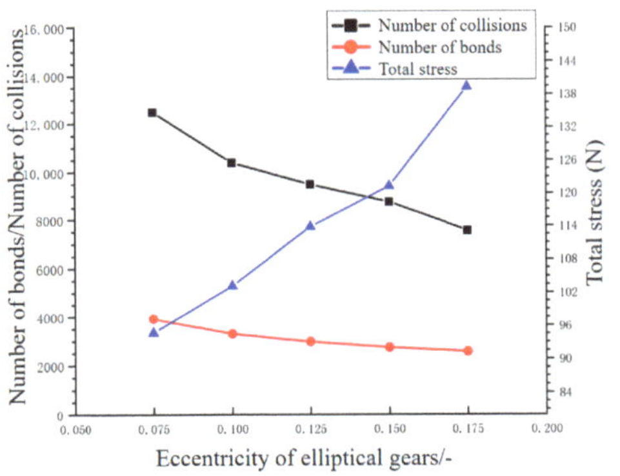

Figure 20. Influence of the eccentricity of elliptical gears on the cleaning effect.

In summary, the rate of rotation of the brush roller, order of the elliptical gears, and eccentricity of the elliptical gears all significantly influence the cleaning effect of tiger nuts. Firstly, with the acceleration or enlarged variation of the rate of rotation of the brush roller, the stress on tiger nuts enlarges while the number of bonds and collisions decreases. Secondly, the stress on tiger nuts is large when the rate of rotation of the brush roller is high or the eccentricity of the elliptical gears is large, which increases the rate of breakage of tiger nuts. Thirdly, as the order of the elliptical gears is increased, the stress on the tiger nuts declines, while the number of bonds and collisions both grow. The cleaning effect of the fourth-order elliptical gears is similar to that of circular gears, without large improvement. Through an overall consideration, 260~300 rpm, 1~3, and 0.1~0.15 are selected as the reference ranges of the rate of rotation of the brush roller, order of the elliptical gears, and eccentricity of the elliptical gears in prototype testing.

5. Bench Tests for Performance Optimization

5.1. Test Conditions

Prototype testing was performed in the Agricultural Engineering Training Centre, College of Mechanical & Electrical Engineering, Henan Agricultural University, Zhengzhou City, China.

Spanish round tiger nuts planted in the Jingjiazhuang test site in Guojia Town, Xinzheng City, Henan Province, China, were used, as shown in Figure 21. The tiger nuts were brown and had a length of 12.46 mm, width of 11.53 mm, and thickness of 10.15 mm. The shape index and water content of the tiger nuts were 91% and 34.58% (the harvested tiger nuts need to be dried until the water content reaches 30–40% before the cleaning process can be carried out), respectively.

Figure 21. Materials used in the prototype testing.

Based on the optimal ranges of the various parameters determined above, a cleaning test platform was built for tiger nuts, as shown in Figure 22. In the tests, the hoisting and feeding devices were used to transport tiger nuts to the inlet of the feeding hopper to ensure continuous feeding. A frequency converter (22KWG3/30KWP3 made in Shenzhen INDVS Technology Co., Ltd., Shenzhen, China) was used to adjust the motor speed to allow the brush roller to rotate at different speeds. The average rate of rotation of the brush roller was measured using a VC6236P non-contact photoelectric tachometer (made in Dongguan Wenxiang Technology Co., Ltd., Dongguan, China). Various test indices were measured after each test.

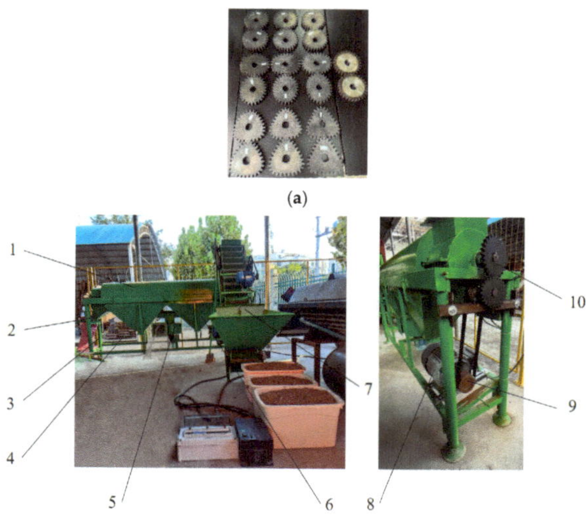

Figure 22. Test bench of the cleaning device and gears with various orders. (**a**) Elliptical gears with various orders and circular gears under different parameters. (**b**) Test bench of the cleaning device. (1) Cleaning chamber; (2) discharge tube; (3) rack; (4) adjustable counterweight; (5) impurity removal blower; (6) settlement and collection tank; (7) hoisting and feeding devices; (8) motor; (9) pulley drive unit; (10) drive unit of high-order elliptical gears.

5.2. Test Indices

By referring to the Enterprise Standard [35], the cleaning efficiency, breakage rate, and impurity rate of tiger nuts were taken as the evaluation standards evincing the operational quality of the cleaning device. The calculation methods of various evaluation indices are described below.

The cleaning efficiency reflects the operational capability of the cleaning device. Tiger nuts, after being cleaned for 5 min, are collected and weighed. Then, the mass of the tiger nuts cleaned by the cleaning device per hour is computed using Equation (13).

$$Q = \frac{12M_1}{1000 \times 1} \tag{13}$$

where Q is the cleaning efficiency (t·h^{-1}); M_1 is the mass of tiger nuts cleaned in 5 min (kg).

The impurity rate refers to the mass fraction of impurities in the cleaned tiger nuts. The impurities include soil, fibrous roots, and shell fragments from the tiger nuts.

The impurity rate of cleaned tiger nuts can be calculated using Equation (14).

$$\eta_z = \frac{M_2}{M_1} \times 100\% \tag{14}$$

where η_z is the impurity rate of tiger nuts (%); M_2 denotes the mass of impurities (kg).

The breakage rate denotes the proportion of broken tiger nuts in the cleaned tiger nuts and is calculated using Equation (15).

$$\eta_p = \frac{M_3}{M_1} \times 100\% \tag{15}$$

where η_p is the breakage rate of tiger nuts (%); M_3 is the mass of broken tiger nuts (kg).

5.3. Box–Behnken Design Using Response Surface Methodology

Design-Expert V8.0.6 was utilized to conduct a Box–Behnken design (BBD) based on the response surface methodology (RSM). A total of 17 groups of three-factor three-level combined tests were designed. To eliminate the errors and ensure the accuracy of the test data, each test was repeated three times. Codes representing the test factors are listed in Table 3, and the specific test schemes and results are summarized in Table 4. A, B, and C separately represent the levels of the rate of rotation of the brush roller, order of the elliptical gears, and eccentricity of the elliptical gears.

Table 3. Codes of the factors.

Group	Rate of Rotation of Brush Roller A/rpm	Order of Elliptical Gears B/-	Eccentricity of Elliptical Gears C/-
−1	260	1	0.100
0	280	2	0.125
1	300	3	0.150

Table 4. Test schemes and results.

Serial Number	Test Factors			Test Indices		
	A	B	C	Cleaning Efficiency Q (t·h^{-1})	Breakage Rate η_p %	Impurity Rate η_z %
1	280.00	3.00	0.150	1.82	0.24	0.40
2	260.00	1.00	0.125	1.61	0.21	0.29
3	280.00	2.00	0.125	1.83	0.17	0.17
4	300.00	3.00	0.125	1.58	0.30	0.39

Table 4. *Cont.*

Serial Number	Test Factors			Test Indices		
	A	*B*	*C*	Cleaning Efficiency *Q* ($t{\cdot}h^{-1}$)	Breakage Rate η_p %	Impurity Rate η_z %
5	280.00	2.00	0.125	1.84	0.17	0.16
6	280.00	1.00	0.150	1.77	0.19	0.34
7	260.00	3.00	0.125	1.62	0.22	0.33
8	300.00	2.00	0.100	1.57	0.30	0.31
9	280.00	2.00	0.125	1.84	0.16	0.17
10	280.00	2.00	0.125	1.85	0.16	0.19
11	300.00	2.00	0.150	1.53	0.28	0.42
12	300.00	1.00	0.125	1.55	0.28	0.35
13	260.00	2.00	0.100	1.59	0.22	0.27
14	280.00	3.00	0.100	1.82	0.21	0.30
15	280.00	1.00	0.100	1.79	0.21	0.31
16	260.00	2.00	0.150	1.63	0.23	0.30
17	280.00	2.00	0.125	1.80	0.15	0.19

5.4. Establishment of Regression Models and Significance Analysis

Design-Expert V8.0.6 was used for analysis of variance of the cleaning efficiency, breakage rate, and impurity rate. Regression models were established for the cleaning efficiency *Q*, breakage rate η_p, and impurity rate η_z. The test results are listed in Table 5.

Table 5. Analysis of variance of the cleaning efficiency, breakage rate, and impurity rate.

Source of Variation	Cleaning Efficiency *Q*		Breakage Rate η_p		Impurity Rate η_z	
	F	*p*	F	*p*	F	*p*
Model	105.60	<0.0001	69.23	<0.0001	83.85	<0.0001
A	23.79	0.0018	159.53	<0.0001	65.65	<0.0001
B	7.08	0.0324	13.02	0.0086	14.15	0.0071
C	0.20	0.6708	0.000	1.0000	61.04	0.0001
AB	0.39	0.5505	0.41	0.5438	0.000	1.0000
AC	6.29	0.0405	3.66	0.0972	10.72	0.0136
BC	0.39	0.5505	10.17	0.0153	8.21	0.0242
A^2	883.56	<0.0001	314.62	<0.0001	161.84	<0.0001
B^2	2.00	0.1998	35.48	0.0006	219.66	<0.0001
C^2	7.30	0.0305	52.78	0.0002	151.33	<0.0001
Lack-of-fit term	0.27	0.8445	0.71	0.5927	0.60	0.6473

Analysis showed that test factors *A* and *B* significantly influence the cleaning efficiency; test factors *A* and *B* also significantly affect the breakage rate; test factors *A*, *B*, and *C* exert significant influences on the impurity rate.

$$\begin{aligned} Q &= 1.83 - 0.028A + 0.015B - 0.02AC - 0.23A^2 - 0.021C^2 \\ \eta_p &= 0.16 + 0.035A + 0.01B + 0.012BC + 0.068A^2 + 0.023B^2 + 0.028C^2 \\ \eta_z &= 0.18 + 0.035A + 0.016B + 0.034C + 0.02AC + 0.018BC + 0.076A^2 + 0.088B^2 + 0.073C^2 \end{aligned} \tag{16}$$

Meanwhile, regression equations for the influence of each factor on the cleaning efficiency *Q*, breakage rate η_p, and impurity rate η_z were obtained, as expressed by Equation (16).

The influence of the interactions of different factors on the cleaning efficiency, breakage rate, and impurity rate was revealed through response surface analysis of the test results. The results are displayed in Figures 23–25.

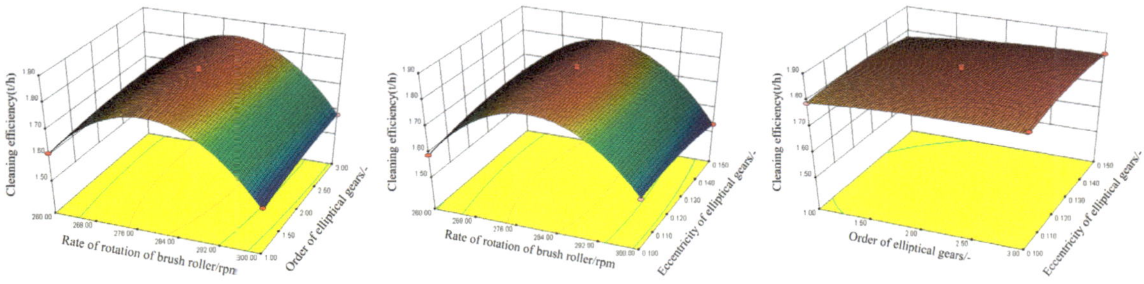

(a) (b) (c)

Figure 23. Response surface diagrams for orthogonal tests of the cleaning efficiency. (**a**) The influence of the rate of rotation of the brush roller and the order of the elliptical gears on the cleaning efficiency. (**b**) The influence of the rate of rotation of the brush roller and the eccentricity of the elliptical gears on the cleaning efficiency. (**c**) The influence of the order and eccentricity of the elliptical gears on the cleaning efficiency.

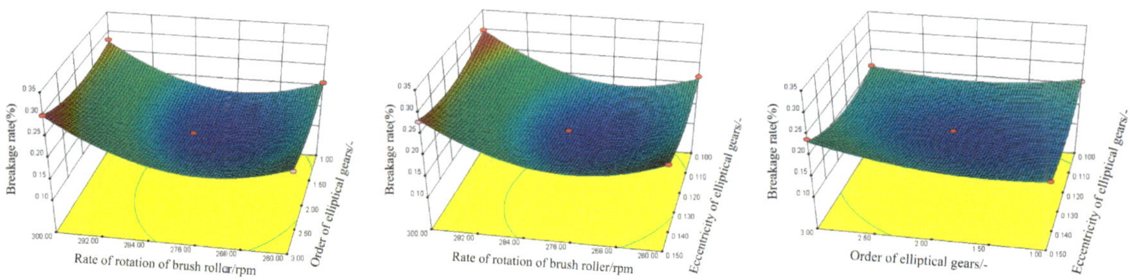

(a) (b) (c)

Figure 24. Response surface diagrams for orthogonal tests of the breakage rate. (**a**) The influence of the rate of rotation of the brush roller and the order of the elliptical gears on the breakage rate. (**b**) The influence of the rate of rotation of the brush roller and the eccentricity of the elliptical gears on the breakage rate. (**c**) The influence of the order and eccentricity of the elliptical gears on the breakage rate.

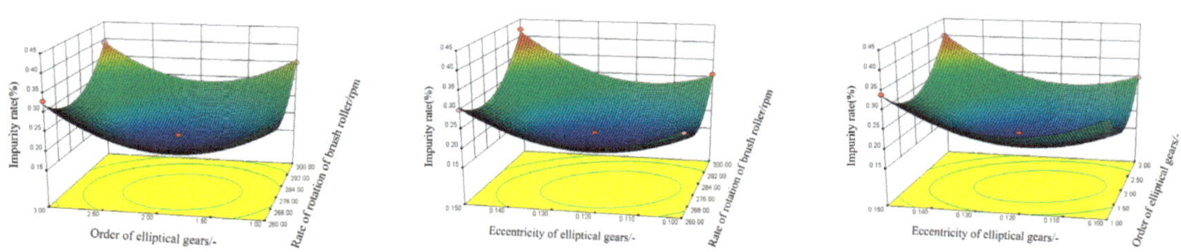

(a) (b) (c)

Figure 25. Response surface diagrams for orthogonal tests of the impurity rate. (**a**) The influence of the rate of rotation of the brush roller and the order of the elliptical gears on the impurity rate. (**b**) The influence of the rate of rotation of the brush roller and the eccentricity of the elliptical gears on the impurity rate. (**c**) The influence of the order and eccentricity of elliptical gears on the impurity rate.

As shown in Figure 23a, when the rate of rotation of the brush roller remains unchanged, the cleaning efficiency of the tiger nuts is improved with the increment of the order of elliptical gears; under a fixed order of elliptical gears, the cleaning efficiency increases, then decreases with the acceleration of the rate of rotation of the brush roller.

Variation of the surfaces indicates that the rate of rotation of the brush roller exerts a more significant influence on the cleaning efficiency.

Figure 23b shows that, at a fixed rate of rotation of the brush roller, the cleaning efficiency of tiger nuts increases, then decreases with the increasing eccentricity of the elliptical gears; when keeping the eccentricity of the elliptical gears constant, the cleaning efficiency first increases, then decreases with the acceleration of the rate of rotation of the brush roller. According to the variation of the surfaces, the rate of rotation of the brush roller more significantly influences the cleaning efficiency of the device as applied to the processing of tiger nuts.

It can be seen in Figure 23c, when keeping the order of the elliptical gears unchanged, the cleaning efficiency of tiger nuts increases, then decreases with the increasing eccentricity; under a fixed eccentricity of the elliptical gears, the cleaning efficiency is improved with the increasing order of the elliptical gears. Variation of the surfaces suggests that the order of elliptical gears more significantly influences the cleaning efficiency.

As shown in Figure 24a, the breakage rate of tiger nuts is shown to decrease, then increase with the increasing order of elliptical gears when keeping the rate of rotation of the brush roller constant; under a fixed order of elliptical gearing, the breakage rate of tiger nuts also decreases, then increases with the acceleration of the rate of rotation of the brush roller. Variation of the surfaces indicates that the rate of rotation of the brush roller more significantly influences the breakage rate of tiger nuts.

It can be seen from Figure 24b that, when keeping the rate of rotation of the brush roller unchanged, the breakage rate of tiger nuts decreases, then increases with the increase in eccentricity; the breakage rate of tiger nuts also shows a similar tendency to decrease, then increase with the accelerated rate of rotation of the brush roller under a fixed eccentricity of the elliptical gears. According to the variation of the surfaces, the rate of rotation of the brush roller exerts a more significant influence on the cleaning efficiency of tiger nuts.

As shown in Figure 24c, when keeping the order of the elliptical gearing constant, the breakage rate of tiger nuts tends to decrease, then increase with the increasing eccentricity; when keeping the eccentricity constant, the breakage rate of tiger nuts also decreases, then increases as the order of the elliptical gears is increased. In accordance with the variation of the surfaces, the order of the elliptical gears more significantly influences the breakage rate of tiger nuts.

Figure 25a shows that, under a constant rate of rotation of the brush roller, the impurity rate of tiger nuts decreases, then increases as the order of the elliptical gears is increased; when keeping the order of the elliptical gears constant, the impurity rate also tends to decrease, then increase with the acceleration of the rate of rotation of the brush roller. The variation of the surfaces shows that the rate of rotation of the brush roller more significantly affects the impurity rate of tiger nuts.

As shown in Figure 25b, when keeping the rate of rotation of the brush roller unchanged, the impurity rate of tiger nuts decreases, then increases as the eccentricity of the elliptical gears is increased; when keeping the eccentricity of the elliptical gears constant, the impurity rate also tends to decrease, then increase with the acceleration of the rate of rotation of the brush roller. The variation of the surfaces shows that the rate of rotation of the brush roller more significantly affects the impurity rate of tiger nuts.

According to Figure 25c, for a given order of the elliptical gears, the impurity rate of tiger nuts decreases, then increases as the eccentricity of the elliptical gears is increased; when keeping the eccentricity of the elliptical gears constant, the impurity rate of tiger nuts also decreases, then increases with the increment in the order of the elliptical gears. Variation of the surfaces indicates that the eccentricity of the elliptical gears exerts a more significant influence on the impurity rate of tiger nuts.

By way of summary, various factors are listed, in descending order of their influence on the cleaning efficiency, as the rate of rotation of the brush roller, order of the elliptical gears, and eccentricity of the elliptical gears; various factors are also listed, in descending order of their influence on the breakage rate, as the rate of rotation of the brush roller, order of the elliptical gears, and eccentricity of the elliptical gears; they are listed, in descending order of their influences on the impurity rate, as the rate of rotation of the brush roller, eccentricity of the elliptical gears, and order of the elliptical gears.

5.5. Parameter Optimization

The Optimization module in Design-Expert V8.0.6 was used to optimize the test factors. The regression equations were solved to maximize the cleaning efficiency while minimizing the breakage and impurity rates.

$$\begin{array}{l} maxY_1(A,B,C) \\ minY_2,Y_3(A,B,C) \\ \begin{cases} 260 \leq A \leq 300 \\ 1 \leq B \leq 3 \\ 0.1 \leq C \leq 0.15 \end{cases} \end{array} \tag{17}$$

Through optimization, the optimal parameter combination is predicted as follows: the rate of rotation of the brush roller is 276.97 rpm, the order of the elliptical gears is 1.94, and the eccentricity of the elliptical gears is 0.122. Under these conditions, the cleaning device designed for use on tiger nuts works best, of which the cleaning efficiency, breakage rate, and impurity rate of tiger nuts are predicted to be 1.83 t·h^{-1}, 0.16%, and 0.17%, respectively.

5.6. Test Verification

The parameters are rounded based on the predicted values and actual working conditions, so the rate of rotation of the brush roller, order of the elliptical gears, and eccentricity of the elliptical gears are set to 280 rpm, 2, and 0.122, respectively, under which conditions, verification tests were conducted. Tiger nuts before and after cleaning are compared in Figure 26. The tests were repeated three times, and the mean average values were recorded. The test results are summarized in Table 6.

(a) (b)

Figure 26. Tiger nuts before and after cleaning. (**a**) Tiger nuts before cleaning; (**b**) tiger nuts after cleaning.

Table 6. Bench test results.

Times of Tests	Cleaning Efficiency Q t·h^{-1}	Breakage Rate η_p %	Impurity Rate η_z %
1	1.85	0.16	0.18
2	1.83	0.15	0.16
3	1.82	0.13	0.14
Average	1.83	0.15	0.16

The test results indicate that the cleaning efficiency, breakage rate, and impurity rate basically agree with the optimization and prediction results and meet the design requirements. Finally, the following are selected as the optimal operation parameters of the cleaning device of tiger nuts: the rate of rotation of the brush roller is 280 rpm, the order of the elliptical gears is 2, and the eccentricity of the elliptical gears is 0.122. Under these conditions, the cleaning efficiency, breakage rate, and impurity rate are 1.83 t·h^{-1}, 0.15%, and 0.16%, respectively.

6. Discussion

A pulsating roll-cleaning device of tiger nuts was designed through theoretical analysis and coupling simulations; combined with the analysis results above, it is confirmed that the rate of rotation of the brush roller mainly determines the cleaning performance, which proves Zhao et al.'s [13] point. On another hand, the work analyzes the effects of the structural and motion parameters of the cleaning device on the cleaning performance of tiger nuts; the deficiencies in the cleaning performance of tiger nuts were investigated:

1. Environmental influence:

The pulsating roll-cleaning method is realized mainly by a variable-speed transmission system consisting of a pair of elliptical gears and appropriately improves the cleaning efficiency and reduces the breakage rate compared to the traditional cleaning method using circular gears. However, the rack is installed on rough ground, and the frequency of vibrations of the cleaning device increases due to the unevenness of the ground. Tiger nuts in the cleaning chamber cannot move axially under the rotation of the brush roller, which reduces the cleaning efficiency and increases the breakage rate of tiger nuts, impairing the efficiency of the process, as evidenced by decreases in the cleaning indices.

2. Random distribution of attitudes of tiger nuts mixed with impurities:

The specific attitudes of large amounts of soil and fibrous roots adhering to the surface of harvested tiger nuts are inconsistent; this differs from the distinct element model of tiger nut–soil aggregate established in virtual simulations. Hence, the time taken to separate impurities including soil and fibrous roots from tiger nuts differs in actual operation, which may influence the cleaning quality of tiger nuts. In the follow-up study, the maximum quality of tiger nuts that the brush roller can withstand can be obtained by flexible treatment of the brush, so as to further improve the cleaning efficiency and obtain better quality tiger nuts.

3. Entanglement of soil and fibrous roots:

When non-uniform quantities of tiger nuts are fed to the hoisting and feeding devices, or much soil is stuck to the tiger nuts and fibrous roots are entangled therewith, the tiger nuts may be accumulated at the bottom of the hoisting hopper and cannot timeously enter the feed-hopper by way of the hoisting device. This prevents tiger nuts in the feed-hopper from continuously and uniformly entering the cleaning chamber, which prolongs the time taken to clean the tiger nuts while also reducing the quality of the cleaning process.

7. Conclusions

1. By adjusting the number of teeth, gear ratio, and deformation coefficient of the elliptical gears and other parameters through the analysis software for the meshing transmission of pitch curves of the elliptical gears with arbitrary orders, the elliptical gear satisfying the transmission stability is obtained, and the pulse variable speed transmission of the brush roller is realized, which ensures the increase in the cleaning efficiency and reduces the breakage rate of tiger nuts.
2. The RecurDyn 2023-EDEM 2022 coupling simulations were used to clarify the influence of key factors on the number of collisions, the number of bonds, and the total stress of tiger nuts during the cleaning process. The optimal ranges of the various parameters of the cleaning device are ascertained as follows: the rate of rotation of the brush roller is 260~300 rpm, the order of the elliptical gears is 1~3, and the eccentricity of the elliptical gears is 0.1~0.15.
3. The results of the bench performance test show that the rate of rotation of the brush roller has the greatest influence on the cleaning efficiency, breakage rate, and impurity rate. Meanwhile, the optimal operation parameters of the cleaning device are as

follows: when the rate of rotation of the brush roller is 280 rpm, the order of the elliptical gears is 2, and the eccentricity of the elliptical gears is 0.122; the cleaning efficiency, breakage rate, and impurity rate are, separately, 1.83 $t \cdot h^{-1}$, 0.15%, and 0.16%. The cleaning effect is favorable under these conditions, meeting the design requirements of the cleaning device.

Author Contributions: Conceptualization, Z.L. and W.W.; methodology, H.H. and D.Y.; software, Z.L. and X.H.; resources, Z.Q. and H.H.; data curation, W.J., Y.M. and D.Y.; writing—original draft preparation, Z.L., B.S. and Y.Z. All authors have read and agreed to the published version of the manuscript.

Funding: This research was funded by the National Key R&D Program of China (2019YFD1002602) and Henan Province Science and Technology Research (222102110330).

Institutional Review Board Statement: Not applicable.

Data Availability Statement: The data used to support the findings of this study are available from the corresponding author upon request.

Acknowledgments: The authors would like to thank their college and the laboratory, as well as gratefully appreciate the reviewers who provided helpful suggestions for this manuscript.

Conflicts of Interest: The authors declare no conflicts of interest.

References

1. Li, P.; Dai, Y.Y.; Wang, W.W. Analysis on the evolution of trade competitiveness of China's main edible oil products. *Cereals Oils* **2019**, *32*, 5–9.
2. Yang, X.D.; Li, Z.Y. Tigernut industry in China: Current status of development, potential and adaptive suggestions. *Chin. J. Oil Crop Sci.* **2022**, *44*, 712–717.
3. Zhang, H.M.; Zhou, Z.; Qu, Z.; Li, Z.J.; Wang, W.Z. Simulation and experiment of sieving process of sieving device for tiger nut harvester. *Agriculture* **2022**, *12*, 2–16. [CrossRef]
4. Qi, J.T.; Gao, J.F.; Chen, S.; Chen, W.H.; Yang, L.Y.; Meng, H.W.; Kan, Z. Parameter optimization of a conveying and separating device based on a five-stage screw and vibrating screen for tiger nut harvesters. *Agriculture* **2024**, *14*, 2–25. [CrossRef]
5. Wang, X.Y.; Sun, J.B.; Xu, Y.C.; Li, X.Q.; Cheng, P.F. Design and experiment of potato cleaning and sorting machine. *Trans. Chin. Soc. Agric. Mach.* **2017**, *48*, 316–322.
6. Ji, L.L.; Yang, H.G.; Xie, H.X.; Yan, J.C.; Wei, H.; Shen, H.Y. Experimental study on potato dry cleaning device. *J. Agric. Mech. Res.* **2022**, *44*, 153–161.
7. Yan, M.L.; Shi, W.H.; Zhou, X.L.; Zhang, Z.D.; Zhang, Y.; Wu, C.L. Research on industry security of China's main oilseeds and vegetable oils from the perspective of import. *Chin. J. Oil Crop Sci.* **2023**, *45*, 643–653. [CrossRef]
8. Branko, V.; Urban, Š. Invasion, distribution and habitat affiliation of *Cyperus esculentus*, a new weed in Slovenia. *Hacquetia* **2021**, *20*, 291–302.
9. Ekow, E.A.; Joshua, A.; Nazir, H.K. Drying kinetics and quality of whole, halved, and pulverized tiger nut tubers. *Int. J. Food Sci.* **2021**, *2021*, 8870001.
10. Avicor, S.W.; Wajidi, M.F.F.; Achoribo, E.S.; Ong, M.T.; Hamzah, S.N. Tiger nut (*Cyperus esculentus*) as a potential phytoinsecticide: Larvicidal activity of crude extracts on Aedes aegypti and *Culex quinquefasciatus* (Diptera: Culicidae). *Trop. Biomed.* **2021**, *38*, 186–191.
11. Gadanya, A.M.; Abubakar, M.Y.; Maigari, F.U.; Mudassir, L.; Abubakar, S.M. Comparative analysis of nutrients content and characterization of oil from two varieties of tiger nut (*Cyperus esculentus*). *Asian J. Res. Biochem.* **2021**, *8*, 11–21. [CrossRef]
12. Qi, J.T.; Pei, M.H.; Kan, Z.; Meng, H.W. Parameter optimization and testing of a conveying and soil-removing device for tiger nut (*Cyperus esculentus*) mechanical harvesting. *Processes* **2022**, *11*, 67. [CrossRef]
13. Zhao, Y.Q.; Li, B.J.; Meng, L.; Wang, L.; Bi, T.; Jian, J.M.; Hou, S.L. Optimal design of mechanical harvesting and separation system for tiger nut. *J. Agric. Mech. Res.* **2022**, *44*, 51–56. [CrossRef]
14. Zhang, H.M.; Zhou, Z.; Lv, Z.J.; Wang, W.Z.; He, H.; Lv, Y.L. Kinematic analysis and testing of vibratory sieving mechanism for *Cyperus esculentus* L. harvester. *J. Chin. Agric. Mech.* **2021**, *42*, 92–99. [CrossRef]
15. Liu, W.L.; Kong, L.Y.; Jiang, C.Y.; Wang, X.Y.; Zhang, L. Design of tiger nut grain harvester. *Agric. Dev. Equip.* **2019**, *10*, 111–112.
16. Wu, X.T.; Wang, G.H. *Non-Circular Gear and Non-Uniform Transmission*; China Machine Press: Beijing, China, 1997.
17. Li, F.S.; Yin, Z.F.; Zhang, Z.L.; Lin, Z.G.; Zhang, Z.D.; Li, H.Y. *Non-Circular Gear and Special Gear Transmission Design*; China Machine Press: Beijing, China, 1983.
18. Chen, H. Simulation analysis in dedusting deposition chamber of the engineering sweeper based on FLUENT. *Equip. Manuf. Technol.* **2019**, 68–71.

19. Wang, Y.X.; Huang, S.T.; Zhang, W.Y.; Qi, B.; Zhou, X.Z.; Ding, Y.Q. Design and experiment of wheat precision seed metering device with staggered hook-tooth. *Trans. Chin. Soc. Agric. Mach.* **2024**, *55*, 142–154.
20. Chen, Y.; Gao, X.X.; Jin, X.; Ma, X.R.; Hu, B.; Zhang, X.L. Calibration and analysis of seeding parameters of *Cyperus esculentus* seeds based on discrete element simulation. *Trans. Chin. Soc. Agric. Mach.* **2023**, *54*, 58–69.
21. Xu, H.B.; Zhang, P.; Zhang, Y.H.; Wang, S.Y.; Yu, Z.Y.; Gu, F.W. Design and experiment of peanut pod-picking and conveying device of multistage tangential flow type peanut combine. *Trans. Chin. Soc. Agric. Mach.* **2023**, *54*, 52–60.
22. Zhang, H.M.; Zhang, C.M.; Li, Z.J.; Ding, L.; Zhu, C.H.; Zhang, J. Simulation analysis and optimization of vibration characteristics of corn finger clip seeding device based on EDEM–RecurDyn Coupling. *Trans. Chin. Soc. Agric. Mach.* **2023**, *54*, 36–46.
23. Yuan, P.P.; Li, H.W.; Huang, S.H.; Jiang, S.; Xu, J.; Lin, H.; Li, R.R. Parameter optimization and experiment of a seed furrow cleaning device for no-till maize seeding. *Agriculture* **2022**, *12*, 2–17. [CrossRef]
24. Zhang, X.J.; Cheng, J.P.; Shi, Z.L.; Wang, M.J.; Fu, H.; Wu, H.F. Simulation and experiment of seed taking performance of swing-clamp type maize precision seed-metering device. *Trans. Chin. Soc. Agric. Mach.* **2023**, *54*, 38–50.
25. Zhang, S.W.; Zhang, R.Y.; Cao, Q.Q.; Zhang, Y.; Fu, J.; Yuan, H.F. Design and experiment of double-deck roller sieve type fruit impurity separation device for tiger nut harvester. *Trans. Chin. Soc. Agric. Mach.* **2023**, *54*, 148–157.
26. Hu, J.P.; Pan, J.; Chen, F.; Yue, R.C.; Yao, M.J.; Li, J. Simulation optimization and experiment of finger-clamping seedling picking claw based on EDEM–RecurDyn. *Trans. Chin. Soc. Agric. Mach.* **2022**, *53*, 75–85.
27. Wang, F.A.; Wen, B.; Xie, X.H.; Xie, K.T.; Guo, S.W.; Zhang, Z.G. Operation mechanism analysis and parameter optimization of conveying and separating device of panax notoginseng harvester. *Trans. Chin. Soc. Agric. Mach.* **2023**, *54*, 201–211.
28. Zhang, S.W.; Zhang, R.Y.; Chen, T.Y.; Fu, J.; Yuan, H.F. Calibration of simulation parameters of mung bean seeds using discrete element method and verification of seed-metering test. *Trans. Chin. Soc. Agric. Mach.* **2022**, *53*, 71–79.
29. He, X.N.; Zhang, X.J.; Zhao, Z.; Shang, S.Q.; Wang, D.W.; Yuan, X.W. Design and test of resistance-reducing excavation device of cyperus edulis based on discrete element method. *Trans. Chin. Soc. Agric. Mach.* **2022**, *53*, 71–79.
30. Yan, B.X.; Wu, G.W.; Fu, W.Q.; Gao, N.N.; Meng, Z.J.; Zhu, P. Influencing factors of corn implantation distribution for high-height planting based on EDEM. *Trans. Chin. Soc. Agric. Mach.* **2020**, *51*, 47–54.
31. Wei, Z.C.; Su, G.L.; Li, X.Q.; Wang, F.M.; Sun, C.Z.; Meng, P.X. Parameter optimization and test of potato harvester wavy sieve based on EDEM. *Trans. Chin. Soc. Agric. Mach.* **2020**, *51*, 109–122.
32. Liao, Y.T.; Liao, Q.X.; Zhou, Y.; Wang, Z.T.; Jiang, Y.J.; Liang, F. Parameters calibration of discrete element model of fodder rape crop harvest in bolting stage. *Trans. Chin. Soc. Agric. Mach.* **2020**, *51*, 73–82.
33. Wang, X.Z.; Yue, B.; Gao, X.J.; Zheng, Z.Q.; Zhu, R.X.; Huang, Y.X. Discrete element simulations and experiments of disturbance behavior as affected by mounting height of subsoiler's Wing. *Trans. Chin. Soc. Agric. Mach.* **2018**, *49*, 124–136.
34. Liu, W.Z.; He, J.; Li, H.W.; Li, X.Q.; Zheng, K.; Wei, Z.C. Calibration of simulation parameters for potato minituber based on EDEM. *Trans. Chin. Soc. Agric. Mach.* **2018**, *49*, 125–135, 142.
35. *QHYS02-2017*; Enterprise Standard for 5QPZ Series of Cleaning and Polishing Machine of Tiger Nut Tubers. Jilin Haoyishou Agricultural Science and Technology Development Co., Ltd.: Changchun, China, 2017.

agriculture

Article

Research on the Jet Distance Enhancement Device for Blueberry Harvesting Robots Based on the Dual-Ring Model

Wenxin Li [1], Hao Yin [1,*], Yuhuan Li [1], Xiaohong Liu [2], Jiang Liu [1] and Han Wang [3]

[1] School of Mechanical and Automotive Engineering, Qingdao University of Technology, Qingdao 266520, China; liwenxin19991013@163.com (W.L.); 18612897966@163.com (Y.L.); liujiang@qut.edu.cn (J.L.)
[2] College of Mechanical and Electrical Engineering, Qingdao Agricultural University, Qingdao 266109, China; lxh964@126.com
[3] College of Mechanical and Vehicle Engineering, Taiyuan University of Technology, Taiyuan 030024, China; wanghan20210827@163.com
* Correspondence: yinhao_vehicle@qut.edu.cn

Abstract: In China, most blueberry varieties are characterized by tightly clustered fruits, which pose challenges for achieving precise and non-destructive automated harvesting. This complexity limits the design of robots for this task. Therefore, this paper proposes adding a jetting step during harvesting to separate fruit clusters and increase the operational space for mechanical claws. First, a combined approach of flow field analysis and pressure-sensitive experiments was employed to establish design criteria for the number, diameter, and inclination angle parameters of two types of nozzles: flat tip and round tip. Furthermore, fruit was introduced, and a fluid–structure coupling method was employed to calculate the deformation of fruit stems. Simultaneously, a mechanical analysis was conducted to quantify the relationship between jet characteristics and separation gaps. Simulation and pressure-sensitive experiments show that as the number of holes increases and their diameter decreases, the nozzle's convergence becomes stronger. The greater the inclination angle of the circular nozzle holes, the more the gas diverges. The analysis of the output characteristics of the working section indicates that the 8-hole 40° round nozzle is the optimal solution. At an air compressor working pressure of 0.5 MPa, force analysis and simulation results both show that it can increase the picking space for the mechanical claw by about 5–7 mm without damaging the blueberries in the jet area. The final field experiments show that the mean distance for Type I (mature fruit) is 5.41 mm, for Type II (red fruit) is 6.42 mm, and for Type III (green fruit) is 5.43 mm. The short and curved stems of the green fruit are less effective, but the minimum distance of 4.71 mm is greater than the claw wall thickness, meeting the design requirements.

Keywords: blueberry; picking robot; jet device; fluid–structure coupling; pressure-sensitive experiments; structural optimization

Citation: Li, W.; Yin, H.; Li, Y.; Liu, X.; Liu, J.; Wang, H. Research on the Jet Distance Enhancement Device for Blueberry Harvesting Robots Based on the Dual-Ring Model. *Agriculture* **2024**, *14*, 1563. https://doi.org/10.3390/agriculture14091563

Academic Editor: Mustafa Ucgul

Received: 10 August 2024
Revised: 2 September 2024
Accepted: 8 September 2024
Published: 9 September 2024

1. Introduction

Blueberries are delicious and nutritious, offering numerous health benefits, which makes them highly popular among consumers [1]. In China, the blueberry planting area and production have both rapidly increased, leading to a thriving planting industry. As a result, China has now become the country with the largest blueberry planting area [2]. The blueberry harvesting process accounts for about 40% of the total workload in the entire production process, making it the most time-consuming and labor-intensive stage [3], which significantly restricts the development of the blueberry industry. Mechanical harvesting can improve efficiency and reduce costs [4]. However, the characteristic of blueberries growing in clusters makes them one of the most challenging fruits and vegetables to harvest using mechanized and automated methods. Therefore, manual harvesting is still the primary method used, leading to low labor efficiency and high labor intensity, which in turn keeps the cost of blueberry cultivation high [5].

To address these challenges, researchers have studied blueberry growth, fruit characteristics, stem mechanical properties, and the connection properties between berries and stems [6–8]. Based on this research, various blueberry harvesting machines have been designed. Currently, the most mature and widely used types are the inertial vibration and comb-brush blueberry harvesters [9]. For example, OSKAR 4WD Plus berry harvesters developed by BEI [10], Littau [11], OXBO, and Poland's JAGODA [12,13] use vibration devices to create contact vibrations with the plants. These vibrations utilize inertial force to separate the stems from the blueberry fruits. Similar harvesting equipment includes the striking-type picking machines [14,15]. These machines primarily use lateral vibrations on the trunk to shake off ripe fruits, making them more suitable for tall, tree-like plants found in regions like North America. However, most blueberries grown in China are low, shrub-like plants, making it difficult to directly adopt this method. Even with single-tooth mechanisms for precise picking, there is a risk of damaging surrounding fruits due to the close clustering of the berries. Pneumatic picking machines [16] use airflow and negative pressure to pick ripe fruits, but they also face the challenge of being unable to distinguish between ripe and unripe fruits, similar to vibration-based equipment.

With the rapid development of artificial intelligence, precise capturing and sorting of ripe and unripe fruits [17], as well as flexible, non-destructive picking, have become major research directions [18,19]. Currently, research on harvesting robots for large single fruits is quite advanced, focusing primarily on fruit recognition and picking path planning. Williams et al. used deep neural networks and stereo matching to detect kiwis [20], enabling robotic picking. Ge et al. proposed a lightweight CNN regression model for full-view coverage of fruit [21]. Li et al. developed a stereo-vision apple recognition and localization algorithm based on multi-task deep learning [22]. Zhou et al. designed a robot that determines apple picking paths using 3D point cloud data [23], achieving a success rate of 82%. Ye et al. developed a lychee picking robot that uses collision-free motion planning [24]. Huang et al. created an autonomous banana-picking robot that employs a corner-turning method for rapid path convergence [25]. For small fruits such as blueberries, cherry tomatoes, and strawberries, the picking process and end-effector selection are particularly important [26]. Liu et al. studied three different cherry tomato picking modes [27], providing a basis for small fruit harvesting. Zhang et al. designed a flexible end effector suitable for cherry tomatoes [28], enabling non-destructive picking. Lemsalu et al. developed a system for real-time recognition of strawberries and flower stems [29], but it performed poorly in dense berry clusters. Cao et al. proposed a strawberry instance segmentation method to enhance recognition and localization capabilities [30]. Tituaña et al. designed a strawberry-picking robot based on the YOLOv4 model [31], achieving a picking success rate of 71.7%. Although there have been some successful cases of harvesting small, soft fruits, most of them still focus on individual fruits.

For blueberries, key issues in precise harvesting include accurate ripeness detection, fruit softness, and the clustering problem, which remains one of the major challenges in the development of blueberry-picking robots. Blueberries have a long maturation period, so a cluster can contain both unripe and ripe fruits. During the harvesting process, the flexible mechanical gripper of a robot cannot directly reach into the fruit cluster to pick the ripe fruits. It requires a separation method to appropriately isolate interfering fruits. For mechanical separation methods, Fu et al. designed a kiwi-picking robot that uses a shielding plate to enclose the fruits [32], allowing the clustered kiwis to be individually separated for precise picking. However, this type of shielding plate design cannot be used for blueberry clusters, and mechanical separation methods are prone to causing fruit damage, especially with small, soft fruits. Therefore, our team studied the morphological and structural characteristics of blueberry fruits and stems, as well as the squeezing and friction characteristics of the fruits [33]. Based on this research, we decided to add a jet mechanism to the picking robot to separate ripe fruits from unripe ones, ensuring that the mechanical gripper can effectively reach the fruits. Before the end-effector clamping in the picking process, the separation of the fruit is interfered with by a momentary jet pass, thus satisfying the extension of the end effector. Therefore, selecting the appropriate nozzle and determining the working distance of the jetting mechanism is crucial. This paper analyzes two different nozzle parameters for the clustered blueberry jetting application. Through simulations and experimental testing, we determined the optimal nozzle structure. We also analyzed the force on the fruit stems under jetting conditions and conducted in-depth simulations and field experiments with the chosen nozzle to observe stem deformation and fruit separation under jetting.

2. Nozzle Simulation and Experiments

2.1. Designing Scheme

As shown in Figure 1, the blueberry-picking robot must complete five basic actions in sequence during each picking cycle: identification, positioning, air jetting, gripping, and separation. First, the recognition device identifies the mature fruit and determines the picking target's location, issuing the corresponding commands. The recognition system assesses the pre-grip force based on the maturity and external characteristics of the fruit to prevent damaging the fruit during gripping. Next, it checks for the presence of interfering fruits; if any are detected, the air jet device separates them from the target fruit. Then, the system determines the spatial position of the picking target by identifying its X, Y, and Z coordinates. The control system moves the robotic arm to the working area, where the air jet device operates to separate the target fruit from interfering fruits. The mechanical claw grips the target fruit, and the end effector's linkage mechanism then separates the fruit from its stem. Finally, the robotic arm follows an optimized path to place and store the blueberries. Since this study's primary goal is to provide an air-jetting solution for separating clustered fruits (using specially designed nozzles to apply high-speed jets to the fruit cluster, allowing for accurate separation and enabling the mechanical claw to reach in), the specific functions of other modules are not discussed. Additionally, major components of the air jet device, such as the air compressor and regulating valves, are installed within the chassis of the picking robot, while the nozzle is connected to the robot's small arm.

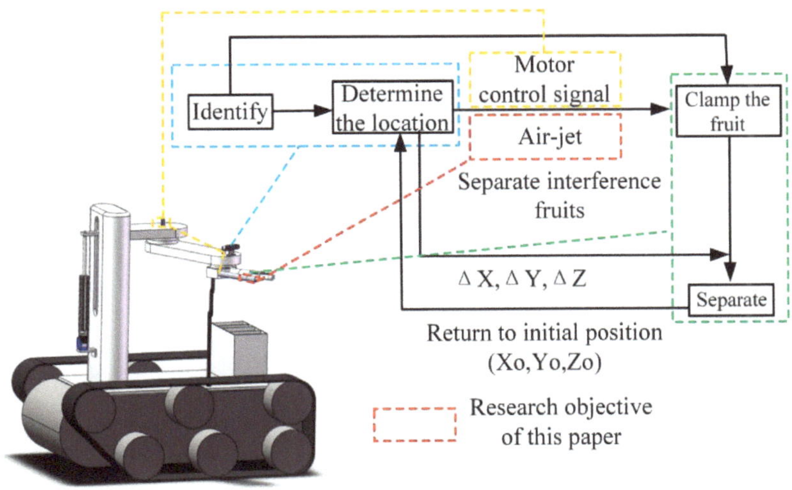

Figure 1. Blueberry picking robot action process.

2.2. Simulation Modeling

2.2.1. Workspace Modeling

This project focuses on the rabbit-eye blueberries grown in greenhouses in northern China, which have a shape similar to a sphere, being either fully rounded or slightly flattened. We selected 300 blueberries and measured their diameter and thickness. The measurement results are shown in Figure 2. The maximum diameter of the blueberries does not exceed 25 mm, and the maximum thickness does not exceed 15 mm. Additionally, for rabbit-eye blueberries, the diameter of the entire fruit cluster typically does not exceed 40 mm. The jetting device needs to separate the target blueberry from the surrounding ones. Therefore, we established an annular model with a weak force at the center and a strong force at the periphery, as shown in Figure 3a. The nozzle output range must cover all sizes of blueberries. Therefore, we selected a weak force area with a diameter of 25 mm at the center of the nozzle's circular working region, and a strong force area with a diameter of no less than 40 mm, with a working distance of 40 mm, as shown in Figure 3a.

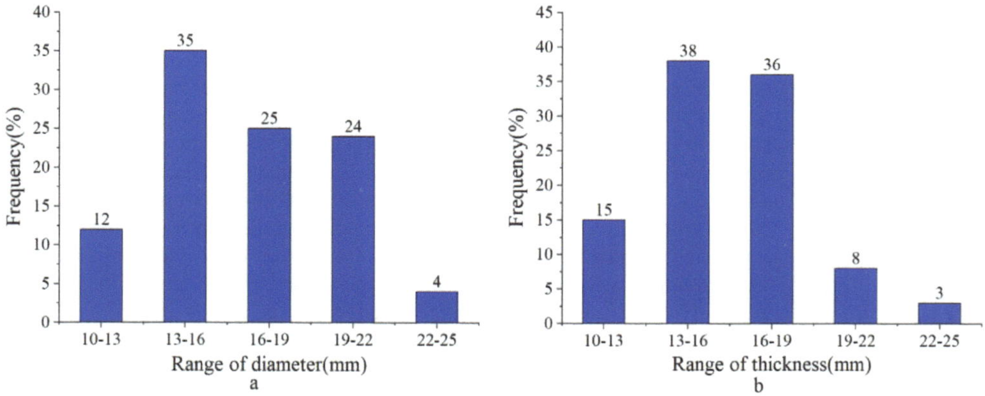

Figure 2. Blueberry size scale diagram. (**a**) Blueberry diameter distribution chart. (**b**) Blueberry thickness distribution chart.

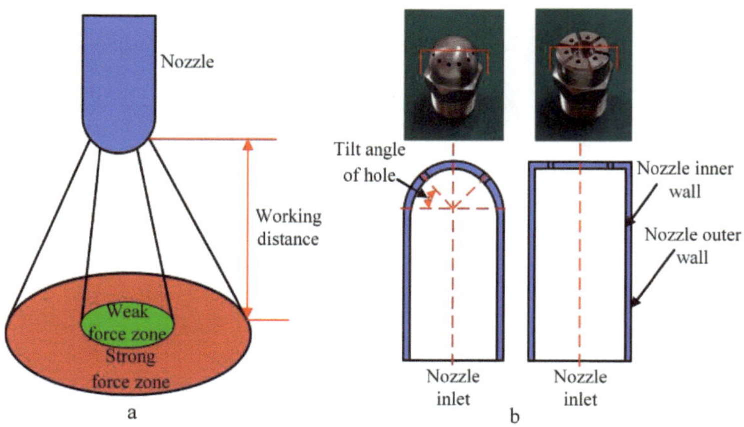

Figure 3. Schematic diagram of nozzle. (**a**): working diagram of nozzle; (**b**): structure diagram of nozzle.

2.2.2. Nozzle Model Establishment

This study aims to effectively separate the target fruit from the interfering fruit and to design dual-ring working area as described earlier. Commonly used single-hole nozzles (cylindrical or conical) do not meet our requirements. Therefore, we have chosen multi-hole nozzles for this study. Considering the difficulty of manufacturing and the need to achieve the dual-ring working area, we focused on studying and applying standard flat-tip and round-tip nozzles. Among them, Figure 3b shows the prototypes and cross-sectional structural models of the flat and round nozzles used. The design parameters include the number of holes, hole diameter, and hole inclination angle. The hole inclination angle is the angle between the centerline of the nozzle shell and the centerline of the hole. Specific data are provided in Table 1.

Table 1. Nozzle design parameters.

Parameters	Notation	Values
Internal diameter	D	11 mm
Length of nozzle	L	23 mm
Number of holes	N	6, 8, 10, 12
Diameter of hole	*d*	1.00, 1.25, 1.5, 1.75, 2.00 mm
Inclination angle of the hole	α	30°, 35°, 40°, 45°

2.2.3. Finite Element Simulation

First, a finite element analysis of the nozzle's flow field was conducted, using the Mosaic method for mesh division as shown in Figure 4. In numerical simulations, a grid independence analysis is often used to balance computational accuracy and time. Six mesh settings were tested using the Mosaic method, with grid numbers of 64,000, 75,000, 88,000, 101,000, 112,000, and 125,000. The effectiveness of each grid setting was evaluated based on the relative deviation between different schemes. Taking the results with 125,000 grids as the reference standard, the relative error in exit velocity and pressure is shown in Figure 5. The results indicate that the relative error between the calculations with 112,000 and 125,000 grids is less than 0.5%, which is within an acceptable range. Therefore, this article will select the 125,000 scheme to set the grid numbers.

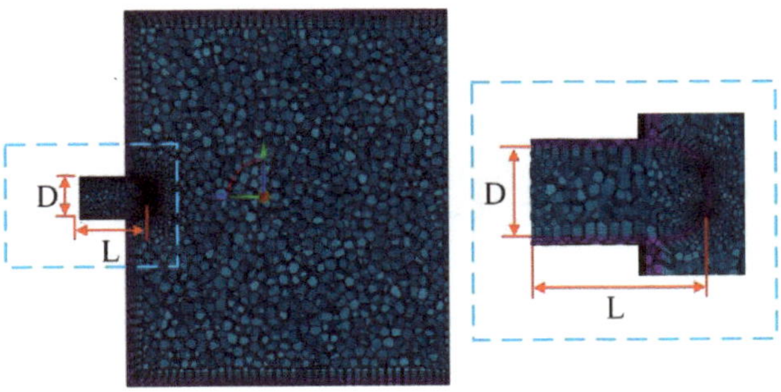

Figure 4. Mesh of the model (the enlarged area is the nozzle size; the blue grid is the fluid domain).

In Figure 4, D represents the inlet diameter, L represents the nozzle length, and the blue mesh denotes the entire fluid domain. Subsequently, numerical simulation methods were used to calculate the Reynolds number for various nozzles based on Equation (1).

$$Re = \frac{\rho v d}{\mu} \tag{1}$$

In the equation, Re represents the Reynolds number, ρ is the air density, v is the average velocity, d is the nozzle diameter, and μ is the air viscosity. Since the minimum Reynolds number for the nozzle is greater than 10,000, the standard k-ε model was chosen, and both complete turbulence and molecular viscosity can be neglected. Given that the Mach number is 0.286, the fluid can be considered incompressible, so the energy equation does not need to be considered [34]. The solution method used is the SIMPLE algorithm, which requires solving the pressure field through the Poisson equation and uses a steady-state solver. To ensure consistency, the same simulation conditions are applied to all structural models once they are determined.

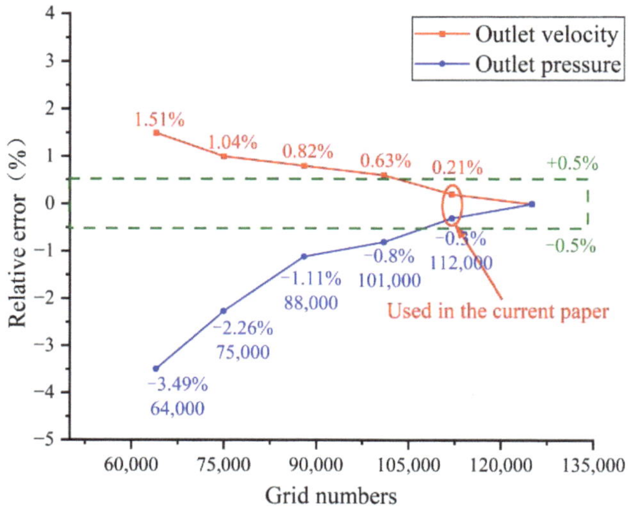

Figure 5. The relative error analysis results of the six grid setting schemes.

Considering the actual output power of the air compressor and the correspondence of the simulation, the output pressure and speed of the air compressor were calibrated. A wind speed meter (Testo 417, Testo, Schwarzwald, Germany) was used to measure the wind speed at the interface between the air compressor and the nozzle. The measurement results are shown in Table 2. At an output pressure of 0.5 MPa, the wind speed at the connection point is approximately 9.73 m/s. Therefore, the simulation was set with an intake speed of 9.73 m/s to match the measured speed, and the outlet environmental pressure was set to one standard atmosphere. All nozzles with different structural parameters were set with the same initial boundary parameters.

Table 2. Interface wind speed at different compressor output pressures.

Air Compressor Pressures (MPa)	Values (m/s)
0.1	6.11
0.2	7.74
0.3	8.56
0.4	9.11
0.5	9.73

2.2.4. Theoretical Analysis of the Working Interface

Based on the mechanical gripper's motion stroke, the working distance in Figure 3a was calculated and set to 40 mm. The nozzle has a multi-hole structure, and the annular area in the working plane involves the coupling effect of the multi-hole airflow. A 2D model of the coupled airflow output characteristics was established, as shown in Figure 6, which includes three stages: the initial section, the basic section, and the dissipation stage.

The output characteristic evaluation indicators include the effective width of the nozzle jet and the sectional velocity of the basic section. The blue dashed line represents the section at the working distance within the basic section, and its length is the effective width. Additionally, the velocity boundaries of the strong and weak force zones should be as circular as possible, and the velocity distribution should be relatively concentrated. In the working area, the speed in the strong force zone must be sufficient to effectively separate the fruits without causing damage, as described by Equation (2).

$$V_w \geq V_e \text{ and } V_{wmax} \leq V_b = min\{V_{b1}, V_{b2}\} \tag{2}$$

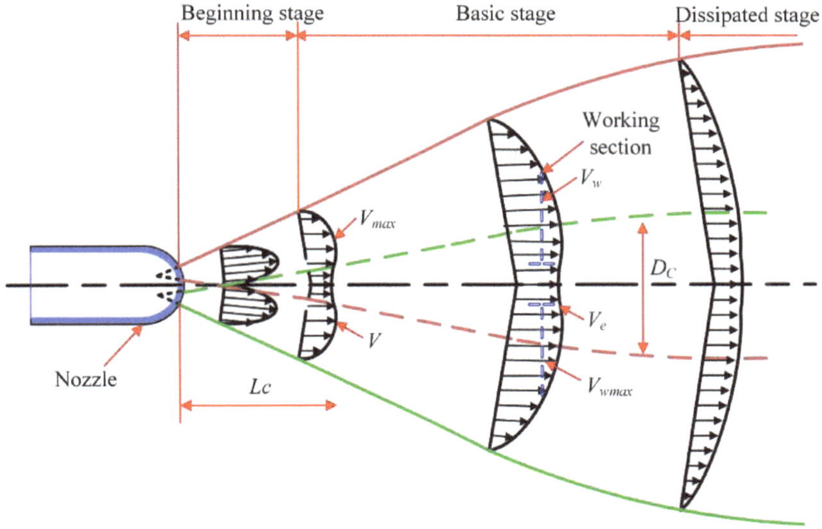

Figure 6. Output characteristics of multi-hole nozzles.

In the equation, V_b is the minimum wind speed that can cause damage to the fruit, V_{b1} is the minimum wind speed that the blueberry skin can withstand, and V_{b2} is the minimum wind speed that the connection between the mature blueberry and the stem can withstand. V_w is the speed in the strong force zone at the working distance, V_{wmax} is the maximum speed at the working distance, and V_e is the effective wind speed, which is the minimum wind speed that can achieve fruit separation. The effective wind speed and the minimum wind speed that can cause damage to the fruit can both be experimentally measured using a wind speed meter. The measurement results are provided in the later section on fruit separation experiments.

In Figure 6, the areas defined by the red and green lines represent the airflow characteristics of two symmetrically positioned holes. The output characteristics of the multi-hole nozzle are influenced by the synergy between the two holes. As shown in Figure 6, the sectional diameter D_C of the airflow intersection region (indicated by the red and green dashed lines) increases with the distance Lc from the nozzle. When Lc is small, D_C increases linearly with Lc. However, when Lc exceeds 6 D, D_C increases more slowly, and beyond 8.5 D, D_C remains almost constant. Compared to a single-hole jet, the dashed line section shows significant contraction.

At different Lc values, the flow velocity V first increases and then decreases, with the peak occurring between 2.5 D and 8.5 D. This range is preferred for the annular model design; thus, in this instance, Lc is set to 40 mm. Notably, the central airflow velocity in the overlapping region is weaker than the velocity at the center of the holes. During actual jetting, when this part is directed at the target fruit, the normal pressure in the central region is small while the pressure in the edge region is high. This forms the mechanical basis for the ideal annular model in Figure 3.

Within the intersecting range of the two-hole airflow, the flow speed increases with distance, starting from zero in the central region and gradually matching the boundary flow speed at the edges. Therefore, the airflow speed in the middle of the multi-hole nozzle gradually increases while the intensity weakens, eventually equaling the boundary speed. Here, V represents the velocity in the nozzle flow field, and V_{max} represents the maximum velocity across different sections of the nozzle flow field. For a single-hole jet, the jet spreads out at a diffusion angle, with the jet radius increasing linearly over the basic stage. The relationship for the jet radius changes in the basic stage is shown in Equation (3).

$$R = 3.4a(x_0 + s) \tag{3}$$

here R represents the jet radius, a represents the exit cross-sectional coefficient, which is determined by the material and manufacturing process, x_0 represents the pole depth, and s represents the turbulence coefficient. Transform the formula to obtain Equation (4):

$$\frac{R}{r_0} = \frac{x_0 + s}{x_0} = 1 + \frac{s}{r_0/tg\beta} = 1 + 3.4a\frac{s}{r_0} \tag{4}$$

here r_0 represents the nozzle radius, β represents the jet diffusion angle. Empirical Equation (5) can be obtained by combining Equation (5):

$$\frac{R}{r_0} = \lambda\left(\frac{as}{r_0} + 0.294\right) \tag{5}$$

here λ represents the outlet sectional coefficient, which is complex for intersecting porous jets. The deformation degree of intersecting jets is generally indicated by φ, as shown in Equation (6). At the initial stage, the value of φ starts to change after the jets intersect. The value of φ is determined by the distance between the nozzles, the inclination angle α of the nozzles, and the diffusion angle β of each jet's outer boundary. Until the dissipation stage, where the primary deformation rate φ becomes constant, the primary deformation rate on any cross-section is equal, making the combined flow resemble a single free jet. At this point, all deformations caused by the intersecting jets disappear.

$$\varphi = \frac{R_x - R}{r_0} \tag{6}$$

here R_x represents the radius of the jet after the intersection. The two-dimensional model can be used to verify the output characteristics at the working distance, and the more accurate and complex spatial coupling model is usually calculated using the finite element simulation method, and its flow field analysis can be compared with the planar model calculation results.

2.3. Simulation Results and Analysis

2.3.1. Analysis of the Influence of Different Parameters of Flat Tip Nozzles

The main influencing parameters of flat-tip nozzles are the number of holes and the hole diameter. Figure 7 shows the output airflow speed under different structural parameters with the same boundary conditions. From the simulation in Figure 7a, we can see that increasing the number of holes while keeping the hole diameter at 1 mm reduces the output speed of the airflow. Under the condition of six holes, the output fluid speed can still maintain 30 m/s at 30 mm. As the number of holes increases, the airflow speed decreases. The designed 12-hole nozzle can maintain 30 m/s at 15 mm. Similarly, as shown in Figure 7b, increasing the hole diameter while keeping the number of holes at six reduces the output speed of the airflow. Under the condition of a 1.25 mm hole diameter, the speed distribution is relatively concentrated, with a larger and more noticeable range of high-speed airflow indicated in red. As the hole diameter increases, the speed output characteristics decrease, and the airflow speed within the working range is less than 20 m/s for a 2 mm hole diameter. The trend of output airflow changing with parameter variation can guide structural selection for different application scenarios. For this study, the primary goal is to separate the fruit. The flat-tip nozzle we need has a very short length between the weak and strong force areas. The output airflow spreads out in a triangular pattern, making it challenging to find a suitable working ring in the flow field. However, the six-hole 1 mm flat-tip nozzle is the best choice among flat-tip nozzles because it has fewer holes and a relatively smaller hole diameter.

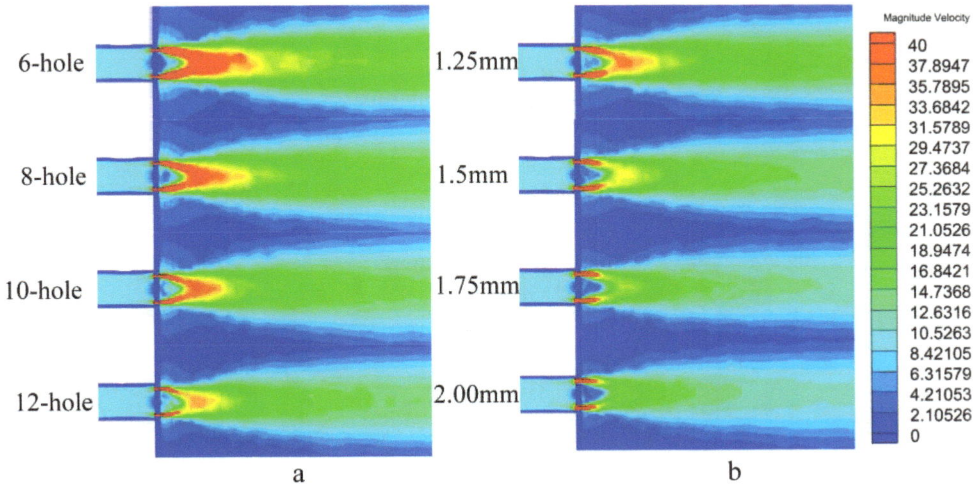

Figure 7. Flat tip nozzles with different parameters. (**a**): different number of holes in 1.00 mm hole diameter; (**b**): different hole diameters for 6-hole configuration.

2.3.2. Analysis of the Influence of Different Parameters on Round Tip Nozzles

The main influencing parameters of round tip nozzles are the number of holes, hole diameter, and hole inclination angle. Figure 8 shows the output airflow speed under different structural parameters of round-tip nozzles with the same boundary conditions. Since the effect of hole diameter on the nozzle's output speed is well-established (the speed continuously decreases as the hole diameter increases, without affecting the shape of the gas output), Figure 8 only displays simulation results for different parameters with a preferable hole diameter of 1 mm.

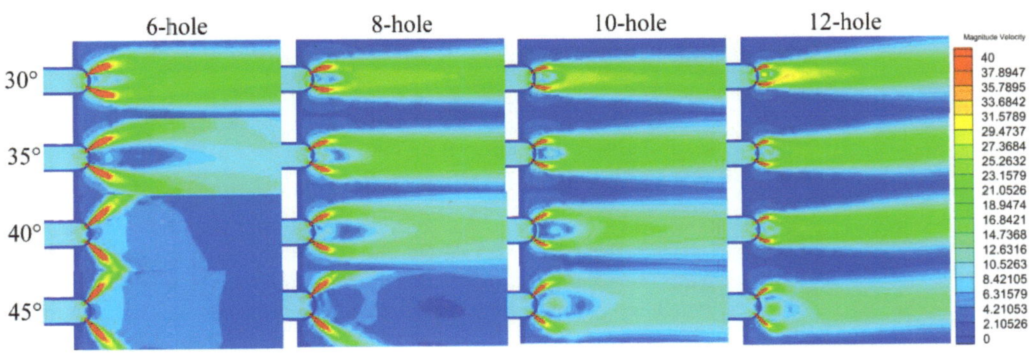

Figure 8. Simulation results of round tip nozzles at different tilt angles and number of holes.

From the simulations in Figure 8, we can see that increasing the inclination angle of the holes, while keeping the number of holes constant, makes the gas output characteristics more divergent. Under the condition of six holes with a 45-degree inclination angle, the gas from two holes on the same cross-section already shoots towards both ends. However, we find that increasing the number of holes can compensate for this loss. Compared to 6 holes, increasing the inclination angle for 8, 10, and 12 holes shows significant improvement. For our work, the goal is to use the annular airflow to retain the target fruit while separating the interfering fruit on the outer ring, so we need to find a suitable working ring. From the simulations, it is clear that the speed output characteristics of 6 holes at 35°, 8 holes at 40°, and 10 holes at 45° are better. However, it is difficult to evaluate their actual effect solely from the cross-sectional view, so we will further analyze and discuss the selected nozzle structures.

2.3.3. Simulation Analysis in the Work Area

To analyze the output characteristics on the working area plane, we selected a 40 mm × 40 mm square cross-section at a distance of 40 mm from the nozzle for comparative analysis of the working area. The simulation results of the nozzles with the better output characteristics mentioned above are shown in Figure 9.

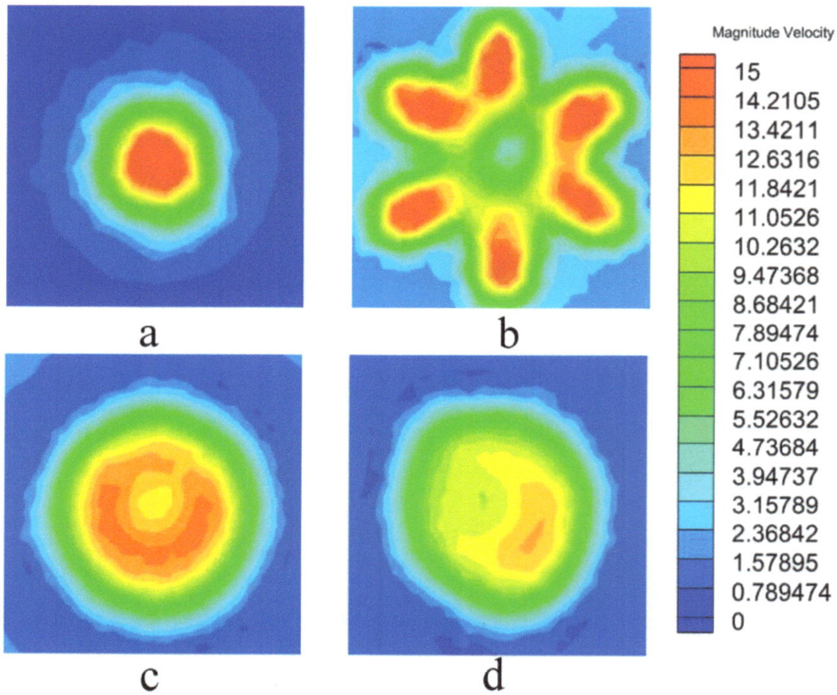

Figure 9. Simulation results for the working section. (**a**): 6-hole flat tip; (**b**): 6-hole 35° round tip; (**c**): 8-hole 40° round tip; (**d**): 10-hole 45° round tip.

As shown in Figure 9a, the average speed of the flat tip six-hole nozzle is very high, the speed characteristic decreases slowly, and the roundness of the plane is good. On the working plane 40 mm away from the nozzle, the simulation results show that the speed at the center exceeds 15 m/s, forming a circle with a diameter of about 10 mm. However, this strong convergence is not favorable for this study as it contradicts our design concept of weak force at the center and strong force at the edges. As shown in Figure 9b, the six-hole 35° nozzle has a clear weak force area at the center and strong force areas around it on the 40 mm working plane. The airflow from each hole is distributed in dots across the working section, arranged sequentially according to the position of the holes. This condition meets the requirements of this study, as the nozzle output characteristics are significantly dispersed with minimal interaction between airflows, though the roundness of the shape is somewhat poor, necessitating further consideration. As shown in Figure 9c, the eight-hole 40° nozzle exhibits a distinct ring distribution of the weak force area at the center and strong force areas around it on the 40 mm working plane. The gas speed at the boundary between the weak and strong force areas is about 14 m/s, and the speed in the central weak force area is about 11 m/s. Additionally, the cohesion of each hole is relatively good. For this study, it meets the need for separating the fruit and branches around the ripe fruit. Finally, the 10-hole 45° nozzle shows a significant decrease in speed output characteristics and earlier diffusion stage due to the larger inclination angle and more holes, resulting in greater interaction between the holes. As shown in Figure 9d, the output airflow speed is lower, and the shape is irregular.

In summary, combining the analysis of output characteristics and working cross-sections with different nozzle parameters, the round-tip nozzle proves more advantageous than the flat-tip nozzle. The analysis of the working cross-section indicates that the 8-hole 40° round nozzle performs the best.

2.4. Pressure Sensing Experiment

2.4.1. Construction of the Nozzle Test Bench

The main components of the nozzle test bench include an air compressor, a pressure regulator, solenoid valves, a field-programmable gate array (FPGA) driver board, nozzles, and a sensor testing control board (Figure 10). The air compressor supplies high-pressure air to the test bench, while the pressure regulator controls the air pressure. The solenoid valves are controlled via the FPGA driver board to regulate the opening and closing of the valves for jetting. Uniformly distributed thin-film sensors are mounted on the panel of the testing platform. After jetting, these sensors transmit signals to the computer through the sensor control board for data collection and processing.

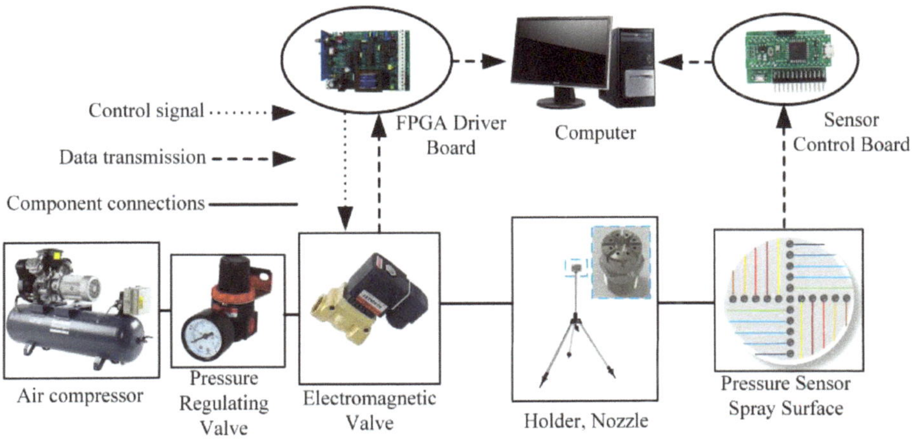

Figure 10. Test bench flow diagram.

2.4.2. Testing and Verification Methods

In this study, the working cross-section is set at a plane 40 mm from the nozzle. The compressor and pressure regulator are controlled to output a pressure of 0.5 MPa. The nozzle is connected to the compressor with a 0.25-inch diameter connection, setting the inlet speed to 9.73 m/s for the pressure sensitivity test. Additionally, during the test, the position of the bracket and sensor panel is kept fixed, and only the nozzles with different parameters are replaced. Each time the nozzle is changed, the compressor is stopped first, then the solenoid valve is closed, and finally, the pressure regulator is turned off. After changing the nozzle, the working conditions are set to the specified settings following the same procedure. Each nozzle experiment is conducted three times, and the computer collects the pressure data through the sensor controller to calculate the average value. By comparing the simulation results and experimenting with different parameters of flat-tip and round-tip nozzles, we analyze the differences between them. To verify the accuracy of the simulation results, a matching chart of simulated wind speed and test pressure for the nozzles is established. Finally, the pressure test board is replaced with blueberry branches and leaves for fruit separation experiments. The differences in mechanical analysis deformation, fluid–structure interaction simulation, and field jetting fruit separation are compared, and the effective separation speed is determined using a wind speed tester.

2.4.3. Pressure Sensitivity Test Results and Analysis

Pressure sensitivity tests were conducted for the relatively better nozzles mentioned above, and the experimental data were recorded with the computer via the sensor control board. The nozzles were labeled as follows: A for the 6-hole flat tip nozzle, B for the 6-hole 35° round tip nozzle, C for the 8-hole 40° round tip nozzle, and D for the 10-hole 45° round tip nozzle.

Under the same simulation settings as mentioned earlier, the air compressor was adjusted to a working pressure of 0.5 MPa. The test environment was not altered, and the same bracket was used to move the pressure sensor panel continuously, recording pressure results. The test procedure consisted of two parts. The first part involved aligning the center of the test panel with the center of the nozzle, maintaining a symmetric position while moving the panel to test the axial pressure variation of the output fluid by changing the distance from the nozzle. The second part involved aligning the center of the test panel with the center of the nozzle, keeping the distance between the panel and nozzle constant while moving the panel laterally to test the pressure variation across the working area concerning the diameter.

The axial pressure variations of the four nozzles are shown in Figure 11. Nozzle A (6-hole flat tip) exhibited the highest axial output pressure. The axial pressure of A initially increased slightly before rapidly decreasing with an increase in distance, reaching a maximum pressure of 106 kPa at 15 mm. At a working distance of 40 mm, the pressure was still 62 kPa, which could cause fruit damage or detachment. The axial output pressures of B (six-hole 35° round tip) and C (eight-hole 40° round tip) were both below 30 kPa, moderate enough to prevent fruit damage or detachment. Their output pressures were similar, with the weakest force located 35 mm from the nozzle, close to the 40 mm working distance mentioned earlier. The axial output pressure of D (10-hole 45° round tip) showed the lowest point of 10 kPa at 25 mm, with a slow increase in pressure afterward. The longitudinal diameter pressures of the four nozzles on the 40 mm working cross-section are shown in Figure 12. It is clear that nozzle A (6-hole flat tip) had the highest pressure of 90 kPa at the center of the circle, decreasing gradually with an increasing diameter, which contradicts our design concept. It is also evident that nozzle D (10-hole 45° round tip) did not meet the requirements, as the gas speed within a 25 mm diameter at the center was similar, failing to achieve the special requirement for fruit separation. The output pressures of B (six-hole 35° round tip) and C (eight-hole 40° round tip) were lowest at the center of the circle, increasing with the diameter. Their pressures started to decrease after reaching the maximum point, indicating a pressure trend from the center to the periphery, which is suitable for the annular working cross-section. The eight-hole 40° round tip nozzle has advantages in terms of the annular working area and pressure peak. The pressure variation trend of the eight-hole 40° round tip nozzle is also clearer. Additionally, Figure 13 shows the corresponding results of test pressure and simulated speed to further demonstrate the relationship between the test and simulation. For nozzle A, an output speed of 34 m/s corresponds to a test pressure of about 88 kPa, while for B, an output speed of 12.5 m/s corresponds to a test pressure of about 36 kPa. For C, an output speed of 17 m/s corresponds to a test pressure of about 45 kPa, and for D, an output speed of 11 m/s corresponds to a test pressure of about 30 kPa. In conclusion, based on the pressure test results, the ranking is as follows: 8-hole 40° round tip > 6-hole 35° round tip > 10-hole 45° round tip > 6-hole flat tip.

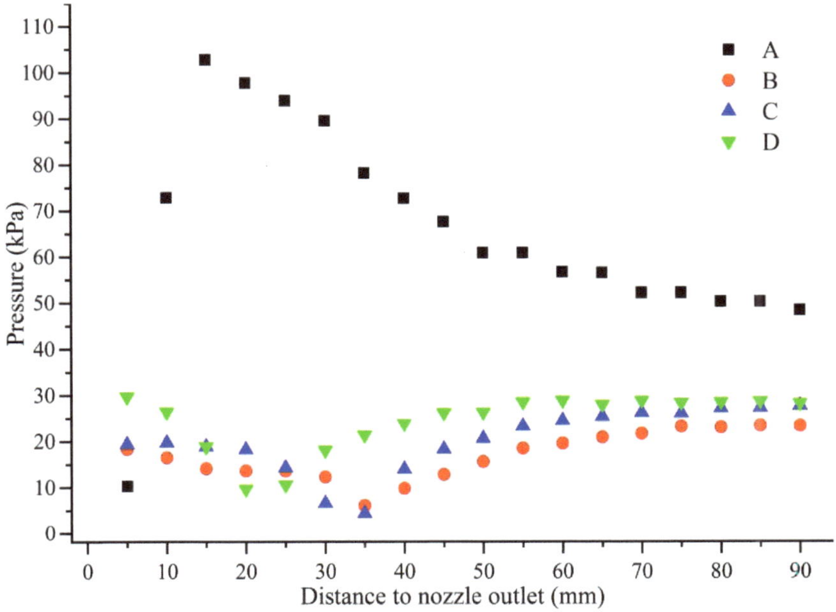

Figure 11. Four types of nozzle axial pressure. A: 6-hole flat tip; B: 6-hole 35° round tip; C: 8-hole 40° round tip; D: 10-hole 45° round tip.

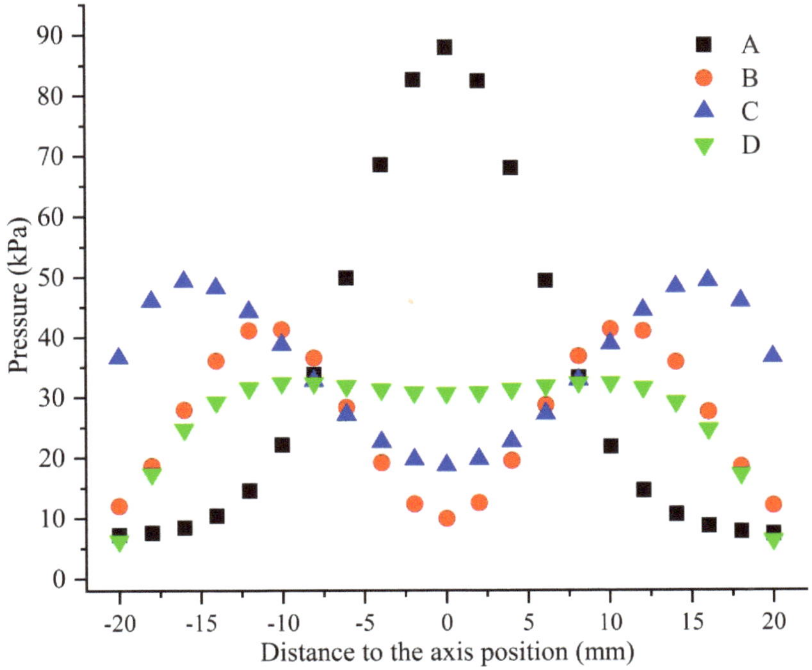

Figure 12. Diameter pressure in the working area of the four nozzles. A: 6-hole flat tip; B: 6-hole 35° round tip; C: 8-hole 40° round tip; D: 10-hole 45° round tip.

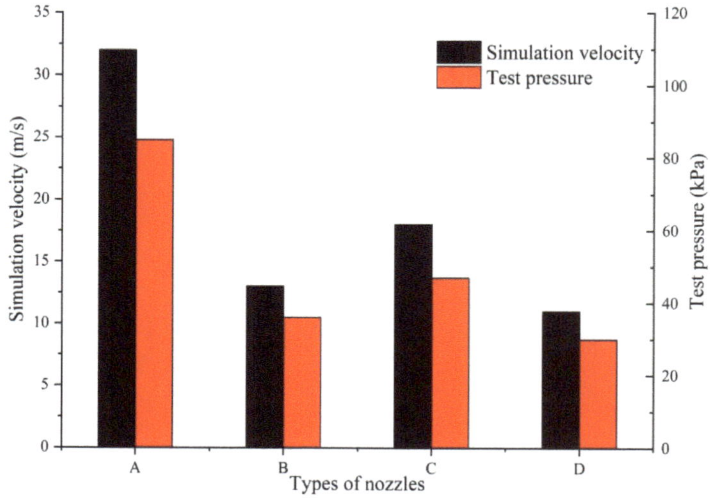

Figure 13. Comparison of test pressure and simulation speed for four nozzles. A: 6-hole flat tip; B: 6-hole 35° round tip; C: 8-hole 40° round tip; D: 10-hole 45° round tip.

3. Simulation and Experimentation of Fruit Stalk

3.1. Mechanical Analysis of Fruit Separation

During the air jet process, the fruit is mainly subjected to the airflow pressure output from the nozzle. The force analysis is shown in Figure 14, where Q represents the distributed load, and G represents its own gravity.

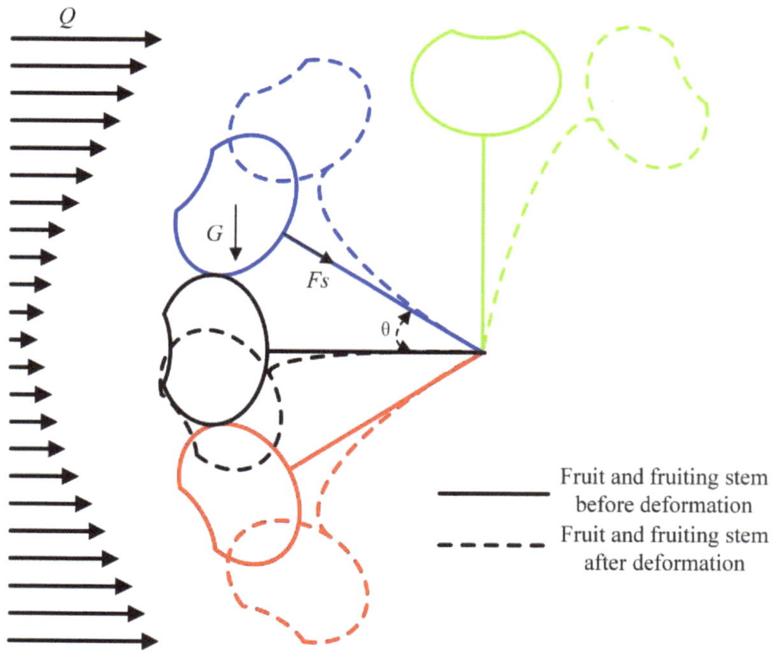

Figure 14. Analysis of forces on fruit stems under airflow. Green indicates the vertical position; black indicates the horizontal position; red and blue indicate the inclined position.

In this study, we disregard the negligible weight of the fruit stem itself. Additionally, the connecting force Fs between the fruit and its stem is considered. The angle between the blueberry stem outside the picking target and the main branch node is defined as θ. The connection between the stem and the branch node is treated as a cantilever beam in this study. Through mechanical simplification, the deflection formula is derived as follows:

$$EIw'' = -M(l) \tag{7}$$

where E is the elastic modulus of the material, I is the moment of inertia of the section, w is the deflection, and M is the relationship between the bending moment M and the distance l from the endpoint to the fruit. Incorporating the boundary conditions of the cantilever beam, Equation (7) is transformed into Equation (8).

$$EIw = \int [\int \left(\int_0^L Qsin\theta l^2 dl \right) dl] dl - \frac{1}{6} GL^3 cos\theta \tag{8}$$

where L is the length of the fruit stalk, and the load Q is fitted using a logistic model based on the output characteristics of the nozzle axis distance. The specific parameters are listed in Table 3, and the model equation is given by formula (9).

$$Q = A_2 + \frac{(A_1 - A_2)}{1 + \left(\frac{x}{x_0}\right)^P} \tag{9}$$

Table 3. Specific parameters of the load model fit.

Parameter	Value
A_1	17.24839
A_2	61.26441
x_0	11.39066
P	2.4368
Reduced Chi-Sqr	0.08925
R square (COD)	0.99925
Adjusted R-squared	0.9991

The relationship between l and x is as follows:

$$x = lsin\theta \tag{10}$$

The parameter l is the direct evaluation criterion for determining the fruit separation effect. Here is a calculation example for the largest fruit: when the fruit peduncle is in a vertical position, the deformation due to the fruit's weight is minimal, and the deformation of the peduncle under full load is approximately 10.87 mm. In a horizontal position, the deformation caused by gravity is about 1.559 mm. When the peduncle length is 15 mm, and the fruit is positioned at a 53° angle, the deformation is approximately 7.247 mm. For the fruit on the lower side, the horizontal deformation of the peduncle is about 1.659 mm. With a remaining gap of approximately 5.588 mm, the minimum gap fully accommodates the mechanical claw insertion.

3.2. Simulation Analysis of Fruit Separation

This section conducts a fluid–structure interaction simulation for the optimized 8-hole 40° nozzle with fruit. The fruit model is placed at a distance of 40 mm from the nozzle's axis on the working cross-section. The simulation uses the maximum fruit as an example, with a peduncle length set to 15 mm. The fruit-peduncle model and its structural and material parameters are detailed in Table 4. The fruit model is imported through the fluid–solid coupling module and uses an unstructured mesh method. This approach has advantages in fluid–solid coupling for the solid domain (force exerted by the fluid on the solid surface).

Smaller and more numerous meshes allow for better resolution of boundaries and analysis of external fluid forces.

Table 4. Material parameters of fruit and fruiting stem.

Parameter	Value
Fruit diameter	25 mm
Fruit thickness	15 mm
Fruit density	1.16 g cm^{-3}
Fruit Poisson's ratio	0.35
Fruit modulus of elasticity	0.225 MPa
Stem length	15 mm
Stem diameter	1.5 mm
Fruiting stem density	38 g cm^{-3}
Fruiting stem Poisson's ratio	0.38
Fruit stem modulus of elasticity	14.2 MPa
Connectivity of fruiting stem to mature fruits	0.17~0.83 N
Connectivity of fruiting stem to immature fruits	1.64~3.67 N

The fluid–structure interaction simulation results for three different fruit orientations (horizontal, vertical, and 53° angle) are shown in Figure 15a–c. These figures clearly illustrate the interaction between the fluid and the fruit. In the software, the end of the peduncle is modeled as a cantilever beam. The airflow from the nozzle exerts forces on the fruit in different orientations, causing deformation. The stress and strain for the horizontal fruit are shown in Figure 15d,g. Since the airflow from the eight-hole nozzle is symmetrically directed, the surrounding pressure is evenly balanced. However, the deformation of the fruit peduncle is downward due to the influence of gravity. The maximum strain is located at the peduncle, where the effect of the airflow load increases after the fruit sags under gravity, resulting in a maximum strain of 0.00496. The total deformation is approximately 1.74 mm, which shows a slight deviation from the theoretical calculation. The stress and strain for the vertical fruit are shown in Figure 15e,h. The stress in the vertical orientation is mainly concentrated in the lower left part of the fruit facing the nozzle. The stress on the back of the fruit and peduncle is relatively small. Therefore, the deformation of the vertical fruit is primarily concentrated at the peduncle, manifesting as an inclination away from the nozzle. As shown in Figure 15f, the maximum strain is 0.0193, and the total deformation is approximately 9.68 mm. Due to the reduced load after bending in the vertical position, combined with the effects of gravity and the pressure difference between the upper and lower parts of the fruit, the deformation deviates from the theoretical calculation.

For the fruit at a 53° angle, the stress and strain are shown in Figure 15f,i. The stress is mainly concentrated on the upper part of the fruit, with other areas having relatively balanced stress. The peduncle deforms upward. As shown in Figure 15i, the maximum strain is 0.014, and the total deformation is approximately 7.12 mm. This is due to the increased load after upward deformation, with higher airflow velocity on the upper side and lower velocity on the lower side, leading to a greater deformation than the theoretical calculation.

Considering the deformation of 1.74 mm in the horizontal case, the minimum gap for the fruit cluster on the lower side is about 5.38 mm, which differs by 0.2 mm from the ideal calculated gap value. However, the results satisfy the requirements for the mechanical claw insertion.

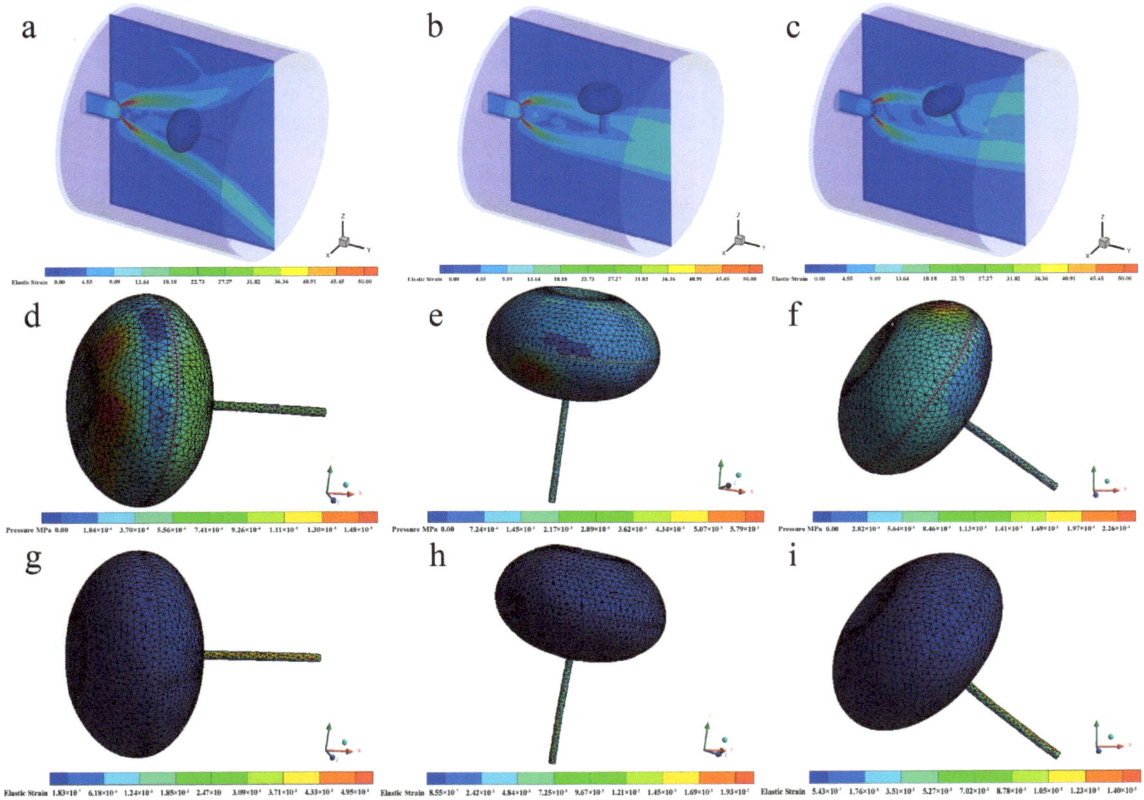

Figure 15. Simulation results for different states of fruit and fruit stem. Simulation of fluid–solid coupling: (**a**) horizontal position; (**b**) vertical position; (**c**) inclination of 53° position. Stress diagram: (**d**) horizontal position; (**e**) vertical position; (**f**) inclination of 53° position. Strain diagram: (**g**) horizontal position; (**h**) vertical position; (**i**) inclination of 53° position.

3.3. Experimental Verification of Fruit Separation

To further validate the accuracy of the simulation and theoretical analysis, we conducted separation experiments on blueberry clusters using the previously optimized eight-hole 40° round nozzle. The results of the separation experiments are shown in Figure 16.

Figure 16 illustrates the interference of fruits at different ripeness stages (green, ripe, and mature) with the target fruit to be picked. To prevent damage to the fruit by the mechanical claw, the gap for fruit separation after spraying should be greater than 40 mm. The nozzle was fixed on a support stand and positioned 40 mm from the target fruit cluster for the spraying experiments. Figure 16a shows the interference of green and ripe fruits with mature fruit, where the red box represents the target fruit to be picked. The separation results are shown in Figure 16b. Due to the hardness of green fruits and their smaller airflow range, but with softer peduncles compared to ripe fruits, the deformation is similar to that of ripe fruits. Under the airflow, the fruits did not fall off, and the gap between fruits measured with a caliper was approximately 6.53 mm. Figure 16c shows the interference of mature fruits with the target ripe fruit, and Figure 16d shows the separation results between mature fruits. With forward air jetting on the target fruit, the side fruit spacing is approximately 5.08 mm. A 0.3 mm error, corresponding to a relative error of 5.9%, arises due to differences between the real blueberry cluster and the simulation. This error is caused by discrepancies in fruit size, stem length, and simulation settings, as well as the fruit's orientation in actual conditions. Despite the errors, the overall results still meet the requirements for the mechanical claw to reach in. In addition to the specific cases shown in the figures, we conducted 30 sets of spraying experiments, with detailed data presented in Table 5. For Type I, the maximum distance was 5.81 mm, the minimum distance was 5.08 mm, and the average distance was 5.41 mm, with a small deviation. For Type II, the maximum distance was 6.42 mm, the minimum distance was 5.62 mm, and the average distance was 5.95 mm, with a small deviation. For Type III, the maximum distance was 5.43 mm, the minimum distance was 4.76 mm, and the average distance was 5.2 mm, with a larger deviation. Despite the short and poorly bent peduncles of green fruits, the minimum distance of 4.71 mm is still greater than the claw thickness, meeting the design requirements.

Figure 16. Results of fruit separation experiments. (**a**) Green and red fruit interfere with target fruit (**b**) Jet results (**c**) Ripe fruit interferes with target fruit (**d**) Jet results.

Additionally, to calibrate the effective wind speed mentioned in the design, a wind speed meter was placed below the fruit cluster during the spraying separation process. With a required fruit separation gap of 4 mm, the nozzle output speed must be at least 9.2 m/s to be considered effective. The spraying device was directed at the back of the fruit, and the wind speed meter was gradually raised with increasing output pressure. When the airflow reached approximately 23.5 m/s, the fruit separated from the peduncle and flew out. With blueberries fixed in a fixture and direct airflow applied at a speed of 23.5 m/s, the fruit skin did not rupture, indicating that the maximum wind speed that the fruit can withstand is 23.5 m/s.

Table 5. Experimental results of different types of disturbances.

No.	Type of Disturbance	Initial Spacing (mm)	Blowing Spacing (mm)
1	II	0.34	5.67
2	II, III	1.40	6.11
3	I, III	0.21	5.42
4	II, III	1.32	5.95
5	II, III	0.38	5.78
6	I	0.00	5.08
7	II	1.68	6.24
8	II	0.71	6.01
9	II, III	1.61	6.53
10	II	0.15	5.88
11	I	0.00	5.34
12	II	0.00	5.62
13	II, III	1.37	6.36
14	I	0.56	5.26
15	II	0.00	5.73
16	I, II	0.00	5.11
17	II	0.17	5.99
18	I, II	1.53	6.89
19	II	1.34	6.42
20	I	1.00	5.81
21	II, III	0.98	6.17
22	I	0.82	5.65
23	III	0.00	4.96
24	II	0.56	6.06
25	I	0.00	5.21
26	II, III	1.21	6.12
27	II	0.43	5.89
28	III	0.00	4.76
29	III	0.28	5.43
30	I	0.00	5.55
Arithmetic mean		0.60	5.766

Ripe fruit type I, red fruit type II, and green fruit type III.

4. Discussion

With the rise in labor costs and advancements in agricultural technology, the mechanization of crop harvesting is undoubtedly a future trend in agriculture. However, blueberry harvesting has lagged in the development of picking robots due to the clustered nature and delicate texture of the fruit.

This study addresses the challenge of accurately picking clustered fruits by proposing the addition of a spraying mechanism in the picking robot. The spraying action increases the spacing between blueberry clusters to facilitate the insertion of the mechanical claw. We compared the output characteristics of different nozzle airflows and assessed the impact of various parameters on airflow. This was validated through pressure sensitivity tests, and we established a relationship between nozzle flow speed and output pressure. We modeled the working area and two-dimensional output flow field for the northern rabbit-eye blueberry used in this study, analyzing effective wind speeds and working ranges. By comparing and analyzing, we selected the most suitable nozzle structure and performed theoretical mechanical analysis. We then imported the fruit peduncle model for fluid–solid coupling simulations and conducted fruit separation experiments at a local plantation to compare the results of different interference fruits and separation experiments. The comparison of mechanical analysis, fluid–solid coupling simulations, and fruit separation experiments provided a more accurate reference for future work.

The spraying device and working area settings are designed for rabbit-eye blueberries, with the working area set based on the maximum fruit size for separation. In actual operations, not all fruits are large. We found that small fruits, due to their high hardness, can be pushed away by the mechanical claw without causing injury to the fruit. In cases where fruit peduncles are short and the fruit hardness is high, the separation gap may not be sufficient. In such cases, if the mechanical claw cannot be inserted, the problem can be improved by increasing the output pressure or using nozzles with more apertures. When harvesting different varieties of blueberries, nozzles or working conditions can be adjusted based on the analysis of specific characteristics to achieve effective fruit separation.

In the future, we will continue to refine the structure of the picking robot. For the spraying device, adjustments will be made based on different varieties, fruit orientations, and densities. Different working conditions will correspond to different nozzle structures and operational settings. Additionally, we will incorporate airflow control into visual recognition systems to address issues where the fruit's front surface is ripe, but the back or sides are not, affecting the robot's visual assessment.

5. Conclusions

This study on blueberry cluster separation through theoretical calculations and fluid-solid coupling simulations has led to the following conclusions:

1. To address the challenge of tightly clustered blueberries that are difficult for mechanical claws to harvest, we proposed a design using a spraying device to separate target fruits from interfering fruits;
2. We established a circular model of the working end face in the spatial flow field. By designing high-pressure and low-pressure zones, we achieved functional separation of the airflow. Finite element analysis shows that the effect of air jet separation is directly related to the nozzle structure type and parameters. A solution using an eight-hole 40° round-head nozzle effectively meets the needs of dispersing clustered blueberries;
3. Fluid–solid coupling simulation indicates that the fruit's orientation significantly affects the air jet separation effect. For example, for fruit with a stem length of 15 mm, the horizontal deformation is about 1.74 mm, the total vertical deformation is approximately 9.68 mm, and the deformation at a 53° horizontal angle is about 7.12 mm, with a minimum gap of about 5.38 mm. Therefore, to achieve more accurate results, it is necessary to consider the mechanical parameters of the fruit stem and the impact of gravity. Additionally, the target fruit in the low-pressure area should be kept as horizontal as possible, and the horizontal orientation angle of the fruit in the high-pressure area should be maximized;
4. Experimental data comparison shows a 10.0% error in theoretical calculations and a 5.9% error in simulation calculations, with the fluid–solid coupling model being more accurate. These errors are due to differences in fruit size, stem length, and simulation parameters, as well as the impact of the fruit's orientation in actual working conditions. Despite these errors, the overall results meet the design requirements;
5. The research methods and calculation models in this study provide a foundational approach for implementing mechanical claw gripping of blueberries and offer insights for separating and harvesting other similar clustered fruits.

Author Contributions: Conceptualization, J.L. and H.Y.; methodology, X.L. and H.Y.; software, W.L.; validation, J.L. and W.L.; formal analysis, Y.L.; investigation, X.L.; resources, J.L. and Y.L.; data curation, W.L.; writing—original draft preparation, W.L.; writing—review and editing, J.L. and H.Y.; visualization, H.W.; supervision, H.W.; project administration, H.Y.; funding acquisition, Y.L. All authors have read and agreed to the published version of the manuscript.

Funding: This research was funded by the Youth Project of Shandong Provincial Natural Science Foundation (ZR2022QE231).

Institutional Review Board Statement: Not applicable.

Data Availability Statement: The data presented in this study are available on request from the corresponding author.

Acknowledgments: The author thanks the Acoustics and Intelligent Control Laboratory of Qingdao University of Technology for the experimental equipment and Wallen Agriculture (Qingdao) for the experimental site.

Conflicts of Interest: The authors declare no conflicts of interest.

References

1. Silva, S.; Costa, E.M.; Veiga, M.; Morais, R.M.; Calhau, C.; Pintado, M. Health promoting properties of blueberries: A review. *Crit. Rev. Food Sci. Nutr.* **2018**, *60*, 181–200. [CrossRef] [PubMed]
2. Li, Y.; Sun, H.; Chen, L. The blueberry industry of China: The past 10 years and the future. *Acta Hortic.* **2017**, *1180*, 531–536. [CrossRef]
3. Margus, A.; Kaarel, S.; Marge, S.; Jüri, O. Mechanical properties of blueberry stems. *Res. Agric. Eng.* **2018**, *65*, 202–208. [CrossRef]
4. Zhou, M.; Sun, H.; Xu, X.; Yang, J.; Wang, G.; Wei, Z.; Xu, T.; Yin, J. Study on the Method and Mechanism of Seedling Picking for Pepper (*Capsicum annuum* L.) Plug Seedlings. *Agriculture* **2024**, *14*, 11. [CrossRef]
5. Cappai, F.; Amadeu, R.R.; Benevenuto, J.; Cullen, R.; Garcia, A.; Grossman, A.; Ferrao, L.F.V.; Munoz, P. High-Resolution Linkage Map and QTL Analyses of Fruit Firmness in Autotetraploid Blueberry. *Front. Plant Sci.* **2020**, *11*, 562171. [CrossRef]
6. Xiao, F.; Wang, H.; Xu, Y.; Shi, Z. A Lightweight Detection Method for Blueberry Fruit Maturity Based on an Improved YOLOv5 Algorithm. *Agriculture* **2024**, *14*, 36. [CrossRef]
7. Niedbala, G.; Kurek, J.; Swiderski, B.; Wojciechowski, T.; Antoniuk, I.; Bobran, K. Prediction of Blueberry (*Vaccinium corymbosum* L.) Yield Based on Artificial Intelligence Methods. *Agriculture* **2022**, *12*, 2089. [CrossRef]
8. Zheng, Z.Q.; Chen, J.H.; Tian, W.; Zhang, C.; An, Z.M. Finite element analysis of blueberry stack damage. *J. Food Process. Eng.* **2021**, *44*, e13840. [CrossRef]
9. Hui, Y.; Liu, H.; Zhang, H.; Wu, Y.; Li, Y.; Li, X.; Wang, D. Application status and development trend of agricultural robot. In Proceedings of the 2018 ASABE Annual International Meeting, ASABE, Detroit, MI, USA, 29 July–1 August 2018; p. 1.
10. Yuan, Y.; Bai, S.; Niu, K. Research progress on mechanized harvesting technology and equipment for forest fruit. *Trans. CSAE* **2022**, *38*, 53–63. [CrossRef]
11. Yu, P.; Li, C.; Takeda, F.; Krewer, G.; Rains, G.; Hamrita, T. Measurement of mechanical impacts created by rotary, slapper, and sway blueberry mechanical harvesters. *Comput. Electron. Agric.* **2014**, *101*, 84–92. [CrossRef]
12. Sarig, Y. Robotics of Fruit Harvesting: A State-of-the-art Review. *J. Agric. Eng. Res.* **1993**, *54*, 265–280. [CrossRef]
13. Erdoğan, D.; Güner, M.; Dursun, E.; Gezer, İ. Mechanical Harvesting of Apricots. *Biosyst. Eng.* **2003**, *85*, 19–28. [CrossRef]
14. Bao, Y.; Yang, J.; Zhao, Y.; Liu, X.; Liang, Z.; Guo, Y. Numerical Simulation of Mechanically Harvested Blueberry Based on EDEM. *J. Harbin Univ. Sci. Technol.* **2020**, *25*, 88–93. [CrossRef]
15. Takeda, F.; Yang, W.Q.; Li, C.; Freivalds, A.; Sung, K.; Xu, R.; Hu, B.; Williamson, J.; Sargent, S. Applying New Technologies to Transform Blueberry Harvesting. *Agronomy* **2017**, *7*, 33. [CrossRef]
16. Gallardo, R.K.; Stafne, E.T.; DeVetter, L.W.; Zhang, Q.; Li, C.; Takeda, F.; Williamson, J.; Yang, W.Q.; Cline, W.O.; Beaudry, R.; et al. Blueberry Producers' Attitudes toward Harvest Mechanization for Fresh Market. *HortTechnology* **2018**, *28*, 10–16. [CrossRef]
17. Tan, K.; Lee, W.S.; Gan, H.; Wang, S. Recognising blueberry fruit of different maturity using histogram oriented gradients and colour features in outdoor scenes. *Biosyst. Eng.* **2018**, *176*, 59–72. [CrossRef]
18. Zhang, J.; Kang, N.; Qu, Q.; Zhou, L.; Zhang, H. Automatic fruit picking technology: A comprehensive review of research advances. *Artif. Intell. Rev.* **2024**, *57*, 54. [CrossRef]
19. Liu, Y.; Zheng, H.; Zhang, Y.; Zhang, Q.; Chen, H.; Xu, X.; Wang, G. "Is this blueberry ripe?": A blueberry ripeness detection algorithm for use on picking robots. *Front. Plant Sci.* **2023**, *14*, 1198650. [CrossRef]
20. Williams, H.A.M.; Jones, M.H.; Nejati, M.; Seabright, M.J.; Bell, J.; Penhall, N.D.; Barnett, J.J.; Duke, M.D.; Scarfe, A.J.; Ahn, H.S.; et al. Robotic kiwifruit harvesting using machine vision, convolutional neural networks, and robotic arms. *Biosyst. Eng.* **2019**, *181*, 140–156. [CrossRef]
21. Ge, Y.; From, P.J.; Xiong, Y. Multi-view gripper internal sensing for the regression of strawberry ripeness using a mini-convolutional neural network for robotic harvesting. *Comput. Electron. Agric.* **2024**, *216*, 108474. [CrossRef]
22. Li, T.; Xie, F.; Zhao, Z.; Zhao, H.; Guo, X.; Feng, Q. A multi-arm robot system for efficient apple harvesting: Perception, task plan and control. *Comput. Electron. Agric.* **2023**, *211*, 107979. [CrossRef]
23. Zhou, Q.; Gao, R.; Ma, Z.; Cao, G.; Chen, J. Research on system integration and control methods of an apple-picking robot in unstructured environment. *Ind. Robot. Int. J. Robot. Res. Appl.* **2024**, *51*, 369–379. [CrossRef]
24. Ye, L.; Duan, J.; Yang, Z.; Zou, X.; Chen, M.; Zhang, S. Collision-free motion planning for the litchi-picking robot. *Comput. Electron. Agric.* **2021**, *185*, 106151. [CrossRef]
25. Huang, P.; Zhu, L.; Zhang, Z.; Yang, C. Row End Detection and Headland Turning Control for an Autonomous Banana-Picking Robot. *Machines* **2021**, *9*, 103. [CrossRef]
26. Bogue, R. Fruit picking robots: Has their time come? *Ind. Robot. Int. J. Robot. Res. Appl.* **2020**, *47*, 141–145. [CrossRef]
27. Liu, J.; Peng, Y.; Faheem, M. Experimental and theoretical analysis of fruit plucking patterns for robotic tomato harvesting. *Comput. Electron. Agric.* **2020**, *173*, 105330. [CrossRef]

28. Zhang, F.; Chen, Z.; Wang, Y.; Bao, R.; Chen, X.; Fu, S.; Tian, M.; Zhang, Y. Research on Flexible End-Effectors with Humanoid Grasp Function for Small Spherical Fruit Picking. *Agriculture* **2023**, *13*, 123. [CrossRef]
29. Lemsalu, M.; Bloch, V.; Backman, J.; Pastell, M. Real-Time CNN-based Computer Vision System for Open-Field Strawberry Harvesting Robot. *IFAC-PapersOnLine* **2022**, *55*, 24–29. [CrossRef]
30. Cao, L.; Chen, Y.; Jin, Q. Lightweight Strawberry Instance Segmentation on Low-Power Devices for Picking Robots. *Electronics* **2023**, *12*, 3145. [CrossRef]
31. Tituaña, L.; Gholami, A.; He, Z.; Xu, Y.; Karkee, M.; Ehsani, R. A small autonomous field robot for strawberry harvesting. *Smart Agric. Technol.* **2024**, *8*, 100454. [CrossRef]
32. Fu, M.; Guo, S.; Chen, A.; Cheng, R.; Cui, X. Design and experimentation of multi-fruit envelope-cutting kiwifruit picking robot. *Front. Plant Sci.* **2024**, *15*, 1338050. [CrossRef] [PubMed]
33. Wang, Z.; Yin, H.; Li, W.; Li, Y.; Liu, J. Research on the Design Method of Blueberry Automatic Harvesting Clamp Force Based on Neural Networks. *Appl. Eng. Agric.* **2024**, *40*, 327–338. [CrossRef]
34. Chen, L.; Feng, Z.; Dong, T.; Wang, W.; Liu, S. Numerical simulation of the internal flow field of a new main nozzle in an air-jet loom based on Fluent. *Text. Res. J.* **2015**, *85*, 1590–1601. [CrossRef]

agriculture

Review

Application of Computer Simulation Technology in the Development of Tractor Transmission Systems

Ya Li [1,†], Xiaohan Chen [1,†], Xiaorong Han [2], Ning Xu [3], Zhiqiang Zhai [4], Kai Lu [5,6], Youfeng Zhu [7] and Guangming Wang [1,*]

1 College of Mechanical and Electronic Engineering, Shandong Agricultural University, Taian 271018, China; liya0683@126.com (Y.L.); jdchenxiaohan@163.com (X.C.)
2 Shandong Weifang Luzhong Tractor Co., Ltd., Weifang 261057, China; hanxiaorong091@126.com
3 Shandong Academy of Agricultural Machinery Sciences, Jinan 252100, China; xuning@saas.ac.cn
4 College of Engineering, China Agricultural University, Beijing 100083, China; zhaizhiqiang@cau.edu.cn
5 State Key Laboratory of Intelligent Agricultural Power Equipment, Luoyang 471003, China; lukai@ytogroup.com
6 Luoyang Tractor Research Institute Co., Ltd., Luoyang 471003, China
7 School of Software & Microelectronics, Peking University, Beijing 102600, China; zhuyoufeng@163.com
* Correspondence: g.wang@sdau.edu.cn
† These authors contributed equally to this work.

Abstract: Tractors are the most important agricultural power machinery. With the development of tractors toward large-scale and electrification, the design of modern tractor transmission systems increasingly relies on computer simulation technology. However, tractor transmission systems, especially power-shift transmissions and CVTs, are highly complex industrial products that involve specialized knowledge from multiple disciplines. Engineers and researchers find it difficult to choose the correct mathematical model and grasp the trend of technological development when applying simulation technology. To address this issue, we conducted a systematic review of the field and carried out the following work: First, the types and basic principles of tractor transmission systems were introduced; Second, the modeling methods and applications of computer simulation technology in the structural optimization, power-shift control, and energy saving of tractor transmission systems were reviewed; Finally, the method of obtaining simulation parameters through system identification was introduced. Although computer simulation technology has been applied in the development of all tractor transmission systems, there are still some issues that need attention, such as the lack of established shift quality evaluation indicators and driving cycles suitable for tractors, which are crucial for the reliability of simulation results but are rarely studied. These issues need to be addressed in future work.

Keywords: tractor; transmission; computer; simulation; agriculture

Citation: Li, Y.; Chen, X.; Han, X.; Xu, N.; Zhai, Z.; Lu, K.; Zhu, Y.; Wang, G. Application of Computer Simulation Technology in the Development of Tractor Transmission Systems. *Agriculture* **2024**, *14*, 1547. https://doi.org/10.3390/agriculture14091547

Academic Editor: Mustafa Ucgul

Received: 25 July 2024
Revised: 30 August 2024
Accepted: 4 September 2024
Published: 6 September 2024

1. Introduction

Tractors are the most important agricultural power machinery, providing power for the vast majority of agricultural production activities such as plowing, seeding, spraying, harvesting, and transportation [1,2]. With the continuous growth of the global population, the demand for increased food production is escalating, thereby driving the development of tractors toward larger scale to achieve higher agricultural production efficiency. Taking China as an example [3], the country's production of large- and medium-power tractors reached 400,000 units in 2022, with ownership exceeding 5.25 million units, gradually replacing smaller-powered tractors. The key components of the tractor are the engine and the transmission. The engine provides power for the tractor but is limited by its own universal characteristics [4]; its power, fuel consumption, and emissions are different at different speeds and torques. The function of the transmission is to adjust the engine torque by changing its own gear ratio so that the engine operates near the high-power or high fuel-efficiency area of the engine fuel consumption map [5,6]. Considering the mature engine technology, modern tractors are usually classified based on the transmission they are equipped with, such as mechanical-shift tractors, power-shift tractors [7,8], and continuously variable transmission (CVT) tractors [9,10]. However, as tractor power increases, the complexity of its transmission system rises, leading to high research and manufacturing costs. Furthermore, with intensifying competition in the high-power tractor market, the product update cycle has been significantly compressed. To address these challenges, an increasing number of researchers and engineers are exploring the application of advanced computer simulation technology in the development of tractor transmission systems.

Currently, simulation technologies related to tractor transmissions can be divided into three categories: fully digital simulation, fully physical simulation, and semi-physical simulation. Full-digital simulation relies entirely on programming or software implementation, with its core task being the construction of mathematical models for the problem under study. Transmission problems, such as the speed characteristics and efficiency characteristics of the powertrain, can usually be solved directly by deducing explicit mathematical equations [11]. For dynamic problems, differential equations must be constructed and solved using numerical methods [12]. In recent years, commercial simulation software has been widely used, such as Matlab/Simulink (the latest version is R2024a), OpenModelica (the latest version is 1.9.0 RC1), SimulationX (the latest version is 2024), AMESim (the latest version is 2404), Easy5 (the latest version is 2024.1), etc. These platforms encapsulate commonly used mathematical models into library components, facilitating the rapid construction of complex mathematical models. Of course, this is not absolute. Software like ADAMS (the latest version is 2024.1) is based on virtual prototyping to solve dynamic problems and does not require library components as support. For complex structural analysis and load simulation problems, it can only be solved by finite element method (FEM) [13] or discrete element method (DEM) [14]. The corresponding commercial simulation software includes Ansys (the latest version is 2024 R2), EDEM (the latest version is 2024), and so on. Full-physical simulation simulates the operating conditions of transmissions or tractors through test benches to obtain the required research results, such as scientific research based on driving simulators [15], loading test benches [16], and power wagons [17]. Semi-physical simulation is a simulation method that falls between fully digital simulation and fully physical simulation. It replaces some physical entities with mathematical models and forms a real-time simulation loop with the remaining physical entities. Common semi-physical simulation systems include Rapid Control Prototypes (RCP) [18] with digitized controllers and Hardware-in-the-Loop (HIL) [19] with digitized controlled objects.

This review primarily discusses the application of computer simulation technology in the development of tractor transmission systems, with a focus on the discussion of modeling methods. The manuscript is structured as follows: Section 1 introduces the background and method of tractor transmission simulation; Section 2 introduces the typical tractor transmission and its working principle; Section 3 introduces the application of simulation technology in transmission structure optimization, power-shift control, and energy saving; Section 4 introduces the system identification method for determining simulation parameters; Section 5 summarizes this review and provides research suggestions for future work.

2. Tractor Transmission System

The transmission system of tractors is significantly different from that of road vehicles. Firstly, the main purpose of road vehicles is to carry people and goods, so they only need a few gears to meet the needs of road driving. By contrast, tractors have complex operating conditions, as depicted in Figure 1 [20]. Different operating conditions correspond to different speeds, requiring tractors to have many more working gears than cars to meet their operational requirements. Secondly, the load of the transmission system of road vehicles mainly includes slop resistance, rolling resistance, and air resistance, while the transmission system of tractors usually has to withstand additional traction resistance generated by the interaction between agricultural machinery and soil [21–23]. Taking plowing as an example, the traction resistance is large and complex, and its value is related to soil conditions, plowing depth, and operating speed, as shown in Figure 2 [24]. For the same reason, modern tractor transmissions need to maintain power transmission when shifting under heavy loads, but road vehicles typically do not have this requirement. Thirdly, the principle of tractor transmission is usually different from that of road vehicles. For example, road vehicle CVTs rely on mechanical friction to transmit power, while tractor CVTs use hydraulic systems to transmit power. Moreover, the efficiency fluctuation of tractor CVT is usually more severe than that of road vehicle CVT, making it difficult to directly apply the control strategy of road vehicle CVT to tractor CVT. Due to the above characteristics of tractor transmission systems, their modeling methods differ from those of road vehicles. Therefore, before introducing the simulation of tractor transmission systems, we need to first understand their transmission principles and characteristics.

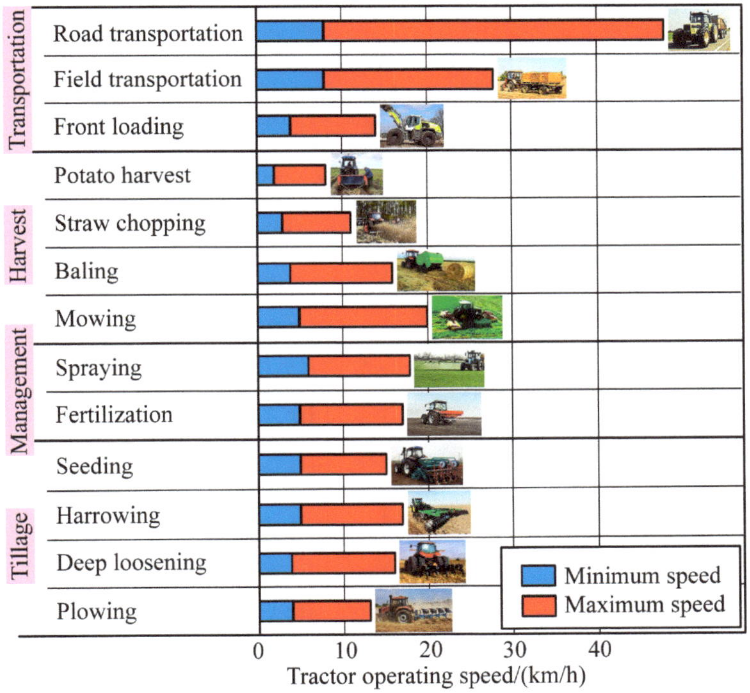

Figure 1. Operating conditions of tractors. (Reproduced from [20]).

As mentioned above, tractor transmissions can be divided into three categories based on the shift method: mechanical-shift transmissions, power-shift transmissions, and CVTs. Tractor mechanical-shift transmission requires manual control of the gear lever to shift gears. Although synchronizers [25,26] were later introduced to assist in shifting, the master clutch still needs to be released to cut off the engine's power transmission during shifting. When the tractor is engaged in heavy load operation, the power interruption during the shift means that the tractor will be quickly braked and stopped by the load, so the driver has to shift by active parking. Frequent start–stop operations not only compromise driving comfort but also diminish tractor operational efficiency.

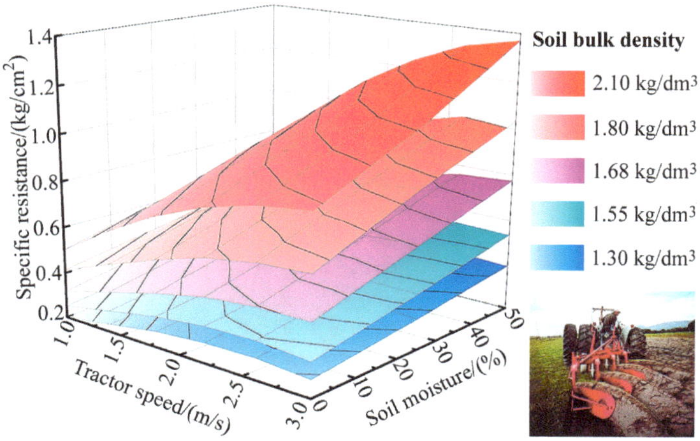

Figure 2. Specific resistance of plowing.

The tractor power-shift transmission equipped with a wet clutch [27,28] can shift without interrupting power. A typical full power-shift system is the e23 transmission developed by John Deere (Moline, IL, USA), which has been equipped on its 7R series tractors, as shown in Figure 3. This transmission has 23 gears for tractor-forward travel, all of which are controlled by wet clutches. To reduce manufacturing costs and improve its operational reliability, semi-power-shift transmissions with only a portion of gears switched by wet clutches have also been widely used [29,30]. For example, the transmission of the LZ2604 tractor developed by YTO (Luoyang, China) is composed of a four-speed power-shift system and a six-range automated mechanical transmission (AMT) system connected in series.

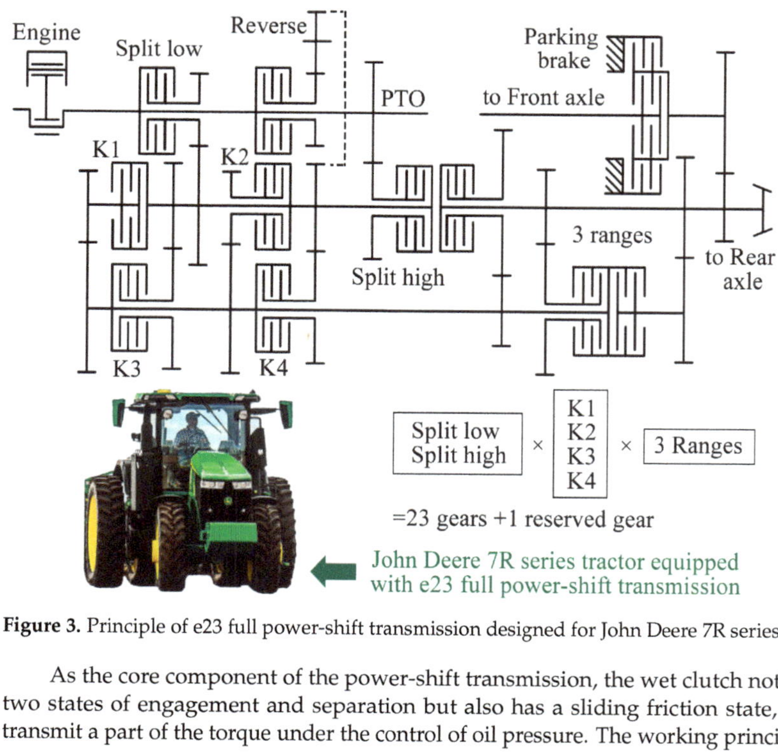

Figure 3. Principle of e23 full power-shift transmission designed for John Deere 7R series tractor.

As the core component of the power-shift transmission, the wet clutch not only has two states of engagement and separation but also has a sliding friction state, so it can transmit a part of the torque under the control of oil pressure. The working principle of the wet clutch is illustrated in Figure 4. When the oil pressure in the clutch cylinder increases, it overcomes the spring force to move the piston to the right, pressing the driving and driven friction plates together, thereby connecting the driving shaft to the driven shaft. On the contrary, when the oil pressure in the clutch cylinder decreases, the spring pushes the piston back to its original position, causing the driving and driven friction plates to release and separate the driving shaft from the driven shaft. Specifically, when the oil pressure is insufficient to fully press the driving and driven friction plates, the clutch will enter a sliding state, and the torque it can transmit depends entirely on the oil pressure. During the shift process, the oil pressure is accurately controlled by the computer, and the torque transfer path can be smoothly switched from the off-going clutch to the on-coming clutch, thereby avoiding power interruption.

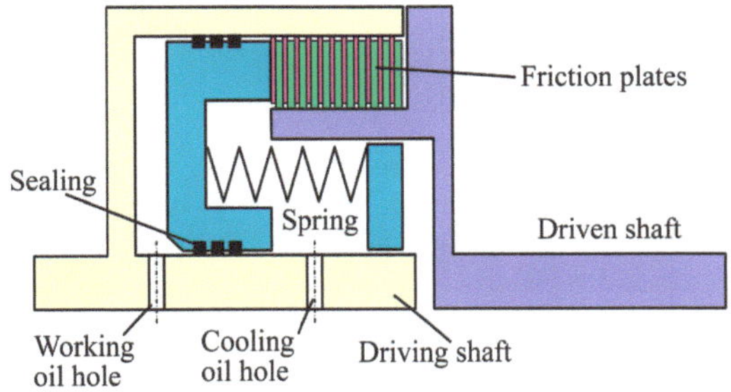

Figure 4. Structure of wet clutch.

The gear-shifting process under computer control greatly reduces the driver's labor intensity and improves driving comfort. Such transmissions typically feature multiple working gears to accommodate diverse operating conditions. For example, the PL2304 tractor produced by ZOOMLION (Changsha, China) has 80 working gears (40 F + 40 R), while the LF2204 tractor produced by YTO (Luoyang, China) has 89 working gears (48 F + 41 R). The increase in the number of gears greatly improves the power performance and fuel economy of tractors, but it also brings difficulties to the design and use of transmissions. Therefore, CVT without gears has become the ideal transmission system for tractors. Over the past century, agricultural engineers have attempted to develop at least six types of tractor CVTs [31]: metal belt (or chain) transmissions, hydrostatic transmissions (HST), electric transmissions, and their power-split versions. Due to performance or cost limitations, except for hydrostatic transmission and hydrostatic power-split transmission, which have been widely commercialized, other tractor CVTs mentioned above have been phased out or have not yet been mass-produced. Tractor HST is a closed system with a variable displacement pump and a hydraulic motor [32]. Although it consumes more energy, its structure is simple, and the cost is lower than its power-split version, thus filling the gap in the latter's market for low-power tractors. The typical application of HST is the L2502 tractor produced by Kubota (Osaka, Japan). Due to space limitations, it will not be elaborated here. Currently, hydrostatic power-split transmission has become the de facto industrial standard for CVT in large and medium power tractors. The world's first hydrostatic power-split tractor transmission was the Vario released by Fendt (Marktoberdorf, Germany) in 1996 [33], as shown in Figure 5. The working principle of Vario is as follows: the engine power enters the carrier of the planetary gear (PG) and is divided into two paths for transmission. One part enters the hydraulic path with swash plate axial piston units as the speed regulating element through the ring gear of PG, and the other part enters the mechanical path through the sun gear of PG. The two parts of power merge at the motor shaft and continue to be transmitted toward the rear axle. Because only a portion of the power flows through the hydraulic system with high energy consumption, this transmission has both the high efficiency of mechanical transmission and the high load driving capability of hydraulic transmission.

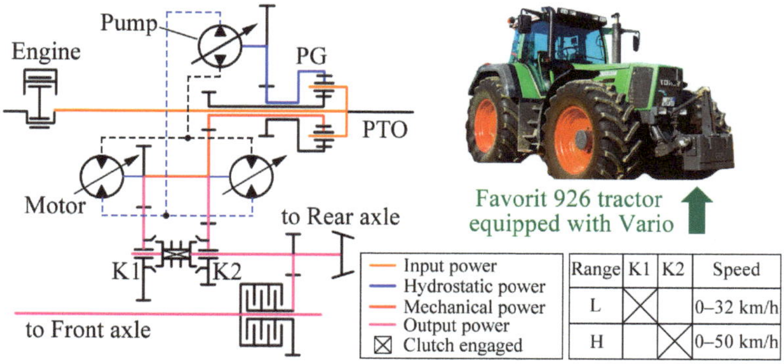

Figure 5. Principle of Vario transmission designed for Fendt Favorit 926 tractor.

It is not difficult to find from the principle of Vario that the most distinctive mechanical component of tractor CVT is the planetary gear. According to the classification method proposed by B.H. Кудрявцев [34], planetary gears can be divided into three basic types: 2Z-X, 3Z, and Z-X-V. Among them, $2Z-X(A)$, $2Z-X(B)$, and $2Z-X(D)$, as well as advanced configurations derived or combined from them, are widely used in CVT tractors, as shown in Figure 6. The structure of $2Z-X(A)$ is very compact, with a transmission efficiency of 97–99%, known as the standard planetary gear, which is used by Fendt Vario; The radial size of $2Z-X(B)$ is small, but the gear ratio range is large, and its transmission efficiency can also reach 97–99%. However, its manufacturing and installation are relatively complex. The typical application of this planetary gear is the dual-clutch power-split transmission developed by CNH (London, UK) [35]. $2Z-X(D)$ has the most compact structure and is therefore adopted by some tractor manufacturers, but in fact, the transmission efficiency of this planetary gear is very low and not suitable for tractor transmission.

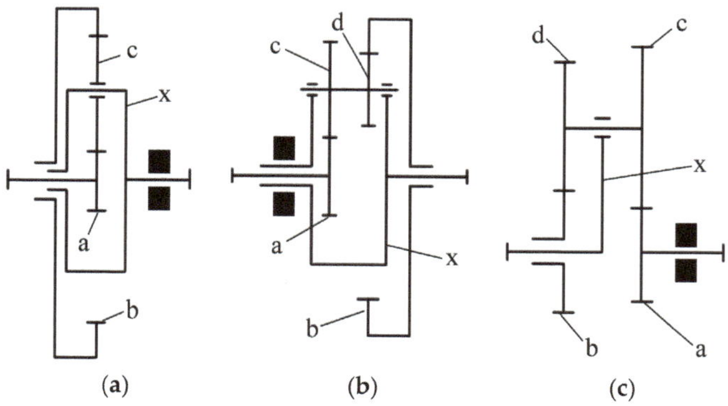

Figure 6. Planetary gears used in CVT tractors. (**a**) $2Z-X(A)$. (**b**) $2Z-X(B)$. (**c**) $2Z-X(D)$. Note: The letters a and b represent the central gear (such as the sun gear or ring gear), x represents the carrier, and c and d represent the planet gears.

In addition to planetary gears, the configuration of the hydrostatic power-split transmission also depends on the structure of its closed transmission chain [36]. Vario is a typical planetary output coupled (OC) transmission, not the mainstream configuration of tractor CVT. To reduce the performance requirements for hydraulic systems, current tractor CVTs commonly use a planetary input coupled (IC) configuration, where the planetary gears are used to collect power rather than distribute it. Different configurations and parameters determine the specific flow pattern of transmission power. When power flows in the opposite direction along the hydraulic or mechanical path, parasitic power is formed, also known as cyclic power [37]. This part of the power does not output externally but only generates internal friction, which is a key consideration in the design of such transmissions.

In fact, there is a relationship between power-shift technology and hydrostatic power-split technology. Modern hydrostatic power-split tractor transmissions usually use multi-range technology [38] to improve transmission efficiency, which divides the speed range of the transmission into several ranges. Due to the limitation of the hydrostatic power portion in each range, the tractor always maintains a high transmission efficiency. For instance, the Class HM8 transmission with seven ranges has a transmission efficiency of up to 89–94% (excluding axles) [39]. Note that the switching of the range is completed by the wet clutch, which means that the hydrostatic power-split tractor transmission is still inseparable from the power-shift technology. However, most hydrostatic power-split tractor transmissions meet the equal-speed shift condition; that is, the gear ratio before and after the shift is constant, which greatly reduces the control difficulty of the power shift.

In recent years, electric tractors have developed rapidly, but they have not yet achieved widespread commercialization. The standard electric tractor is made up of batteries, motors, and transmission devices in series [40,41], which is the most promising tractor transmission system, such as the H_2 Dual Power tractor produced by CNH (London, UK) (Figure 7). However, due to very high manufacturing costs and short operating time, this "direct with battery" (i.e., pure electric) tractor has had difficulty being accepted by the market in the short term. On the basis of the solution mentioned above, the "direct with engine" (i.e., series hybrid) solution that uses an engine-driven generator (i.e., range extender) to charge the battery can extend the tractor's operating time, but its transmission efficiency is low, and its performance advantages are not obvious. The most likely to enter the market in the short term is the power-split version of electric tractors (i.e., electric CVT or eCVT). Replacing the variable displacement pump and hydraulic motor with a generator and an electric motor can roughly understand the working principle of this transmission, but its electrical energy can be recovered, which is a common advantage of all-electric tractors.

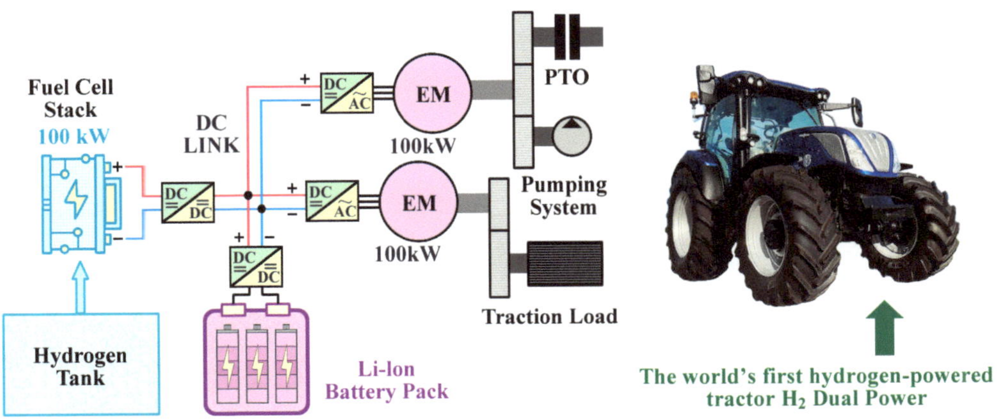

Figure 7. Principle of electric transmission system designed for CNH H_2 Dual Power tractor.

Finally, we summarized the types and application status of various tractor transmissions, as shown in Table 1, which details the advantages and disadvantages of each type of tractor transmission system. It is not difficult to see that mechanical-shift transmissions will gradually be phased out. Power-shift transmissions and hydrostatic power-split transmissions are the most advanced tractor transmission systems that have been commercially available in the world today. The electric transmission system directly connected to the battery has long-term potential, but before its technology matures, the electric transmission system connected to the engine through power splitting is expected to become a transitional tractor transmission technology.

Table 1. Types and application status of tractor transmission systems.

Type	Sub-Type	Principle of Speed Regulation	Comments	Application in Tractors
Mechanical	Mechanical	Shift with sliding gears, engagement sleeves, or synchronizers.	• Low manufacturing cost and high transmission efficiency. • Low operational efficiency.	Suitable for tractors of various powers, such as the 5-754 tractor produced by John Deere (Moline, IL, USA) that uses engagement sleeves for shifting and the MH2204 tractor produced by Wuzheng (Rizhao, China) that uses synchronizers for shifting. Note that this type of tractor has gradually been phased out in many developed countries.
Power-shift	Full power-shift	Shift with wet clutches.	• High transmission efficiency and operational efficiency. • High manufacturing cost and difficulty in controller design.	Suitable for large power tractors, such as the 9RT590 tractor produced by John Deere (Moline, IL, USA) and the Puma2304 tractor produced by CNH (London, UK).
	Semi power-shift	Combination of power-shift and mechanical-shift.	• Medium manufacturing costs and high transmission efficiency. • Medium operating efficiency.	Suitable for large and medium power tractors, such as the LZ2604 tractor produced by YTO (Luoyang, China) and the 8W2804 tractor produced by Deutz-Fahr (Treviglio, Italy).
CVT	Metal belt or chain (direct)	Change the radius of the traction force.	• High transmission efficiency and operating efficiency, and low manufacturing cost. • Low load driving capability.	Although engineers have developed a large number of tractor prototypes, such as the Munich Research Tractor released by TUM (München, Germany) in 1988, they have never been accepted by the market. However, this type of transmission is widely used in road vehicles.
	Metal belt or chain (power-split)		• High transmission efficiency but lower than its direct version; High operating efficiency and medium manufacturing cost. • Low load driving capacity but higher than its direct version.	
	Hydrostatic (direct)	Change the unit's displacement.	• High load driving capability and operating efficiency. • Medium manufacturing cost, low transmission efficiency, and high noise level.	Suitable for small power tractors, such as the L2502 HST tractor produced by Kubota (Osaka, Japan) and the Solis26 HST tractor produced by ITL (Hoshiarpur, India). Note that this transmission system can also be used for large and medium power tractors, but its performance is not as good as its power-split version.
	Hydrostatic (power-split)		• High load driving capability, transmission efficiency, and operating efficiency. • High manufacturing cost and noise level.	Suitable for large and medium power tractors, such as the P7000 tractor produced by Lovol (Weifang, China) and the 1000 Vario tractor produced by Fendt (Marktoberdorf, Germany).
	Electrical (direct with engine)	Change the frequency of current or electric flux.	• High operating efficiency. • High manufacturing cost and low transmission efficiency.	Upcoming. Some tentative applications: Belarus 3023 tractor produced by Minsk Tractor Works (Minsk, Belarus), and HB2204 tractor produced by YTO (Luoyang, China).
	Electrical (direct with battery)		• High operating efficiency. • Very high manufacturing cost and short operation time.	Upcoming, with long-term potential. Some tentative applications: SESAM tractor produced by John Deere (Moline, Illinois, USA), and e100 Vario tractor produced by Fendt (Marktoberdorf, Germany).
	Electrical (power-split)		• High transmission efficiency and operating efficiency. • High manufacturing cost.	Upcoming, the most promising next-generation tractor CVT. Some tentative applications: 8R 410 tractor produced by John Deere (Moline, IL, USA).

3. Application of Simulation in Tractor Transmission Development

Simulation technology is usually used to solve complex transmission and dynamics problems in the development of tractor transmissions. This review focuses on three issues: structural optimization, power-shift control, and energy saving.

3.1. Structural Optimization

In the field of structural optimization, static analysis, modal analysis, acoustic analysis, thermal analysis, etc., are usually required for tractor transmissions, and the main method used is the Finite Element Method (FEM). These analyses typically have a similar methodological process: preparing geometric models, preprocessing, solving, and post-processing. In this type of research, commercial finite element software such as Ansys (the latest version is 2024 R2) is widely used, but for the analysis of simple structures, it can also be achieved through programming. There is a significant amount of research in this field. For example, Mujiburrahman et al. [42] developed a shell model of a tractor transmission based on Solidworks and conducted finite element analysis of stress and deformation using Ansys. In this study, the authors optimized the structural parameters of the transmission, such as web thickness, gear box housing depth, and gearbox housing thickness, and compared the stress of the shell based on E-Glass and gray cast iron. The results show that the latter has better mechanical properties. Dong et al. [43] conducted similar optimization work. The authors constructed a shell model of a tractor transmission under Creo and conducted finite element analysis on it using Ansys. However, in addition to verifying the structural strength, this study also analyzed the modal characteristics of the structure to ensure that the natural frequency of the transmission casing is far away from the engine frequency and the meshing frequency of the gears in the transmission system, thereby avoiding resonance. Hartmann et al. [44] constructed a vibration acoustic model, including the transmission casing and cab on finite element mesh, and proposed a hybrid method of Dynamic Energy Analysis (DEA)–Statistical Energy Analysis (SEA) for analyzing tractor acoustic response. Sergienko et al. [45,46] studied the frictional behavior of tractor oil-cooled friction multidisc clutches. The authors constructed dynamic differential equations describing the sliding process of the clutch driving and driven friction plates, as well as thermodynamic equations for the friction heating clutch process, and solved them using Runge Kutt method and finite element method, respectively, to obtain the formation law of the stationary temperature field of the clutch.

In the past, FEM methods only provided visual simulation results, and engineers needed to manually improve and optimize the structure and parameters of tractor transmission systems under the guidance of cloud maps. With the advancement of technology, most finite element software, such as Ansys, provide automatic optimization programs and can even perform topology optimization on structures. The optimization methods involved include Response Surface Methodology (RSM), Genetic Algorithm (GA), etc. Although these automated operations are not mentioned in the above literature, they can significantly improve work efficiency in the development of tractor transmissions. The general process of using FEM for the structural design and optimization of tractor transmission is shown in Figure 8.

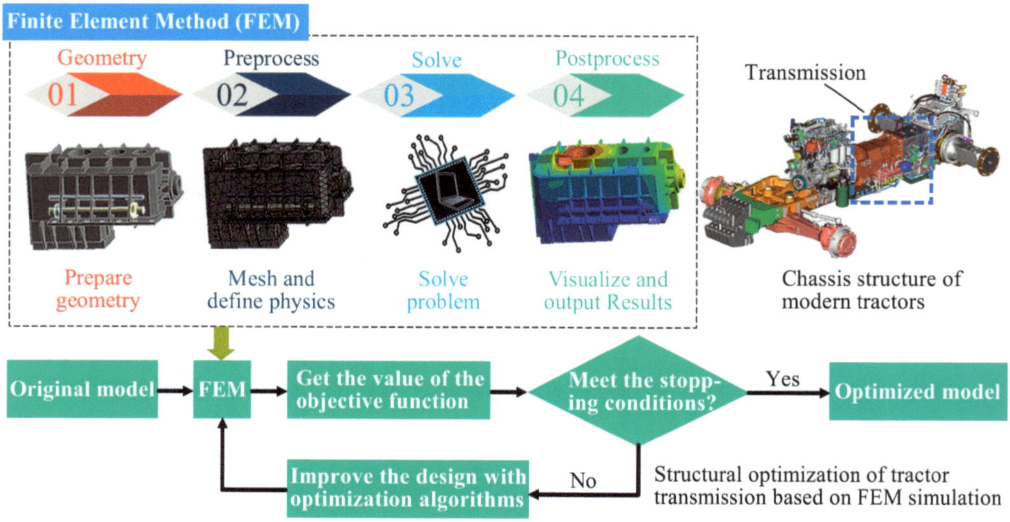

Figure 8. General process of using FEM for the structural optimization of tractor transmissions. (Reproduced from [35,47]).

3.2. Power-Shift Control

The primary task of tractor power-shift research is to determine the evaluation method for shift impacts. In road vehicles, peak acceleration J_1 and jerk J_2 are the most commonly used indicators for evaluating shift quality and are also widely used in tractor research. Their expressions are

$$J_1 = Max(dV_t/dt) \tag{1}$$

$$J_2 = Max\left(d^2V_t/dt^2\right) \tag{2}$$

where, V_t represents the speed of the tractor, and its first and second derivatives with respect to time dV_t/dt and d^2V_t/dt^2 represent the rate of change in tractor speed and acceleration, respectively. Therefore, the peak acceleration J_1 and jerk J_2 reflect the impact of the shift process on the tractor's movement, which directly affects driving comfort.

However, the validity of these two indicators is still widely debated today. For instance, Winkel et al. [48] conducted research on this issue based on a simulated driving test involving 23 drivers and redefined the driving comfort of road vehicles as a function of acceleration, jerk, and direction. In fact, Duncan and Wegscheid [49] from John Deere conducted similar research on tractors for field operations as early as the 1980s. This work was conducted on the Vehicle Operations Simulator (VOS) of the Deere & Company Technical Center's human factors laboratory, where 18 experienced drivers rated the shift quality under different peak accelerations, jerks, and background vibrations, as shown in Figure 9. The findings indicate that the peak acceleration during shift can better reflect the driver's subjective evaluation of the tractor shift quality, but jerk is not suitable for tractors used in field operations. The conclusion of this study is very important because it fully considers the impact of background vibration on tractor driving comfort. Unfortunately, this early work has not received sufficient attention in recent years.

Figure 9. Research on tractor shift quality based on vehicle operations simulator. (Reproduced from [49]).

In the study of tractor power-shift, in addition to the subjective feelings of the driver, the energy loss of the wet clutch, the core component of the shift system, needs to be considered [50]. It can be evaluated by peak sliding friction power J_3 and sliding friction work J_4:

$$J_3 = Max\left(T_{cf}\omega_{dc}\right) \tag{3}$$

$$J_4 = \int_{t_a}^{t_b} T_{cf}|\omega_{dc}|dt \tag{4}$$

where the product of the torque T_{cf} and slip speed ω_{dc} of the clutch $T_{cf}\omega_{dc}$ is the power consumed by the clutch during the sliding process, and its integral over time $\int_{t_a}^{t_b} T_{cf}|\omega_{dc}|dt$ is the accumulated energy loss, which will be dissipated in the form of frictional heat. Therefore, the peak sliding friction power J_3 and sliding friction work J_4 reflect the impact of the shift process on the clutch. Note that these two indicators are important factors affecting the service life of the clutch. In extreme cases, the friction heat released by excessive energy loss may quickly burn the clutch.

Other common indicators for evaluating gear shift quality include speed fluctuation, torque fluctuation, and customized, comprehensive indicators, as detailed in the studies conducted by Chen et al. [51], Wang et al. [52], and Tanelli et al. [53], which will not be elaborated on here.

We have summarized the indicators commonly used in previous studies to evaluate the shift quality of tractors, as shown in Table 2. It is not difficult to find that the rationality of most indicators has not been confirmed, which poses difficulties for the scientific evaluation of simulation results. John Deere's research clearly indicates that some evaluation indicators for the shift quality of road vehicles cannot be directly used in tractors. Therefore, research in this field needs to be strengthened. Currently, we recommend using peak acceleration and sliding friction work in simulations to evaluate the shift quality of tractors.

Table 2. Summary of evaluation indicators for tractor shifting dynamics simulation.

Evaluation Indicator	Expression	Type of Shift Quality Indicated	Comments
Speed fluctuation.	$V_\infty - V_{min}$	Impact on the driver.	• The relationship between this indicator and tractor driving comfort needs to be confirmed. • Power interruption (i.e., $V_{min} = 0$) will affect the evaluation of shift quality.
Peak acceleration.	$Max(dV_t/dt)$	Impact on the driver.	• John Deere has confirmed the close relationship between this indicator and tractor driving comfort under field background vibration. • Suitable for evaluating the shift quality of tractors during field operations.
Jerk.	$Max\left(d^2V_t/dt^2\right)$	Impact on the driver.	• John Deere has already denied the validity of this indicator under field background vibration, but it is widely recognized in evaluating the shift quality of road vehicles. • Suitable for evaluating the shift quality of tractors during road transportation.
Torque fluctuation.	T_{max}/T_∞ or $T_{max} - T_\infty$	Impact on the entire transmission system.	• The relationship between this indicator and clutch service life needs to be confirmed. • Excessive T_∞ will affect the sensitivity of this evaluation indicator.
Peak sliding friction power of clutch.	$Max\left(T_{cf}\omega_{dc}\right)$	Impact on the clutch.	• The relationship between this indicator and clutch service life needs to be confirmed.
Sliding friction work of clutch.	$\int_{t_a}^{t_b} T_{cf}\lvert\omega_{dc}\rvert dt$	Impact on the clutch.	• It directly reflects the clutch energy loss.
Customized evaluation indicator.	-	-	Not recommended for use without experimental support.

After determining the evaluation index, the core task of the power-shift control technology is to optimize the oil pressure curve of the wet clutch to reduce the shift impact. The simplest clutch pressure control method is to use a solenoid directional valve [54], which typically has a hydraulic circuit, as shown in Figure 10. When the solenoid valve is energized, the clutch cylinder connects to the main oil circuit, causing its pressure to rise from zero to the main line pressure to engage the clutch. On the contrary, when the solenoid valve is powered off, the clutch cylinder connects to the oil tank, causing its oil pressure to drop from the main line pressure to zero to separate the clutch. Whether the clutch is engaged or separated, its oil pressure curve varies in segments. When the clutch is engaged, its working oil pressure rises in two stages to the termination pressure:

Phase 1: As the oil pressure increases, when it is sufficient to overcome the spring force and static friction force, the piston begins to move, and its discriminant equation is as follows:

$$p_c A_{cp} > k_{spr} x_{s0} + F_{fst} \tag{5}$$

where the left side of the equation represents the thrust of hydraulic oil acting on the piston, while the two terms on the right side of the equation, respectively, represent the reaction force of the spring and the static friction force that prevents the piston from moving.

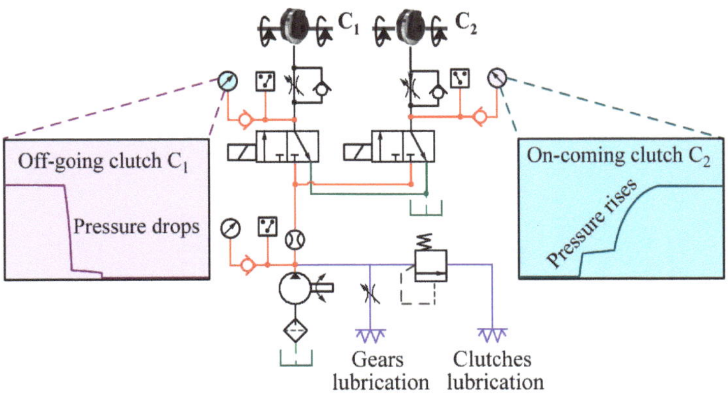

Figure 10. Power-shift system based on solenoid reversing valve.

Before the piston moves to the end stop, the clutch oil pressure is mainly used to overcome the spring force and sliding friction, the value of which is determined by the following equation:

$$p_c A_{cp} = k_{spr}(x_{s0} + x_s) + F_{fsl} + m_{cp}(dv_{cp}/dt) \tag{6}$$

where the left side of the equation represents the thrust of hydraulic oil acting on the piston, while the three terms on the right side of the equation respectively represent the reaction force of the spring after compressing the displacement x_s, the sliding friction force that prevents the piston from moving, and the inertial force of the piston movement.

Wang et al. [55] confirmed through experiments that the movement speed of the piston is approximately constant, i.e., $dv_{cp}/dt \approx 0$. The sliding friction force is usually a constant value and satisfies $F_{fri} \leq F_{fst}$. Therefore, after experiencing the first rapid increase, the clutch pressure only increases slightly with the change of spring displacement.

Phase 2: When the piston moves to the end stop, it stops moving. At this point, except for the part used to overcome the spring force, the remaining part of the oil pressure is used to press the driving and driven friction plates to engage the clutch:

$$F_{cN} = p_c A_{cp} - k_{spr}(x_{s0} + x_{s,\ Max}) \tag{7}$$

where the left side of the equation represents the normal force acting on friction plates, while the two terms on the right side of the equation respectively represent the thrust of hydraulic oil acting on the piston and the reaction force of the spring after compressing the stroke x_s.

The maximum torque that the clutch can transmit is [56,57]

$$T_{cf,\ max} = F_{cN} \times 2n_{sf}\mu_c\left(r_{co}^3 - r_{ci}^3\right) / \left[3\left(r_{co}^2 - r_{ci}^2\right)\right] \tag{8}$$

Once the load torque of the clutch is higher than $T_{cf,\ max}$, its driving and driven friction plates enter a slipping state, and the torque transmitted by it no longer increases. The above equation is widely used in the modeling of power-shift systems, and the key to solving it is to obtain the value of clutch oil pressure p_c in advance. It can be obtained through experimental methods or simulation. Chen et al. [58] constructed a power-shift model based on solenoid directional valves and wet clutches under AMESim and obtained clutch pressure curves consistent with experimental results through simulation.

The separation process of the clutch can also be divided into two stages, which are the reverse of the clutch engagement process. Once the clutch pressure decreases, the driving and driven friction plates immediately enter a slipping state, and subsequent piston movement does not affect the shift quality.

Continuing to discuss the above equation, Kim et al. [59] pointed out in their study of power-shift tractors that the friction coefficient of the clutch μ_c is not a constant but a variable related to the slip speed between the driving and driven friction plates, as shown in Figure 11. Their tractor shift dynamics model constructed under EASY5 achieved high computational accuracy and was experimentally validated under multiple operating conditions.

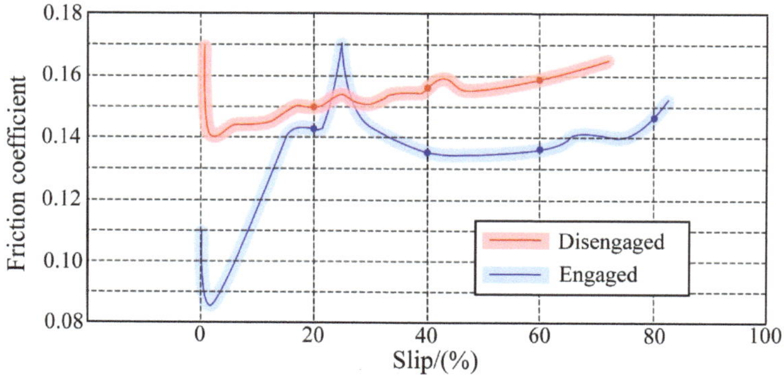

Figure 11. The friction coefficient of the wet clutch (Reproduced from [59]).

The clutch model under SimulationX also fully considers this point and provides the following equation to describe the actual friction coefficient of the clutch [60]:

$$\mu_c(\omega_{dc}) = \mu_{cst} - (\mu_{cst} - \mu_{csl}\mathrm{M}_1)tanh(\mathrm{M}_2\omega_{dc}) + \mathrm{M}_3\omega_{dc} \quad (9)$$

where the reference values of the static friction coefficient μ_{cst}, sliding friction coefficient μ_{csl} and constant M_1, M_2 and M_3 are shown in Table 3.

Table 3. Variable values related to clutch friction coefficient [60].

Friction Materials	μ_{cst}	μ_{csl}	M_1	M_2	M_3
steel–steel	0.10 ... 0.12	0.05 ... 0.07	0.95	0.3	0.0003
steel–sintered-bronze	0.11 ... 0.13	0.07 ... 0.09	0.85	0.08	0.0004
steel–organic	0.30 ... 0.40	0.30 ... 0.40	0.85	0.05	0
steel–paper	0.10 ... 0.12	0.10 ... 0.14	0.85	0.15	−0.0003

Baumann and Klittich [61] constructed a more detailed tractor shifting dynamics model under AMESim, which considered the effect of centrifugal force generated by the oil in the clutch cylinder during rotation on the normal force F_{cN} mentioned above, as shown in Figure 12.

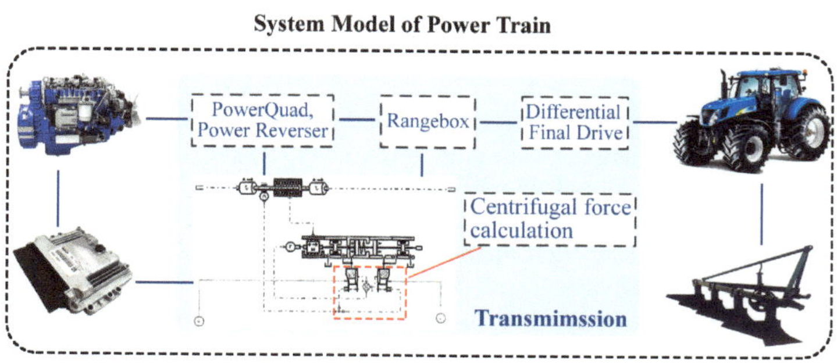

Figure 12. AMESim model of tractor power-shift system considering centrifugal force generated by the oil in the clutch cylinder (Reproduced from [61]).

For the above power-shift system with the solenoid directional valve as the core component, the most effective way to avoid the power interruption of the tractor and improve its shift quality is to control the action sequence of the two clutches, that is, the on-coming clutch should act before the off-going clutch. In addition, the clutch pressure and oil filling flow also have a great influence on the shift quality.

This control method has significant limitations and is mainly suitable for power-shift systems that meet equal-speed shift conditions or have small speed differences between adjacent gears. For instance, Chen et al. [62] analyzed the shift quality of a tractor equipped with a series hydrostatic power-split transmission based on SimulationX (Figure 13). By optimizing the flow, pressure, timing of the clutch, and pump displacement, the shift impact of the tractor is greatly reduced. Zhu et al. [63] discussed the shift quality issue of a CVT tractor involving six clutch switches. The authors also constructed the tractor shift dynamics model under SimulationX and improved the shift quality by optimizing the timing.

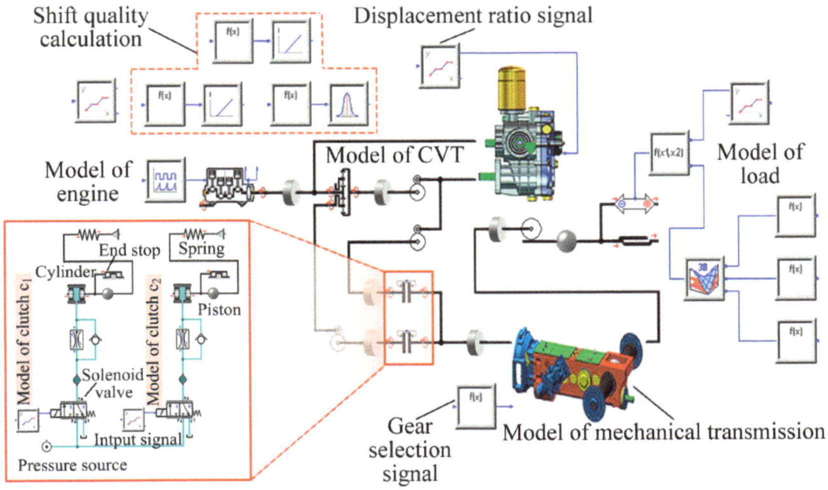

Figure 13. SimulationX model of CVT tractor that meets equal-speed shift conditions. (Reproduced from [62]).

For most power-shift systems, the proportional reducing valve is more commonly used than the solenoid directional valve. The proportional valve can accurately adjust clutch pressure through input current, thereby significantly reducing shift impact. The shift process based on proportional pressure control typically consists of two phases: the torque phase and the inertia phase. The torque phase aims to increase the torque of the on-coming clutch and decrease the torque of the off-going clutch to zero without changing the gear ratio of the transmission, while the task of the inertia phase is to change the gear ratio of the transmission to the target gear value.

Goetz et al. [64] analyzed the tractor power-shift process involving two groups of clutch switching based on Matlab/Simulink. The operation sequence of upshift is the torque phase first and then the inertia stage. The control strategy and simulation results are shown in Figure 14.

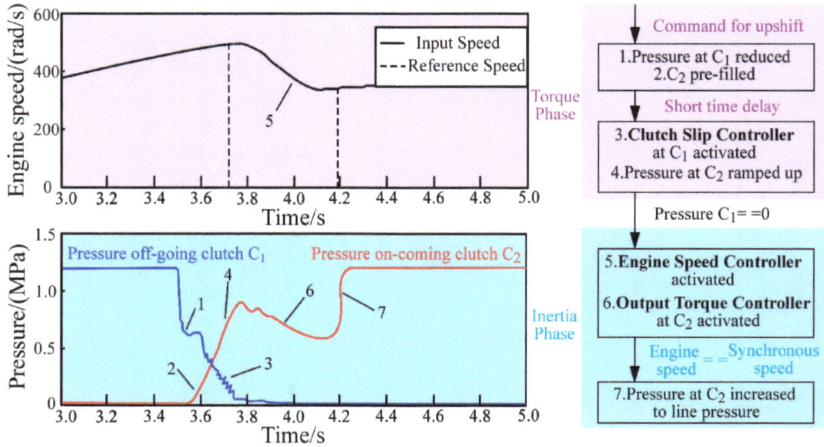

Figure 14. Control strategy for tractor upshift. (Reproduced from [64]).

In the torque phase: firstly, reduce the oil pressure of the off-going clutch C_1 to make it close to the slipping state, and at the same time, increase the oil pressure of the on-coming clutch C_2 to charge the oil cylinder and eliminate the piston clearance; secondly, while increasing the pressure of clutch C_2 along the slope, the pressure of clutch C_1 is reduced to zero by the clutch slip controller. In the inertia phase, the engine speed controller gradually reduces the engine speed to the target gear value. At the same time, the output torque controller adjusts the clutch oil pressure to make the output torque transition to the target gear value; finally, it increases the oil pressure of clutch C_2 to the termination value.

The downshift sequence begins with the inertia phase, followed by the torque phase. The control strategy and simulation results are illustrated in Figure 15. In the inertia phase: firstly, reduce the oil pressure of the off-going clutch C_1 to approach a slipping state; secondly, increase the engine speed to the target gear value through the engine speed controller; finally, increase the oil pressure of the on-coming clutch C_2 to charge its cylinder and eliminate piston clearance. In the torque phase, while increasing the pressure of clutch C_2 along the slope, the pressure of clutch C_1 is reduced to zero through a clutch slip controller; finally, it increases the oil pressure of clutch C_2 to the termination value.

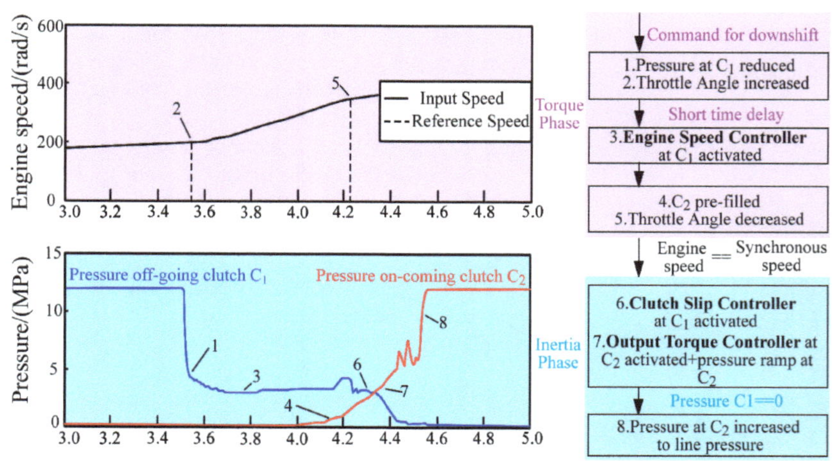

Figure 15. Control strategy for tractor downshift. (Reproduced from [64]).

Note that the research of Goetz et al. involves the control of engine speed, which needs to consider the speed governing characteristics of the engine in modeling. The engine speed is a function of throttle opening and engine torque:

$$n_e = f(\alpha_e,\ T_e) \tag{10}$$

In addition, the engine model also needs to consider the moment of inertia of its flywheel to accurately describe its dynamic behavior, which will not be elaborated on here.

On this basis, Li et al. [65] investigated a more complex shifting process for tractors involving coordinated control of four wet clutches. The authors developed a mathematical model of the tractor powertrain in Matlab/Simulink. They proposed an upshift control strategy consisting of three phases (torque phase–inertia phase–torque phase) and a downshift control strategy also comprising three phases (inertia phase–torque phase–inertia phase), as detailed in [65].

In some engineering applications, fixed proportional valve current curves or clutch pressure curves can be used to implement shift control, as shown in Figure 16. For this control mode, it is necessary to optimize the time and amplitude (i.e., control parameters) of each node point of the control curve.

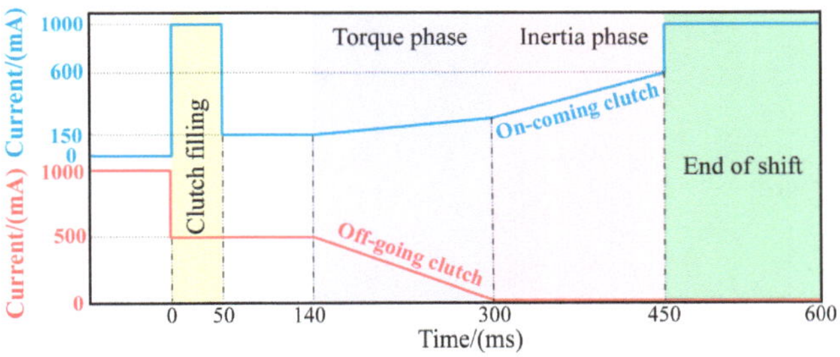

Figure 16. The control current curve of the proportional valve.

There is a lot of research in this field. For example, Xia et al. [66] constructed a mathematical model of a tractor power-shift transmission based on differential equations under Matlab/Simulink. This study assumes that the tractor performs an upshift operation while working in the field at a speed of 6.02 km/h, and the corresponding engine throttle opening during the shift is 50%. For this specific working condition, the authors optimized the clutch control parameters using the Pattern Search Algorithm (PSA) with the weighted values of jerk, sliding friction work, and torque fluctuation as the objective function. In their subsequent study [67], the authors further optimized the clutch termination pressure of power-shift tractors based on the minimum optimal control theory.

Wang et al. [68] studied the power-shift problem of a tractor equipped with a CVT that does not meet the equal-speed shift conditions. Although only two clutches are involved in the shift process, the tractor needs to complete range switching at a speed of 4 km/h under heavy load, and at the same time, the hydraulic motor needs to quickly complete reverse rotation during a shift. The authors constructed the simulation model of the tractor transmission system under SimulationX, optimized the control parameters corresponding to the pressure curve of the clutch based on the proposed high-dimensional visualization method, and obtained satisfactory shift quality.

Xu [69] studied the shift quality of a CVT tractor involving four clutch switching. The authors constructed the dynamic model of the transmission system under AMESim, and proposed a shift control strategy based on range bridging. By adjusting the on-off state of the four clutches, the authors found a temporary intermediate range between the two adjacent ranges. For this shifting process, the transmission first shifts from the initial range to the intermediate range, followed by a rapid transition from the intermediate range to the final range, during which all node parameters on each clutch pressure curve are orthogonally optimized. Simulation results indicate that this approach improves shift quality by minimizing the difference in gear ratios before and after the direct shift.

In fact, to ensure the robustness of gear shift control, utilizing a fixed command current curve or voltage curve does not imply the abandonment of closed-loop control The general practice in industry is to control the input current of the proportional valve through PID. In recent years, more researchers have proposed direct tracking control of the command oil pressure curve. Relevant simulation studies can be found in the works of Li et al. [70], Li et al. [71], and Li et al. [72]. In these studies, the research of Zhang et al. [73] deserves attention. The authors not only introduced the concept of digital twins in the pressure tracking control but also used Rapid Control Prototypes (RCP) to complete the semi-physical simulation task.

According to the above analysis, it is very difficult to establish a sufficiently accurate model of the tractor power-shift transmission system, but not all modeling details are necessary. We have summarized the model options that need to be considered in the study of tractor-shifting dynamics, which can serve as a reference for modeling, as shown in Table 4.

Table 4. Summary of modeling options to be considered in tractor power-shift simulation.

Model	Sub-Model	Modeling Options	Comments
Power	Engine	Speed control.	Three models are available for selection: • Constant speed: It is suitable for simulating tractor CVTs that meet the equal-speed shift conditions. • Speed regulation using a closed-loop controller: It is suitable for simulating general power-shift transmissions. • Speed regulation using engine map: It is also suitable for simulating general power-shift transmissions. However, the accuracy of controlling engine speed through throttle is lower than that of closed-loop controllers.
		Fuel consumption map.	• Not a mandatory option.
		Crankshaft inertia.	• It must be considered when using variable engine speeds.
Transmission and axle	Wet clutch	Friction coefficient.	• The vast majority of models mentioned in previous studies use fixed friction coefficients, while those validated through multi-condition experiments use variable friction coefficients. Therefore, we recommend using variable friction coefficients in precise modeling. The calculation of the friction coefficient can refer to Equation (9) and Table 3.
		Centrifugal force of oil.	• Although some studies have considered the influence of centrifugal force on the shape of clutch pressure curves, no further shift tests have been conducted. Since most fully validated models do not take this into account, it is not a mandatory choice.
	Gear or PG	Meshing efficiency.	• In general, gears and PG have high efficiency, so energy loss can be ignored in simple transmission systems. Otherwise, meshing efficiency should be taken into account.
	Shaft	Stiffness and damping.	• Only when observing torsional vibration is it necessary to consider stiffness and damping in the model.
		Moment of inertia.	• It must be considered.
	Pump or motor	Mechanical efficiency and volumetric efficiency.	• Hydraulic pumps and hydraulic motors are commonly used in CVTs. Due to the low efficiency of hydraulic systems, it is necessary to consider their efficiency, especially their volumetric efficiency.
		Dynamics of the displacement control unit.	• It can be regarded as a second-order system and represented by a difference equation, as detailed in [60].

Table 4. *Cont.*

Model	Sub-Model	Modeling Options	Comments
Load	Theoretical load	Constant load.	• A constant load can be used in qualitative analysis, such as comparing the effects of parameters on shift quality.
		Load with noise.	• To analyze the robustness of the shift control strategy, noise can be added to the load model.
	Dynamic load	Mass of tractor unit.	• It must be considered.

3.3. Energy Saving

The tractor transmission is accompanied by energy loss when transmitting power, which is commonly expressed as transmission efficiency. This performance indicator must be considered in transmission design; otherwise, it will significantly affect the tractor's power performance and fuel economy. For traditional mechanical-shift tractors, power-shift tractors, and HST tractors, the direction of power flow in the transmission system is clear, making it easier to estimate their energy and fuel consumption. Relevant simulation studies can refer to the work of Kolator [74], Oyelade and Oni [75], Siddique et al. [76], and Li et al. [77]. For hydrostatic power-split transmissions, the presence of swash plate axial piston units, differential planetary gears, closed transmission chains, and parasitic power poses difficulties for its energy consumption calculation. In addition, the energy consumption modeling of the electric transmission system differs from the existing tractor transmission system and requires consideration of new components such as batteries and motors. Therefore, this study focuses on the simulation of energy consumption of tractor hydrostatic power-split transmission and tractor electric transmission systems.

One effective method to obtain the efficiency of the tractor hydrostatic power-split transmission is to separately calculate the speed and torque of all its transmission shafts. For swash plate axial piston units, the speed and torque equations for its variable displacement pump and fixed displacement motor are as follows [78]:

$$\varepsilon D_{p,max}\omega_p = D_m\omega_m + C_s\Delta p\left(D_{p,max} + D_m\right)/\mu \tag{11}$$

$$\Delta p D_m = T_m + f_m\omega_m + C_{fm}\Delta p D_m \tag{12}$$

For standard planetary gears, the rotational speeds of its basic components (i.e., sun gear, ring gear, and carrier) meet the following requirements:

$$n_a + k_p n_b - (1+k)n_x = 0 \tag{13}$$

According to the above equation, once the rotational speed of the two basic components is determined, the rotational speed of the remaining component can be calculated. In addition, considering energy loss, the torque between the three basic components satisfies the following relationship:

$$T_a + T_b + T_x = 0 \tag{14}$$

If the power flows from the sun gear of the PG conversion mechanism to the ring gear,

$$T_a = -T_x/\left(1 + k_p\eta_{ab}\right) \tag{15}$$

If the power flows from the ring gear of the PG conversion mechanism to the sun gear,

$$T_a = -T_x/\left(1 + k_p/\eta_{ab}\right) \tag{16}$$

For ordinary gear transmission, the speed equation is

$$n_B = -n_A/i_{AB} \tag{17}$$

If power flows from gear A to gear B, the torque equation is

$$T_A = T_B/(i_{AB}\eta_{AB}) \tag{18}$$

If power flows from gear B to gear A, the torque equation is

$$T_A = T_B/(i_{AB}/\eta_{AB}) \tag{19}$$

After obtaining the rotational speed and torque of the input shaft and output shaft, the transmission efficiency of the transmission can be expressed as

$$\eta_{CVT} = n_{out}T_{out}/(n_{in}T_{in}) \tag{20}$$

The difficulty of this analysis method is in determining the direction of the power flow because in the presence of parasitic power, the power may flow reversely along the mechanical path or hydraulic path of the transmission. Currently, the direction of power flow can be determined by symbolic analysis or formulaic methods, which can be referred to in the research of Zhang et al. [79], Wang et al. [80], and Xu [81].

The above efficiency calculation method requires separately calculating the speed and torque of each transmission shaft, and the flow of power follows a strict causal relationship, making it easy to connect as an independent transmission model with the engine model and axle model. Therefore, this modeling method is not only used for efficiency analysis but can also be extended to study the energy consumption and fuel economy of the entire tractor through model extension. If it is only used for analyzing and optimizing the energy consumption of the transmission, the efficiency formula of the transmission can also be directly derived.

For the closed transmission system shown in Figure 17, its transmission efficiency can be derived through the meshing power method [34]. If the input shaft is γ, its efficiency is

$$\eta_{\gamma\delta} = 1 - \left[k\psi_{x1}\left(z_{b2}/z_{a2} + \left|i_{\gamma\delta}\right|\right) + kz_{b2}\psi_{\alpha\delta}/z_{a2} + z_{b3}\psi_{\beta\delta}/z_{a3}\right] / \left[(1+k)\left|i_{\gamma\delta}\right|\right] \tag{21}$$

If the input shaft is δ, its efficiency is

$$\eta_{\delta\gamma} = 1/\eta_{\gamma\delta} \tag{22}$$

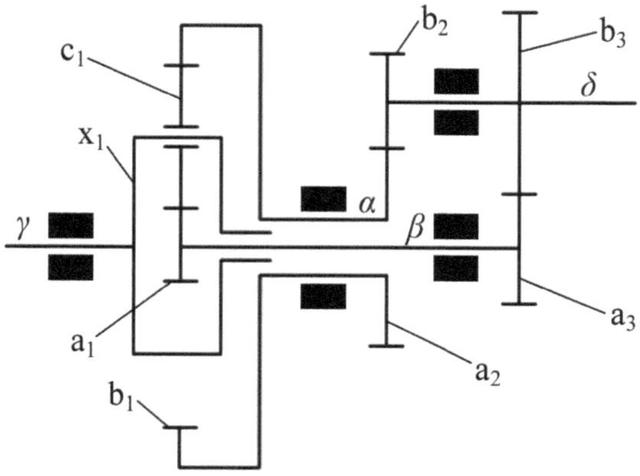

Figure 17. Closed planetary gear transmission.

For this closed transmission system, replacing any one set of gear pairs in path $\alpha - \delta$ or $\beta - \delta$ with swash plate axial piston units can use the above equation to calculate the transmission efficiency of a tractor CVT based on standard planetary gear. However, this method requires separately solving for the loss coefficient or efficiency of the swash plate axial piston units. Liu [82] provided the efficiency equation for this subsystem:

$$\eta_H = \eta_{p,v}\eta_{p,m}\eta_{m,v}\eta_{m,m} \tag{23}$$

where

$$\eta_{p,v} = 1 - 60C_s\Delta p/(\varepsilon\mu n_p) \tag{24}$$

$$\eta_{p,m} = 1/\left[1 + C_v\mu n_p/(60\varepsilon\Delta p) + C_f/\varepsilon + 2\pi T_l/(\Delta p\varepsilon D_{p,max})\right] \tag{25}$$

$$\eta_{m,v} = 1/[1 + 60C_s\Delta p/(\mu n_m)] \tag{26}$$

$$\eta_{m,m} = 1 - C_f - C_v\mu n_m/(60\Delta p) \tag{27}$$

The meshing power method was widely used in the early research of CVT tractors, but it requires specifying the pressure difference between the inlet and outlet of the pump during calculation. Δp is determined by the load of the transmission, which poses difficulties for the extended application of this mathematical model. Similar efficiency calculation methods include the M.A. Крейнес method, Sadako Tsung-Ching method, and their derivative methods, as detailed in the research of Rao [34], Zhang et al. [83], and Zhang et al. [84].

One application of energy consumption calculation is the analysis and optimization of parameters in transmissions. For instance, Wang et al. [80] constructed an energy consumption model for a tractor CVT based on Simpson planetary gear set according to Equations (11)–(20). By using a global visualization method to match the gear ratio of the gear pair on the pump shaft and the standing ratios of the planetary gears, the transmission efficiency of the CVT was improved under all operating conditions. Xia et al. [85] calculated the transmission efficiency of the swash plate axial piston units based on Equations (23)–(27) and constructed the simplest energy consumption model for a tractor CVT while ignoring mechanical energy loss. Based on this model, the authors used the NSGA-II multi-objective genetic algorithm to match the standing ratios of Simpson planetary gears, thereby optimizing multiple tractor performance indicators, including transmission efficiency.

Zhang et al. [86] also used the NSGA-II algorithm to optimize the CVT, but the power confluence mechanism of the transmission is a standard planetary gear. Liu et al. [38] constructed an energy consumption model for a CVT tractor in two steps: firstly, the authors obtained the efficiency values of the transmission under different operating conditions based on the efficiency equation and stored them in a 2D table; secondly, the authors constructed a tractor model based on AMESim and inserted the table data storing the efficiency map as a transmission model into the tractor model, thereby obtaining a complete simulation model for analyzing the tractor's transmission performance. İnce et al. [87] conducted a similar study, revealing the impact of power flow and the Willis gear ratio (i.e., the reciprocal of standing ratio) of planetary gears on CVT transmission efficiency.

In recent years, the relationship between the transmission configuration and energy consumption has gradually attracted attention and has been considered in related transmission optimization research. For instance, Rossetti et al. [88] developed a Multi-objective Particle Swarm Optimizer (MOPSO) algorithm in Matlab and constructed an energy consumption model for a hydrostatic power-split CVT in AMESim. Through the co-simulation of these two software platforms (Figure 18), transmission parameters were optimized for both planetary input-coupled and planetary output-coupled configurations, resulting in an efficient and compact transmission design. In subsequent studies, the authors conducted more extensive research based on the Direct Search algorithm [89]. They explored all possible configurations of a single standard planetary gear and systematically optimized the transmission efficiency of tractor CVTs based on this approach. Zhao [90] also considered the influence of configuration when optimizing the energy consumption of tractor CVTs.

In addition to the CVT configuration based on standard planetary gear, the author also compared and analyzed the configuration based on Simpson planetary gear.

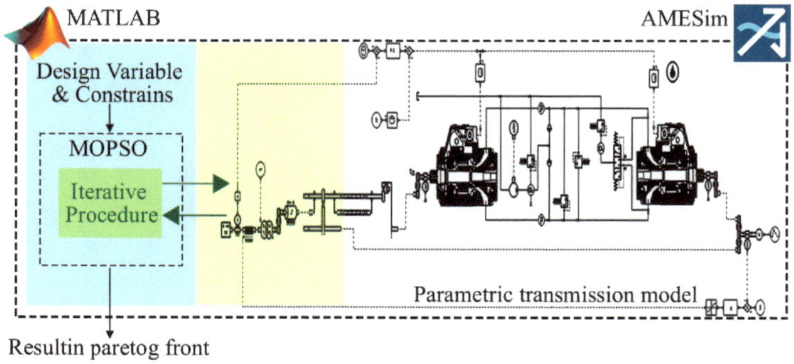

Figure 18. Co-simulation model of CVT based on Matlab and AMESim. (Reproduced from [88]).

Another application of energy consumption calculation is to analyze the fuel economy and emissions of CVT tractors. İnce and Güler [91] constructed energy consumption models for hydrostatic power-split CVT tractor and mechanical-shift tractor in Matlab and compared the fuel economy of the two tractors in 8 different scenarios through simulation. The results showed that tractors equipped with CVTs saved 8.2% of fuel compared to traditional tractors. Rossetti et al. [92] constructed energy consumption models for hydrostatic power-split transmission and power-shift transmission under AMESim, and they obtained further research conclusions than those of İnce et al., that CVT can not only save fuel but also effectively reduce emissions. The above studies were conducted under specific operating conditions. Wang et al. [80] simulated the efficiency and fuel consumption of the tractor under all field operating conditions (4–20 km/h) based on the constructed energy consumption model. They found that the fuel economy of the CVT tractor varies under different operating conditions, but the highest fuel consumption can be saved by about 20% compared to mechanical-shift tractors. This study also considered the issue of emissions, and with appropriate control strategies, the fuel efficiency and emissions problems of tractors can be solved simultaneously. Note that the following equation was used to represent fuel consumption in this study:

$$g_t = g_e / \eta_{CVT} \tag{28}$$

where, g_e is the fuel consumption of the engine, which can be obtained by querying the engine fuel consumption map through engine speed and torque. In simulation, interpolation or fitting can also be performed on the raw data in the engine fuel consumption map to obtain the mathematical expression of g_e:

$$g_e = f(n_e, T_e) \tag{29}$$

On this basis, g_t in Equation (28) is defined as the fuel consumption of the "engine-transmission" system, which physically means the mass of fuel consumed by the engine per unit of work output of the transmission. Considering that the efficiency of a hydrostatic power-split transmission varies within a relatively large range and that tractors are power machines designed for output power, this indicator better reflects the energy consumption level of tractors than engine fuel consumption. Currently, this indicator has been widely accepted and used in this field, as detailed in the studies of Rossetti and Macor [93], Liang et al. [94], and Zhang et al. [95].

Next, we will review the modeling issues of tractor electric transmission systems. Note that electric tractors still require a transmission system, and their modeling method is consistent with traditional tractors, but additional consideration needs to be given to the modeling of batteries and motors.

For fuel cells, the stack voltage is determined by the following equation [96]:

$$U_{stack} = n_{cell}(U_{ner} - U_{act} - U_{ohm} - U_{con}) \tag{30}$$

where

$$U_{ner} = U_{scell} - 0.5R_g T_{stack}/C_F \ln\left(p_{H_2O}/p_{O_2}^{0.5} p_{H_2}\right) \tag{31}$$

$$U_{act} = 0.5R_g T_{stack} \log(I_{dens}/I_0)/(C_F \tau_c) \tag{32}$$

$$U_{ohm} = R_{ohm} I_{dens} \tag{33}$$

$$U_{con} = 0.5R_g T_{stack} \log(1 - I_{dens}/I_{lim})/C_F \tag{34}$$

For lithium-ion batteries, their open circuit voltage U_{OCV} is a function of their state of charge (SOC) [97]:

$$U_{OCV} = f(SOC) \tag{35}$$

The relationship between U_{OCV} and SOC can be determined through experiments, and the value of SOC can be estimated using the Ampere-hour counting method [98,99]:

$$SOC_k = SOC_0 - \left[\int_0^k (\eta_c I_k - S_d)dt\right] T_{sp}/C_n \tag{36}$$

where

$$I_k = 0.5\left(U_{OCV} - \sqrt{U_{OCV}^2 - 4R_{bat}P_{bat}}\right)/R_{bat} \tag{37}$$

$$P_{bat} = \begin{cases} n_{mot} T_{mot} \eta_{mot}/9550 & Charge \\ n_{mot} T_{mot}(9550\eta_{mot}) & Drive \end{cases} \tag{38}$$

Note that the relationship between battery resistance R_{bat} and SOC can also be obtained through experiments.

The battery voltage at this time is

$$U_{bat} = U_{OCV} - R_{bat} I_k \tag{39}$$

Correspondingly, the power of the motor depends on its operating state [100]:

$$P_{mot} = \begin{cases} n_{mot} T_{mot}/(9550\eta_{mot}), & Driving \\ n_{mot} T_{mot} \eta_{mot}/9550, & Generating \end{cases} \tag{40}$$

In the above equations, the efficiency of the motor can be obtained by querying the motor efficiency map through its speed and torque. In addition, it can also be represented as a function by interpolating or fitting the raw data of the motor efficiency map:

$$\eta_{mot} = f(n_{mot}, T_{mot}) \tag{41}$$

As mentioned above, the modeling method for the mechanical transmission system of electric tractors can refer to power-shift transmissions and hydrostatic power-split transmissions. For example, the principle of the coupler of eCVT is similar to that of hydrostatic power-split transmission, and the efficiency calculation method of the differential planetary gear set for both is the same. Therefore, the modeling problem of the mechanical transmission system of electric tractors will not be elaborated here.

Based on the above equations, combined with the modeling method of the mechanical transmission system, a complete energy consumption model of the electric tractor can be constructed. As an example, Figure 19 shows an electric tractor model constructed under SimulationX (Version 4.0) [101].

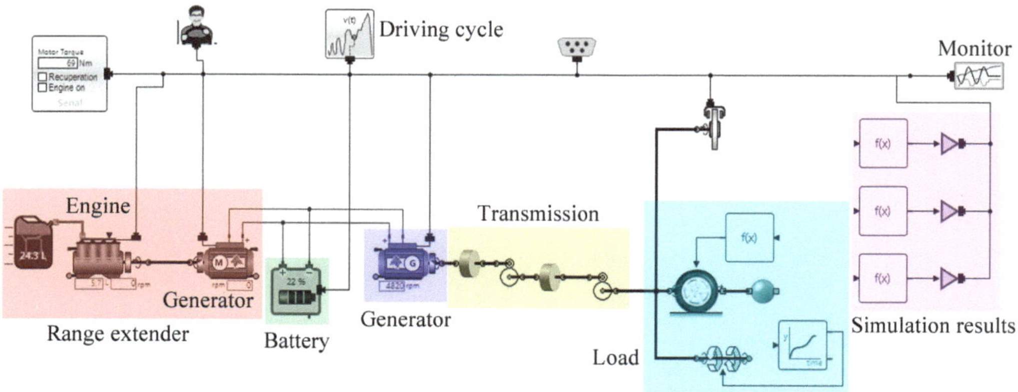

Figure 19. Simulation model of electric tractor based on SimulationX. (Reproduced from [101]).

Currently, researchers have analyzed the energy consumption and energy management strategies for various electric tractors based on constructed simulation models. For the "direct with engine" electric tractor (i.e., series hybrid tractor), Medževeprytė et al. [102] constructed mathematical models for a series hybrid tractor and a CVT tractor equipped with a hydrostatic power-split transmission (ZF ECCOM 3.5) in Matlab (Version R2021b) and AVL Cruise (Version 2016), respectively. The authors compared the performance of two tractors under field operation and road transportation conditions, demonstrating that electric tractors have higher fuel efficiency. The authors also proposed a series hybrid tractor based on a Compressed Natural Gas (CNG) engine and constructed its energy consumption model during soil cultivation in Matlab/Simscape. The simulation found that the transmission energy consumption of the tractor is highly dependent on the efficiency of the engine operating point, and the tractor requires a larger capacity of energy storage to extend its working time under heavy load conditions [103]. For the "direct with battery" electric tractor (i.e., pure electric tractor), Mao et al. [104] constructed a transmission system model of a 36.8 kW pure electric tractor under OpenModelica, analyzed its traction performance and speed range, and obtained simulation results consistent with experiments. Li et al. [105] proposed an energy management strategy based on optimal system efficiency for a pure electric tractor with dual-motor power coupling. The authors constructed a mathematical model of the tractor in Matlab/Simulink (Version R2018a) and analyzed its energy consumption under two working conditions: plowing and rotary tillage. This study reveals the influence of transmission system parameters on the efficiency of electric tractors. Based on similar operating conditions, Wang et al. [106] conducted a similar study on the efficiency of dual-motor power-coupled electric tractors under HIL. For the "power-split" electric tractor (i.e., eCVT tractor), Mocera et al. [107] constructed energy consumption models for an eCVT tractor and a traditional tractor separately in Matlab/Simulink. The simulation results showed that the peak power capability of eCVT tractor under the given energy management strategy was comparable to that of a traditional tractor, but it had better fuel economy than the latter. Rossi et al. [108] conducted similar comparative studies on eCVT tractors of different configurations on the same simulation platform, further confirming the potential application of eCVT transmission systems in tractors.

Regardless of the type of tractor transmission system, the load model is crucial in its energy consumption simulation, as tractors have complex operating conditions. The load of a tractor includes slop resistance, rolling resistance, air resistance, and traction resistance, where the expressions for the first three are [35,109]:

$$F_{slop} = M_t g\sin \propto_{slope} \tag{42}$$

$$F_{roll} = R_{roll}\left[M_t g\cos \propto_{slope} - 0.5\rho_{air}S_x C_z(V_t + V_{wind})|V_t + V_{wind}|\right] \tag{43}$$

$$F_{air} = 0.5\rho_{air}S_x C_x(V_t + V_{wind})|V_t + V_{wind}| \tag{44}$$

The traction resistance of tractors is usually large and difficult to calculate accurately, and in the past, it could only be obtained through experiments. Taking plowing operation as an example, the most classic study on plowing load is the resistance data published by Wicha in 1956 [24]. According to this study, plowing load is related to soil bulk density, soil moisture, and operating speed (Figure 2), and these data are given in the form of specific resistance, allowing it to be used as a universal load model for plowing simulation:

$$F_{traction} = \kappa_s g abn_{pl} \tag{45}$$

where abn_{pl} represents the total cross-sectional area of the soil block cut by the plow, and $\kappa_s g$ represents the plowing traction resistance per unit cross-sectional area.

With the development of computer simulation technology, some of the publicly available load models in recent years have been obtained through the discrete element method. Specific references can be made to the research of Makange et al. [110], Tong et al. [111], Zhai et al. [112], and Kim et al. [113]. In addition, there are also some load data obtained through theoretical prediction models, which can be referred to in the studies of Čermák and Mitáš [114] and Kim et al. [115].

At the end of this section, we also summarized the model options that need to be considered in tractor energy consumption research, which can serve as a reference for modeling, as shown in Table 5.

Table 5. Summary of modeling options to be considered in tractor energy consumption simulation.

Model	Sub-Model	Modeling Options	Comments
Power and energy	Engine	Speed control.	Three models are available for selection: • Constant speed: It is suitable for simulating transmission energy consumption—for example, simulating the CVT efficiency map at a specified input speed. • Speed regulation using a closed-loop controller: It is suitable for simulating energy management strategies that require simultaneous control of engine speed and gear ratio. • Speed regulation using engine map: It is suitable for simulating energy management strategies that only save fuel by adjusting the gear ratio.
		Fuel consumption map.	• It must be considered when simulating fuel economy and used to query fuel consumption at a specified engine speed and torque.
		Crankshaft inertia.	• It is not a mandatory option in simple energy consumption calculations, but the impact of dynamic processes on tractor fuel consumption should be considered in driving cycles.
	Battery	-	• Its modeling method refers to Equations (30)–(39).
	Motor or Generator	-	• Its modeling method refers to Equations (40) and (41).

Table 5. *Cont.*

Model	Sub-Model	Modeling Options	Comments
Transmission and axle	Wet clutch	Frictional behavior.	Not a mandatory option.
	Gear or PG	Meshing efficiency.	• It must be considered, and the meshing efficiency of gears and PG can usually be regarded as a constant. • In power-split systems, meshing efficiency is used to accurately calculate the torque of each basic component of PG. Some studies did not consider the efficiency of PG to simplify calculations, but its rationality needs to be confirmed. • In precise energy consumption simulation, it is usually necessary to consider the efficiency of bearings supporting gears or PG.
	Shaft	Stiffness and damping.	Not a mandatory option.
		Moment of inertia.	• It is not a mandatory option in simple energy consumption calculations, but the impact of dynamic processes on tractor fuel consumption should be considered in driving cycles.
	Pump or motor	Mechanical efficiency and volumetric efficiency.	• It must be considered. In some literature, fixed efficiency values have been used, but we still recommend using measured data or accurate energy consumption models to run simulations.
		Dynamics of the displacement control unit.	Not a mandatory option.
Load	Theoretical load	Slop resistance.	• It must be considered when the tractor is operating on sloping ground.
		Rolling resistance.	• It must be considered.
		Air resistance.	• It must be considered when the tractor has a cab with a large windward area and is driving at high speeds.
		Traction resistance.	• It must be considered when the tractor is towing agricultural machinery for operation.
	Dynamic load	Mass of tractor unit.	• It is not a mandatory option in simple energy consumption calculations, but the impact of dynamic processes on tractor fuel consumption should be considered in driving cycles.

4. Parameters in Modeling

Although we have provided many applications for simulating tractor transmission systems, we have not mentioned the methods for constructing mathematical models and determining their parameters. For most of the studies mentioned above, the simulation models are based on equations. The reality is that many parameters of these equations are unknown, and for some complex dynamic problems, we cannot even successfully derive their mathematical equations. For these issues, we need to minimize the number of unknown parameters as much as possible. For example, the structural parameters of the wet clutch can usually be directly provided by the manufacturer, while the moment of inertia of the shaft and gear can be automatically settled by virtual prototyping software such as Solidworks according to the material. If this still cannot solve the problem, the parameters can be obtained through system identification methods. We take the modeling of the swash plate axial piston units of a tractor CVT as an example to illustrate the general process of this method, as shown in Figure 20. The input of this system is the PWM duty cycle of the electromagnetic coil current of the variable displacement pump, and the output is the transmission ratio (i.e., the ratio of the motor speed to the pump speed). Its transfer function has the following form:

$$G(s) = \left(b_m s^m + b_{m-1} s^{m-1} + \ldots + b_1 s + 1\right) / \left(a_n s^n + a_{n-1} s^{n-1} + \ldots a_1 s + 1\right), \ (n \geq m) \tag{46}$$

The values of m and n can be determined based on prior knowledge. On this basis, an excitation experiment was designed: a PWM step signal was applied to the pump on the test bench, and the transmission ratio curve data of the system response $h^*(t)$ was measured through a speed sensor. In the identification calculation of this system, the area method is used to obtain the parameters to be estimated. Specifically, the ith area of the step response mentioned above is [116]

$$A_i = \int_0^{\infty} [1 - h^*(t)](-t)^{i-1}/(i-1)!dt + \sum_{j=0}^{i-2} A_{i-j-1} \int_0^{\infty} [1 - h^*(t)](-t)^i/j!dt \tag{47}$$

The parameters to be estimated can be calculated using the following equations:

$$\begin{bmatrix} b_1 \\ b_2 \\ \vdots \\ b_m \end{bmatrix} = \begin{bmatrix} A_n & A_{n-1} & \cdots & A_{n-m+1} \\ A_{n+1} & A_n & \cdots & A_{n-m+2} \\ \vdots & \vdots & \vdots & \vdots \\ A_{n+m-1} & A_{n+m-2} & \cdots & A_n \end{bmatrix} \begin{bmatrix} A_{n+1} \\ A_{n+2} \\ \vdots \\ A_{n+m} \end{bmatrix} \tag{48}$$

$$\begin{bmatrix} a_1 \\ a_2 \\ \vdots \\ a_m \end{bmatrix} = \begin{bmatrix} 1 & 0 & \cdots & 0 & 0 \\ A_1 & 1 & \cdots & 0 & 0 \\ \vdots & \vdots & \vdots & \vdots & \vdots \\ A_{n-1} & A_{n-2} & \cdots & A_1 & 1 \end{bmatrix} \begin{bmatrix} b_1 \\ b_2 \\ \vdots \\ b_m \\ 0 \end{bmatrix} + \begin{bmatrix} A_1 \\ A_2 \\ \vdots \\ A_n \end{bmatrix} \tag{49}$$

This study was completed by one of the authors of this review, Wang et al., in 2013, as detailed in [117]. Note that the results of system identification usually require experimental verification, and the expression of mathematical models can be transfer functions and state space models, polynomial models, etc., which will not be described in detail here.

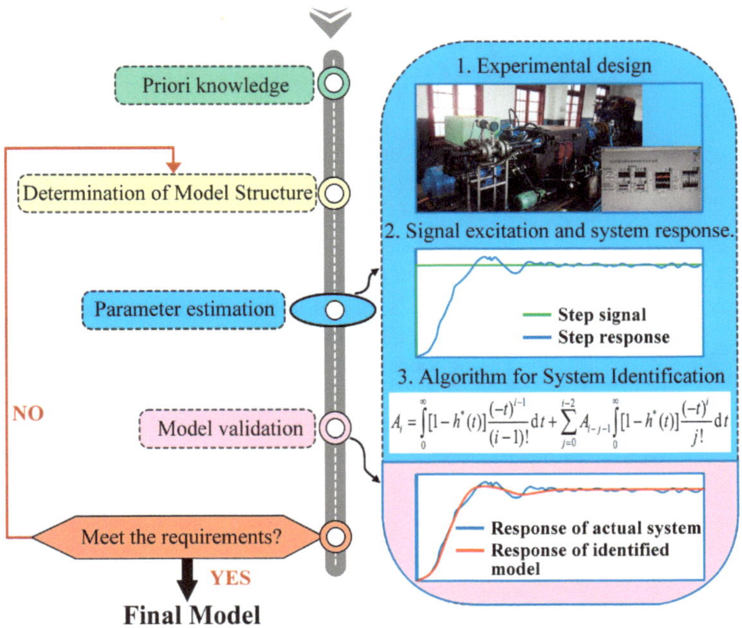

Figure 20. The general process and case study of obtaining tractor transmission parameters through system identification.

In addition to the above example, there are numerous practical applications of system identification in obtaining tractor transmission parameters. Cheng and Lu [118] obtained unknown parameters of the tractor hydrostatic system model through an improved simulated annealing algorithm based on experimental data and theoretical models. On this basis, the authors [119] further proposed a system identification method based on a prior-model-heuristic-intelligent-optimization algorithm and obtained the response model of the tractor CVT accordingly. Musialek et al. [120] identified the transfer function of the clutch and designed a PI controller that meets the control requirements based on this model. Similar research can refer to the work of Tørdal et al. [121] and Cheng et al. [122].

5. Summary and Development Recommendations

With the advancement of tractor technology, modern tractor transmission systems are no longer the physical objects of "what you see is what you get", which makes computer simulation technology play an important role in the development of their transmission systems. Simulation can help us optimize the structure of the tractor transmission system, analyze the influence and mechanism of various parameters on transmission performance, and guide us in proposing effective control strategies. However, tractor transmissions are highly complex electromechanical-hydraulic integrated systems involving numerous disciplines and diverse modeling methods. It is difficult for researchers to quickly construct the knowledge system required for simulation from limited literature, let alone grasp technological development trends. To address these issues, the work we have done is as follows:

(1) A systematic introduction was given to the types and working principles of tractor transmission systems. Tractor transmissions not only come in a wide variety of forms but also have different operating conditions and principles from road vehicles. Understanding this knowledge is the foundation for correctly constructing a tractor transmission system model.
(2) The modeling method of the tractor transmission system was introduced from three aspects: structural optimization, power-shift control, and energy saving. Specifically, key mathematical equations and modeling options were provided as references for correct modeling. On this basis, the applications of computer simulation technology in these key fields were reviewed separately.
(3) The method of obtaining simulation parameters through system identification was introduced, and the implementation process was explained through case studies. The practical application of this method in the development of tractor transmission systems was also listed.

This review also reveals some problems in the current simulation application of tractor transmission systems, as follows:

(1) In the simulation research of tractor power-shift transmission, the vast majority of evaluation indicators for shift quality lack an experimental basis. Some of the existing indicators come from the evaluation system of road vehicles, while others are customized by researchers. The experimental results from John Deere clearly indicate that some evaluation indicators for the shift quality of road vehicles are not applicable to tractors used in field operations, which seriously undermines the scientific foundation of research in this field.
(2) The vast majority of tractor simulation work relies on load models. However, tractors not only lack standard driving cycles but also systematic load datasets. This situation poses challenges to the performance evaluation of different tractor transmission systems and control strategies.

In future work, the problems encountered in the above simulations should be taken seriously and resolved. More evaluation indicators for tractor shift quality should be confirmed through experiments, and more tractor load datasets should be made public.

These efforts will bring new opportunities for the application of computer simulation technology in the development of tractor transmission systems.

Author Contributions: Conceptualization, G.W.; investigation, Y.L., X.C., X.H. and Y.Z.; resources, X.H., N.X., Z.Z. and K.L.; writing—original draft preparation, Y.L., X.C. and G.W.; writing—review and editing, X.H., N.X. and G.W.; visualization, X.C.; supervision, G.W.; project administration, G.W.; funding acquisition, G.W. All authors have read and agreed to the published version of the manuscript.

Funding: This research was funded by the Shandong Province Agricultural Machinery R&D, Manufacturing, Promotion and Application Integration Pilot Project, grant number NJYTHSD-202302, and the Shandong Provincial Natural Science Foundation, grant number ZR2020QE163.

Data Availability Statement: The original contributions presented in the study are included in the article, further inquiries can be directed to the corresponding author.

Conflicts of Interest: Author Xiaorong Han was employed by the company Shandong Weifang Luzhong Tractor Co., Ltd., Author Kai Lu was employed by the company Luoyang Tractor Research Institute Co., Ltd. The remaining authors declare that the research was conducted in the absence of any commercial or financial relationships that could be construed as a potential conflict of interest.

Nomenclature

CVT	Continuously variable transmission.
AMT	Automated mechanical transmission.
FEM	Finite element method.
DEM	Discrete element method.
RCP	Rapid Control Prototypes.
HIL	Hardware-in-the-Loop.
HST	Hydrostatic transmission.
PG	Planetary gear.
OC	Planetary output coupled.
IC	Planetary input coupled.
PSA	Pattern search algorithm.
PSGA	Particle swarm genetic algorithm.
NSGA-II	Non-dominated sorting genetic algorithm II.
MOPSO	Multi-objective Particle Swarm Optimizer.
SOC	State of charge.
J_1, J_2, J_3, J_4	Peak acceleration, jerk, peak sliding friction power, and sliding friction work.
V_t	Tractor velocity.
t, t_a, t_b	Time, shift start time, shift start time.
T_{cf}, $T_{cf,\,max}$	Actual torque and maximum torque transmitted by the clutch.
ω_{dc}	Slip speed of clutch.
p_c	Clutch pressure.
k_{spr}	Spring stiffness.
x_{s0}, x_s, $x_{s,\,Max}$	Initial displacement, displacement increment, and stroke of clutch piston.
F_{fst}, F_{fsl}	Static and sliding friction forces of clutch piston.
k_p	Standing ratio of PG.
η_{ab}, η_{AB}, $\eta_{\gamma\delta}$, η_{CVT}	Transmission efficiency of PG, meshing gears, closed PG, and CVT.
n_A, n_B	Speed of input gear and output gear.
T_A, T_B	Torque of input gear and output gear.
i_{AB}	Gear ratio of meshing gears.
n_{in}, n_{out}	Input and output speeds of the transmission.
T_{in}, T_{out}	Input and output torques of the transmission.
ψ_{x1}, $\psi_{\alpha\delta}$, $\psi_{\beta\delta}$	Loss coefficients in closed PG.
z_{a2}, z_{b2}, z_{a3}, z_{b3}	Number of gear teeth in enclosed PG.
n_{cell}	Number of cells in stack.
U_{stack}, U_{ner}, U_{act}, U_{ohm}, U_{con}, U_{scell}	Stack voltage, nernst voltage, activation loss voltage, ohmic loss voltage, concentration loss voltage, and standard cell voltage.
R_g	Ideal gas constant.
T_{stack}	Stack temperature.
C_F	Faraday constant.
p_{H_2O}, p_{O_2}, p_{H_2}	Partial pressures of water, oxygen, and hydrogen.
I_{dens}, I_0, I_{lim}	Current density, reaction exchange current density, and maximum current density.
τ_c	Charge transfer coefficient.
R_{ohm}	Membrane resistance.
U_{OCV}	Open circuit voltage.
SOC_k, SOC_0	Current and initial values of SOC.
η_c	Coulombic efficiency.
I_k	Load current.
S_d	Self-discharging rate.
k	Current time.

A_{cp}	Effective area of clutch piston.	T_{sp}	Sampling period.
m_{cp}	Mass of clutch piston.	C_n	Nominal capacity of battery.
v_{cp}	Movement speed of clutch piston.	R_{bat}	Battery resistance.
F_{cN}	Normal force acting on clutch friction plates.	P_{bat}	Battery power.
n_{sf}	Number of clutch friction surfaces.	P_{mot}	Motor power.
$\mu_c, \mu_{cst}, \mu_{csl}$	Actual friction coefficient, static friction coefficient, and sliding friction coefficient of clutch friction plates.	n_{mot}	Motor speed.
r_{ci}, r_{co}	Inner and outer radii of clutch friction plates.	T_{mot}	Motor torque.
M_1, M_2, M_3	Constants related to materials.	η_{mot}	Efficiency of motor.
n_e	Engine speed.	$F_{slop}, F_{roll}, F_{air}, F_{traction}$	Slop resistance, rolling resistance, air resistance, and traction resistance.
T_e	Engine torque.	M_t	Mass of tractor unit.
α_e	Throttle opening.	g	Acceleration of gravity.
g_e	Fuel consumption of engine.	$\propto_{slope}$	Slope angle.
g_t	Fuel consumption of "engine-transmission" system.	R_{roll}	Rolling friction coefficients.
ε	Displacement ratio of pump to motor.	ρ_{air}	Air density.
$D_{p,max}, D_m$	Rated displacement of pump and motor.	S_x	Tractor active area for air resistance.
ω_p, ω_m	Speed of pump and motor.	C_x, C_z	Drag coefficient for translation on longitudinal axis and vertical axis.
C_s	Leakage coefficient.	V_{wind}	Wind velocity.
Δp	Pressure difference between motor inlet and outlet.	κ_s	Plowing specific resistance.
μ	Dynamic viscosity of hydraulic oil.	a, b	Plowing depth and width of a single plow bottom.
T_m	Torque of motor shaft.	n_{pl}	Number of plow bottoms.
f_m	Viscous damping coefficient of motor.	$G(s)$	Transfer function.
C_{fm}	Friction loss coefficient of motor.	$h^*(t)$	The step response of the system.
n_a, n_b, n_x	Speed of sun gear, ring gear and carrier of PG.	$a_{x\in\{1, 2, ..., m\}}$, $b_{x\in\{1, 2, ..., m\}}$	Parameters to be estimated.
T_a, T_b, T_x	Torque of sun gear, ring gear and carrier of PG.	A_i	The ith area of the step response.

References

1. Calcante, A.; Oberti, R.; Tangorra, F.M. Definition of linear regression models to calculate the technical parameters of Italian agricultural tractors. *J. Agric. Eng.* **2023**, *54*, 1525. [CrossRef]
2. Chakraborty, A.; Arunachalam, K. Test methodology for validating tractor structural part under multiaxial loading. *Proc. Inst. Mech. Eng. Part D-J. Automob. Eng.* 2023, *in press*. [CrossRef]
3. National Bureau of Statistics. Available online: https://data.stats.gov.cn (accessed on 22 July 2024).
4. Chen, F.; Jiang, S.; Xie, X.; Chen, L.; Lan, Y. Engine universal characteristic modeling based on improved ant colony algorithm. *Int. J. Agric. Biol. Eng.* **2015**, *8*, 26–35. [CrossRef]
5. Zhu, Z.; Yang, Y.; Wang, D.; Cai, Y.; Lai, L. Energy Saving Performance of Agricultural Tractor Equipped with Mechanic-Electronic-Hydraulic Powertrain System. *Agriculture* **2022**, *12*, 436. [CrossRef]
6. Siddique, M.A.A.; Baek, S.-M.; Baek, S.-Y.; Jeon, H.-H.; Kim, Y.-J.; Kim, Y.-S.; Kim, W.-S.; Lee, D.-H.; Lim, R.-G.; Kim, T.-J. Application of auto power shift (APS) controller for minimizing fuel consumption and performance evaluation of an agricultural tractor. *Comput. Electron. Agric.* **2023**, *214*, 108279. [CrossRef]
7. Li, B.; Pan, J.; Li, Y.; Ni, K.; Huang, W.; Jiang, H.; Liu, F. Optimization method of speed ratio for power-shift transmission of agricultural tractor. *Machines* **2023**, *11*, 438. [CrossRef]
8. Qian, Y.; Wang, L.; Lu, Z. Design and Optimization of Power Shift Tractor Starting Control Strategy Based on PSO-ELM Algorithm. *Agriculture* **2024**, *14*, 747. [CrossRef]
9. Herzog, D.; Schlosser, J.F.; Bertinatto, R.; de Farias, M.S.; Casali, L. Transmission systems in agricultural tractors marketed in Brazil. *Cienc. Rural* **2021**, *51*, e20200514. [CrossRef]
10. Ahn, S.; Choi, J.; Kim, S.; Lee, J.; Choi, C.; Kim, H. Development of an integrated engine-hydro-mechanical transmission control algorithm for a tractor. *Adv. Mech. Eng.* **2015**, *7*, 1687814015593870. [CrossRef]
11. Wang, J.; Xia, C.; Fan, X.; Cai, J. Research on transmission characteristics of hydromechanical continuously variable transmission of tractor. *Math. Probl. Eng.* **2020**, *2020*, 6978329. [CrossRef]
12. Xiang, Y.; Li, R.; Brach, C.; Liu, X.; Geimer, M. A novel algorithm for hydrostatic-mechanical mobile machines with a dual-clutch transmission. *Energies* **2022**, *15*, 2095. [CrossRef]

13. Chen, Y.; Mou, X. Natural characteristics analysis and experimental test of spiral bevel gear pair in the tractor transmission system. *Adv. Mech. Eng.* **2021**, *13*, 16878140211037120. [CrossRef]
14. Pue, J.D.; Lamandé, M.; Schjønning, P.; Cornelis, W.M. DEM simulation of stress transmission under agricultural traffic Part 3: Evaluation with field experiment. *Soil Tillage Res.* **2020**, *200*, 104606. [CrossRef]
15. Watanabe, M.; Sakai, K. Identifying tractor overturning scenarios using a driving simulator with a motion system. *Biosyst. Eng.* **2021**, *210*, 261–270. [CrossRef]
16. Zhang, X.; Zhou, Z. Speed control strategy for tractor assisted driving based on chassis dynamometer test. *Int. J. Agric. Biol. Eng.* **2021**, *14*, 169–175. [CrossRef]
17. Yan, X.; Tao, Q.; Xu, L. Dynamic loading system of an agricultural vehicle power wagon based on complex loading. *IEEE Access* **2020**, *8*, 196667–196674. [CrossRef]
18. Yin, Y. Study on the Synchronous Ranges Shifting Stability of Hydro-Mechanical Continuously Variable Transmission. Master's Thesis, Henan University of Science and Technology, Luoyang, China, 2017.
19. Raikwar, S.; Wani, L.J.; Kumar, A.S.; Rao, M.S. Hardware-in-the-Loop test automation of embedded systems for agricultural tractors. *Measurement* **2019**, *133*, 271–280. [CrossRef]
20. Meyer, H.J. *Fahrzeuggetriebetechnik Getriebe in Traktoren*; Technical Report for Tag der Wissenschaftlichen Aussprache; Technische Universität Berlin: Berlin, Germany, 2009.
21. Contessotto, E.E.; Botta, G.F. Power and draft required by chisel plow on soils under no-tillage in the Argentinean flat Pampas. *Rev. Fac. Cienc. Agrar.* **2020**, *52*, 102–110.
22. Ortopan, M.; Simonovic, V.; Tasic, N.; Veg, E.; Zlatanovic, I.; Gubeljak, N. Analyzing site-specific tractor draft force in different passes during plowing. *Teh. Vjesn.* **2024**, *31*, 228–232. [CrossRef]
23. Badua, S.A.; Sharda, A.; Flippo, D.; Ciampitti, I.A. Real-Time Gauge Wheel Load Variability of a Row-Crop Planter During Field Operation. *Trans. ASABE* **2018**, *61*, 1517–1527. [CrossRef]
24. Wicha, A. *Maschinen und Geräte für die Bodenbearbeitung*; Fachbuchverlag: Leipzig, Germany, 1956.
25. Han, G.; Ahn, D.-V.; Kwon, D.; Kim, H.-S.; Park, Y.-J.; Lee, J.W. Development of a real-time tractor model for gear shift performance verification. *Agriculture* **2023**, *13*, 2133. [CrossRef]
26. Byeon, S.-J.; Kim, S.-J; Park, J.-K.; Park, Y.-J.; Lee, J.-W. Parameter study for establishing a synchronizer control strategy in tractor dual-clutch transmission. *Agriculture* **2024**, *14*, 218. [CrossRef]
27. Wu, B. Friction Coefficient of Wet Clutches as a Function of Service Mileage. *Lubricants* **2024**, *12*, 147. [CrossRef]
28. Shin, S.; Choi, S.B. Control-oriented modeling of wet clutch friction considering thermal dynamics. *Mechatronics* **2024**, *99*, 103146. [CrossRef]
29. Han, H.-W.; Han, J.-S.; Chung, W.-J.; Kim, J.-T.; Park, Y.-J. Prediction of synchronization time for tractor power-shift transmission using multi-body dynamic simulation. *Trans. ASABE* **2021**, *64*, 1483–1498. [CrossRef]
30. Li, B.; Sun, D.; Hu, M.; Liu, J. Automatic starting control of tractor with a novel power-shift transmission. *Mech. Mach. Theory* **2019**, *131*, 75–91. [CrossRef]
31. Renius, K.T.; Resch, R. Continuously variable tractor transmissions. In Proceedings of the Agricultural Equipment Technology Conference, Louisville, KY, USA, 14–16 February 2005.
32. Liu, Z.; Zhang, G.; Chu, G.; Niu, H.; Zhang, Y.; Yang, F. Design matching and dynamic performance test for an HST-based drive system of a hillside crawler tractor. *Agriculture* **2021**, *11*, 466. [CrossRef]
33. Bukashkin, A.Y.; Dobretsov, R.Y.; Galyshev, Y.V. Split transmission of tractor with automatic gearbox. *Procedia Eng.* **2017**, *206*, 1728–1734. [CrossRef]
34. Rao, Z. *Design of Planetary Gear Transmission*, 2nd ed.; Chemical Industry Press Co., Ltd.: Beijing, China, 2019.
35. Renius, K.T. *Fundamentals of Tractor Design*; Springer Nature: Cham, Switzerland, 2020.
36. Yu, J.; Cao, Z.; Cheng, M.; Pan, R.Z. Hydro-mechanical power split transmissions: Progress evolution and future trends. *Proc. Inst. Mech. Eng. Part D* **2019**, *233*, 727–739. [CrossRef]
37. Xia, Y.; Sun, D. Characteristic analysis on a new hydro-mechanical continuously variable transmission system. *Mech. Mach. Theory* **2018**, *126*, 457–467. [CrossRef]
38. Liu, F.; Wu, W.; Hu, J.; Yuan, S. Design of multi-range hydro-mechanical transmission using modular method. *Mech. Syst. Signal Proc.* **2019**, *126*, 1–20. [CrossRef]
39. Blumenthal, U.; Hansen, T.; Hartmann, R.O. Claas CVT Development for Different Applications in Commercial Vehicles. In Proceedings of the International Off-Highway & Powerplant Congress & Exposition, Milwaukee, WI, USA, 11–13 September 2000.
40. Chen, Y.-C.; Chen, L.-W.; Chang, M.-Y. A design of an unmanned electric tractor platform. *Agriculture* **2022**, *12*, 112. [CrossRef]
41. Liu, J.; Xia, C.; Jiang, D.; Sun, Y. Development and testing of the power transmission system of a crawler electric tractor for greenhouses. *Appl. Eng. Agric.* **2020**, *36*, 797–805. [CrossRef]
42. Mujiburrahman, K.; Saravanakumar, S.; Satheesh kumar, K.; Carciyo kaviya, J.J.; Krishnaraj, R. Design and analysis of E-glass gear box housing in tractor and optimization of its design parameters. *Mater. Today: Proc.* **2022**, *49 Pt 8*, 3696–3704. [CrossRef]
43. Dong, S.; Li, S.; Fu, S.; Wang, K. Finite element analysis and optimization of tractor gearbox body under various kinds of working conditions. *Sci. Rep.* **2022**, *12*, 17386. [CrossRef]
44. Hartmann, T.; Morita, S.; Tanner, G.; Chappell, D.J. High-frequency structure- and air-borne sound transmission for a tractor model using Dynamical Energy Analysis. *Wave Motion* **2019**, *87*, 132–150. [CrossRef]

45. Sergienko, V.P.; Tseluev, M.Y.; Kupreev, A.V. Numerical simulation of thermal conditions in the oil-cooled friction multidisc clutch of a power-intensive wheel tractor. Part 1. Problem formulation. *J. Frict. Wear* **2010**, *31*, 261–269. [CrossRef]
46. Sergienko, V.P.; Tseluev, M.Y.; Kupreev, A.V. Numerical simulation of the thermal conditions in the multidisk oil-cooled friction clutch of a power-intensive wheel tractor. Part 2. Study results. *J. Frict. Wear* **2010**, *31*, 334–340. [CrossRef]
47. Ganesha, B.; Umesh, K.N.; Setty, G.R.; Naik, S. Dynamic vibration analysis of gear box casing using ANSYS software. *Int. J. Innov. Res. Sci. Eng. Technol.* **2017**, *6*, 13354–13362.
48. Winkel, K.N.; Irmak, T.; Happee, R.; Shyrokau, B. Standards for passenger comfort in automated vehicles: Acceleration and jerk. *Appl. Ergon.* **2023**, *106*, 103881. [CrossRef]
49. Duncan, J.R.; Wegscheid, E.L. Determinants of off-road vehicle transmission 'shift quality'. *Appl. Ergon.* **1985**, *16*, 173–178. [CrossRef] [PubMed]
50. Ahn, H.; Park, Y.; Kim, S.; Choi, C. Theoretical calculations and experimental studies of power loss in dual-clutch transmission of agricultural tractors. *Agriculture* **2023**, *13*, 1225. [CrossRef]
51. Chen, Y.; Cheng, Z.; Qian, Y. Research on Wet Clutch Switching Quality in the Shifting Stage of an Agricultural Tractor Transmission System. *Agriculture* **2022**, *12*, 1174. [CrossRef]
52. Wang, J.; Xia, C.; Fan, X.; Cai, J. Research on the Influence of Tractor Parameters on Shift Quality, Based on Uniform Design. *Appl. Sci.* **2022**, *12*, 4895. [CrossRef]
53. Tanelli, M.; Panzani, G.; Savaresi, S.M.; Pirola, C. Transmission control for power-shift agricultural tractors: Design and end-of-line automatic tuning. *Mechatronics* **2011**, *21*, 285–297. [CrossRef]
54. Bao, M.; Ni, X.; Zhao, X.; Li, S. Research on the HMCVT gear shifting smoothness of the four-speed self-propelled cotton picker. *Mech. Sci.* **2020**, *11*, 267–283. [CrossRef]
55. Wang, G.; Zhang, X.; Zhu, S.; Zhang, H.; Tai, J.; Nguyen, V. Shift performance of tractor hydraulic power-split continuously variable transmission. *Trans. CSAM* **2015**, *46*, 7–15.
56. Li, A.; Qin, D.; Guo, Z. Adaptive torque control of wet dual clutch based on dynamic friction coefficient estimation. *Mechatronics* **2024**, *100*, 103175. [CrossRef]
57. Qian, Y.; Wang, L.; Cheng, Z.; Zhao, Y.; Wang, X.; Lu, Z. Research on optimal oil filling control strategy of wet clutch in agricultural machinery. *Actuators* **2022**, *11*, 315. [CrossRef]
58. Chen, Y.; Qian, Y.; Lu, Z.; Zhou, S.; Xiao, M.; Bartos, P.; Xiong, Y.; Jin, G.; Zhang, W. Dynamic characteristic analysis and clutch engagement test of HMCVT in the high-power tractor. *Complexity* **2021**, *2021*, 8891127. [CrossRef]
59. Kim, D.C.; Kim, K.U.; Park, Y.J.; Huh, J.Y. Analysis of shifting performance of power shuttle transmission. *J. Terramech.* **2007**, *44*, 111–122. [CrossRef]
60. SimulationX Online Help. Available online: https://doc.simulationx.com/4.6/1033/Content/Libraries/PowerTransmission/Couplings/DiscClutch.htm (accessed on 10 July 2024).
61. Baumann, J.; Klittich, M. Dynamic system simulation for model based design of tractor power train components. *VDI-Berichte* **2008**, *2045*, 19–26.
62. Chen, X.; Zhao, Y.; Liu, K.; Wang, G.; Song, Y.; Chakrabarti, P. Shifting quality analysis of unmanned tractor equipped with series hydro-mechanical transmission. *Trans. Can. Soc. Mech. Eng.* **2024**, *48*, 221–237. [CrossRef]
63. Zhu, Z.; Gao, X.; Pan, D.; Zhu, Y.; Cao, L. Study on the control strategy of shifting time involving multigroup clutches. *Math. Probl. Eng.* **2016**, *2016*, 9523251. [CrossRef]
64. Goetz, M.; Levesley, M.C.; Crolla, D.A. Dynamics and control of gearshifts on twin-clutch transmissions. *Proc. Inst. Mech. Eng. Part D J. Automob. Eng.* **2005**, *219*, 951–963. [CrossRef]
65. Li, B.; Sun, D.; Hu, M.; Zhou, X.; Liu, J.; Wang, D. Coordinated control of gear shifting process with multiple clutches for power-shift transmission. *Mech. Mach. Theory* **2019**, *140*, 274–291. [CrossRef]
66. Xia, G.; Yu, X.; Tang, X.; Zhao, L.; Sun, B. Power shift control of high-power tractor based on pattern search. *Appl. Eng. Agric.* **2020**, *36*, 565–578. [CrossRef]
67. Xia, G.; Wang, Y.; Tang, X.; Zhao, L.; Hu, J. Double-clutch power shift quality optimization based on optimal control theory. *Appl. Eng. Agric.* **2021**, *37*, 677–689. [CrossRef]
68. Wang, G.; Song, Y.; Wang, J.; Xiao, M.; Cao, Y.; Chen, W.; Wang, J. Shift quality of tractors fitted with hydrostatic power split CVT during starting. *Biosyst. Eng.* **2020**, *196*, 183–201. [CrossRef]
69. Xu, Z. Design and Power Shift Analysis of Hydrostatic Power Split Transmission with Single Planetary Gear Set. Master's Thesis, Shandong Agricultural University, Taian, China, 2023.
70. Li, A.; Qin, D.; Guo, Z.; Xia, Y.; Lv, C. Wet clutch pressure hysteresis compensation control under variable oil temperatures for electro-hydraulic actuators. *Control Eng. Pract.* **2023**, *141*, 105723. [CrossRef]
71. Li, J.; Dong, H.; Han, B.; Zhang, Y.; Zhu, Z. Designing Comprehensive Shifting Control Strategy of Hydro-Mechanical Continuously Variable Transmission. *Appl. Sci.* **2022**, *12*, 5716. [CrossRef]
72. Li, S.; Wu, C.; Sun, Z. Design and implementation of clutch control for automotive transmissions using terminal-sliding-mode control and uncertainty observer. *IEEE Trans. Veh. Technol.* **2016**, *65*, 1890–1898. [CrossRef]
73. Zhang, Y.; Du, Y.; Mao, E.; Song, Z.; Chen, D.; Zhu, Z. Pressure control method of wet clutch in high-powered tractor based on digital twin. *J. Mech. Eng.* **2023**, *59*, 268–279. [CrossRef]
74. Kolator, B.A. Modeling of Tractor Fuel Consumption. *Energies* **2021**, *14*, 2300. [CrossRef]

75. Oyelade, O.A.; Oni, K.C. Modelling of tractor fuel consumption for ploughing operation in a sandy loam soil. *AMA-Agric. Mech. Asia Afr. Lat. A.* **2021**, *52*, 44–52.
76. Siddique, M.A.A.; Baek, S.-M.; Baek, S.-Y.; Kim, W.-S.; Kim, Y.-S.; Kim, Y.-J.; Lee, D.-H.; Lee, K.-H.; Hwang, J.-Y. Simulation of fuel consumption based on engine load level of a 95 kw partial power-shift transmission tractor. *Agriculture* **2021**, *11*, 276. [CrossRef]
77. Li, B.; Sun, D.; Hu, M.; Zhou, X.; Wang, D.; Xia, Y.; You, Y. Automatic gear-shifting strategy for fuel saving by tractors based on real-time identification of draught force characteristics. *Biosyst. Eng.* **2020**, *193*, 46–61. [CrossRef]
78. Guo, Z.; Yuan, S.; Jing, C.; Wei, C. Modeling and simulation of shifting process in hydraulic machinery stepless transmission based on AMESim. *Trans. Chin. Soc. Agric. Eng.* **2009**, *25*, 86–91.
79. Zhang, Z.; Li, X.; Peng, Z.; Jing, C.; Yang, S.; Chen, Z. Full-condition efficiency model and power loss analysis of hydro-mechanical CVT. *Trans. CSAE* **2024**, *40*, 48–58. [CrossRef]
80. Wang, G.; Zhao, Y.; Song, Y.; Xue, L.; Chen, X. Optimizing the fuel economy of hydrostatic power-split system in continuously variable tractor transmission. *Heliyon* **2023**, *9*, e15915. [CrossRef]
81. Xu, L. Study on Characteristics of Hydro-Mechanical Continuously Varlable Transmission of Tractor. Ph.D. Thesis, Xi'an University of Technology, Xi'an, China, 2007.
82. Liu, F. Design and Characteristics Research on the Hydro-menchanical Continuously Variable Transmission of High-power Tractors. Master's Thesis, Nanjing Agricultural University, Nanjing, China, 2015.
83. Zhang, G.; Wang, K.; Xiao, M.; Zhou, M. HMCVT steady state transmission efficiency based on HST-EGT torque ratio. *Trans. CSAM* **2021**, *52*, 533–541.
84. Zhang, J.; Liu, X.; Wei, H.; Liu, M.; Huang, W.; Yan, X. Study on shifting performance of tractor multi-clutch under different engagement rules. *Agriculture* **2024**, *14*, 254. [CrossRef]
85. Xia, Y.; Sun, D.; Qin, D.; Zhou, X. Optimisation of the power-cycle hydro-mechanical parameters in a continuously variable transmission designed for agricultural tractors. *Biosyst. Eng.* **2020**, *193*, 12–24. [CrossRef]
86. Zhang, Q.; Sun, D.Y.; Qin, D.T. Optimal parameters design method for power reflux hydro-mechanical transmission system. *Proc. Inst. Mech. Eng. Part D* **2019**, *233*, 585–594. [CrossRef]
87. İnce, E.; Güler, M.A. Design and analysis of a novel power-split infinitely variable power transmission system. *J. Mech. Des.* **2019**, *141*, 54501. [CrossRef]
88. Rossetti, A.; Macor, A. Multi-objective optimization of hydro-mechanical power split transmissions. *Mech. Mach. Theory* **2013**, *62*, 112–128. [CrossRef]
89. Rossetti, A.; Macor, A.; Scamperle, M. Optimization of components and layouts of hydromechanical transmissions. *Int. J. Fluid Power* **2017**, *18*, 123–134. [CrossRef]
90. Zhao, Y. Research on Energy Consumption Law of Transmission System of Hydro-Mechanical Transmission Tractors. Master's Thesis, Shandong Agricultural University, Taian, China, 2024.
91. İnce, E.; Güler, M.A. On the advantages of the new power-split infinitely variable transmission over conventional mechanical transmissions based on fuel consumption analysis. *J. Clean. Prod* **2020**, *244*, 118795. [CrossRef]
92. Rossetti, A.; Macor, A.; Benato, A. Impact of control strategies on the emissions in a city bus equipped with power-split transmission. *Transport. Res. Transp. Environ.* **2017**, *50*, 357–371. [CrossRef]
93. Rossetti, A.; Macor, A. Control strategies for a powertrain with hydromechanical transmission. *Energy Proc.* **2018**, *148*, 978–985. [CrossRef]
94. Liang, L.; Li, Q.; Tan, Y.; Yang, Z. Multi-parameter optimization of HMCVT fuel economy using parameter cyclic algorithm. *Trans. CSAE* **2023**, *39*, 48–55. [CrossRef]
95. Zhang, M.; Wang, J.; Wang, J.; Guo, Z.; Guo, F.; Xi, Z.; Xu, J. Speed changing control strategy for improving tractor fuel economy. *Trans. CSAE* **2020**, *36*, 82–89. [CrossRef]
96. Martini, V.; Mocera, F.; Somà, A. Numerical investigation of a fuel cell-powered agricultural tractor. *Energies* **2022**, *15*, 8818. [CrossRef]
97. Zhang, R.; Xia, B.; Li, B.; Cao, L.; Lai, Y.; Zheng, W.; Wang, H.; Wang, W. State of the art of lithium-ion battery SOC estimation for electrical vehicles. *Energies* **2018**, *11*, 1820. [CrossRef]
98. Baek, S.-Y.; Kim, Y.-S.; Kim, W.-S.; Baek, S.-M.; Kim, Y.-J. Development and verification of a simulation model for 120 kw class electric AWD (All-Wheel-Drive) tractor during driving operation. *Energies* **2020**, *13*, 2422. [CrossRef]
99. Zhang, L.; Liu, W.; Qi, B. Energy optimization of multi-mode coupling drive plug-in hybrid electric vehicles based on speed prediction. *Energy* **2020**, *206*, 118126. [CrossRef]
100. Zhang, L.; Liu, W.; Qi, B. Innovation design and optimization management of a new drive system for plug-in hybrid electric vehicles. *Energy* **2019**, *186*, 115823. [CrossRef]
101. Baek, S.Y.; Kim, W.S.; Kim, Y.S.; Kim, Y.J.; Park, C.G.; An, S.C.; Moon, H.C.; Kim, B.S. Development of a simulation model for an 80 kw-class electric All-Wheel-Drive (AWD) tractor using agricultural workload. *J. Drive Control* **2020**, *17*, 27–36.
102. Medževeprytė, U.K.; Makaras, R.; Lukoševičius, V.; Kilikevičius, S. Application and efficiency of a series-hybrid drive for agricultural use based on a modified version of the world harmonized transient cycle. *Energies* **2023**, *16*, 5379. [CrossRef]
103. Medževeprytė, U.K.; Makaras, R.; Lukoševičius, V.; Keršys, A. Evaluation of the working parameters of a series-hybrid tractor under the soil work conditions. *Teh. Vjesn.* **2022**, *29*, 45–50. [CrossRef]

104. Mao, Y.; Wu, Y.; Yan, X.; Liu, M.; Xu, L. Simulation and experimental research of electric tractor drive system based on Modelica. *PLoS ONE* **2022**, *17*, e0276231. [CrossRef] [PubMed]
105. Li, T.; Xie, B.; Li, Z.; Li, J. Design and optimization of a dual-input coupling powertrain system: A case study for electric tractors. *Appl. Sci.* **2020**, *10*, 1608. [CrossRef]
106. Wang, S.; Wu, X.; Zhao, X.; Wang, S.; Xie, B.; Song, Z.; Wang, D. Co-optimization energy management strategy for a novel dual-motor drive system of electric tractor considering efficiency and stability. *Energy* **2023**, *281*, 128074. [CrossRef]
107. Mocera, F.; Martini, V. Numerical performance investigation of a hybrid eCVT specialized agricultural tractor. *Appl. Sci.* **2022**, *12*, 2438. [CrossRef]
108. Rossi, C.; Pontara, D.; Falcomer, C.; Bertoldi, M.; Mandrioli, R. A Hybrid–Electric Driveline for Agricultural Tractors Based on an e-CVT Power-Split Transmission. *Energies* **2021**, *14*, 6912. [CrossRef]
109. LMS Imagine. *AMESim User's Guides*; LMS Imagine: Roanne, France, 2013.
110. Makange, N.R.; Ji, C.; Torotwa, I. Prediction of cutting forces and soil behavior with discrete element simulation. *Comput. Electron. Agric.* **2020**, *179*, 105848. [CrossRef]
111. Tong, J.; Jiang, X.; Wang, Y.; Ma, Y.; Li, J.; Sun, J. Tillage force and disturbance characteristics of different geometric-shaped subsoilers via DEM. *Adv. Manuf.* **2020**, *8*, 392–404. [CrossRef]
112. Zhai, S.; Shi, Y.; Zhou, J.; Liu, J.; Huang, D.; Zou, A.; Jiang, P. Simulation optimization and experimental study of the working performance of a vertical rotary tiller based on the discrete element method. *Actuators* **2022**, *11*, 342. [CrossRef]
113. Kim, Y.-S.; Siddique, M.A.A.; Kim, W.-S.; Kim, Y.-J.; Lee, S.-D.; Lee, D.-K.; Hwang, S.-J.; Nam, J.-S.; Park, S.-U.; Lim, R.-G. DEM simulation for draft force prediction of moldboard plow according to the tillage depth in cohesive soil. *Comput. Electron. Agric.* **2021**, *189*, 106368. [CrossRef]
114. Čermák, M.; Mitáš, S. Theoretical analysis of the forces and torques acting on a tractor during ploughing. *J. Terramech.* **2021**, *96*, 23–27. [CrossRef]
115. Kim, W.; Kim, Y.; Kim, Y. Development of prediction model for axle torque of agricultural tractors. *Trans. ASABE* **2020**, *63*, 1773–1786. [CrossRef]
116. Fang, C. *Process Identification*; Tsinghua University Press: Beijing, China, 2003.
117. Wang, G.; Zhu, S.; Wang, S.; Shi, L.; Ni, X.; Ouyang, D. Speed ratio control of tractor hydraulic mechanical CVT. *Trans. CSAE* **2013**, *29*, 17–23.
118. Cheng, Z.; Lu, Z. Model modification and parameter identification of tractor hydraulic transmission system characteristics. *Trans. CSAE* **2022**, *38*, 33–41. [CrossRef]
119. Cheng, Z.; Lu, Z. System response modeling of HMCVT for tractors and the comparative research on system identification methods. *Comput. Electron. Agric.* **2022**, *202*, 107386. [CrossRef]
120. Musialek, I.; Kesy, A.; Olszak, A.; Choi, S.B. Parameter estimation of transfer function of viscous clutch with electrorheological fluid and torque control. *J. Intell. Mater. Syst. Struct.* **2024**, *35*, 410–423. [CrossRef]
121. Tørdal, S.S.; Klausen, A.; Bak, M.K. Experimental system identification and black box modeling of hydraulic directional control valve. *Model. Identif. Control* **2015**, *35*, 225–235. [CrossRef]
122. Cheng, Z.; Lu, Z.; Dai, F. Research on HMCVT efficiency model based on the improved SA algorithm. *Math. Probl. Eng.* **2019**, *2019*, 2856908. [CrossRef]

agriculture

Article

Design and Experiment of Planting Mechanism of Automatic Transplanter for Densely Planted Vegetables

Jiawei Shi, Jianping Hu *, Jing Li, Wei Liu, Rencai Yue, Tengfei Zhang and Mengjiao Yao

Jiangsu Provincial Key Laboratory of Hi-Tech Research for Intelligent Agricultural Equipment, Jiangsu University, Zhenjiang 212013, China; 2112116016@stmail.ujs.edu.cn (J.S.); 2211916011@stmail.ujs.edu.cn (J.L.); mario_liu@ujs.edu.cn (W.L.); 2111916008@stmail.ujs.edu.cn (R.Y.); 2112016001@stmail.ujs.edu.cn (T.Z.); 2112216018@stmail.ujs.edu.cn (M.Y.)

* Correspondence: hujp@ujs.edu.cn; Tel.: +86-511-88797338

Abstract: The planting mechanism of existing transplanters cannot meet the agronomic requirements of planting densely planted vegetables with multiple rows, small plant spacing, and small row spacing. In order to solve this current problem, an eight-row duckbill planting mechanism driven by a motor and a cylinder was designed. According to the agronomic guidance and mechanism design requirements for transplanting densely planted vegetable seedlings, this paper analyzes the working principle of the planting mechanism, establishes its kinematic theoretical model, and determines the structural parameters of the driving device and opening and closing device in the planting mechanism. Aimed at the problem of large planting resistance when eight-row planting end effectors of the planting mechanism are planting at the same time, based on the existing research, three duckbill planting end effectors with double incisions, four incisions, and conical structures were selected, and the planting process was simulated using an EDEM 2022-RecurDyn 2024 coupling simulation. The single-factor analysis method and the interactive factor Box–Behnken response surface analysis method were used. It is concluded that the duckbill end effector with double incisions has the smallest planting resistance, and the rationality of the mechanism design is preliminarily verified. A planting resistance measurement platform was built based on the STM32 platform and HX711 module, and a planting resistance test of the duckbill planting end effector was carried out to verify the correctness of the planting mechanism simulation results. The planting mechanism performance test was carried out, and the test results showed that the planting qualification rate of the prototype reached 96.62%, the planting spacing variation coefficient was only 3.55%, and the planting efficiency reached about 7135 plants/h, which met the agronomic requirements of small plant spacing and small row spacing for densely planted vegetables and verified the feasibility and practicality of the planting mechanism.

Keywords: planting mechanism; densely planted vegetables transplantation; duckbill planting end effector; coupling simulation; performance test

Citation: Shi, J.; Hu, J.; Li, J.; Liu, W.; Yue, R.; Zhang, T.; Yao, M. Design and Experiment of Planting Mechanism of Automatic Transplanter for Densely Planted Vegetables. *Agriculture* **2024**, *14*, 1357. https://doi.org/10.3390/agriculture14081357

Academic Editor: Mustafa Ucgul

Received: 25 July 2024
Revised: 9 August 2024
Accepted: 12 August 2024
Published: 14 August 2024

1. Introduction

Vegetable transplanting is an important technology for improving crop yield and quality. Transplanting technology is conducive to alleviating the impact of seasons on vegetable planting, ensuring the consistency of vegetable growth conditions and development, facilitating the stable growth of vegetables, and improving land utilization and the multiple cropping index [1–7]. Densely planted vegetables are mainly pak choi. Because pak choi are planted in multiple rows on a single ridge, and the row and plant spacing is relatively small (10–15 cm), there is currently no suitable automatic transplanter. It mainly relies on manual planting, which has a low operation efficiency and high labor intensity [8–11].

European and American countries with mature vegetable transplanting mechanization technology have developed large-scale automatic transplanters that integrate mechanical, electrical, hydraulic, and control technologies [12–15]. For example, the five-row automatic transplanter Futura, from the company Ferrari in Italy, has a transplanting row spacing of 450–750 mm and a minimum planting spacing of 60 mm. The planting mechanism of this type of automatic transplanter has a large structure and is mainly used for multi-row vegetable seedling transplanting operations with a large plant spacing and large row spacing. It does not meet the agronomic requirements of a small plant spacing and small row spacing for densely planted vegetables. Japan's automatic transplanters are small-sized equipment mainly used for two-row operations, with a transplanting spacing of 260–800 mm and a transplanting row spacing of 450–650 mm, which does not meet the agronomic requirements of densely planted vegetables with multiple rows, small plant spacing, and small row spacing [16–19]. Up to now, the research and development of transplanters in Japan is mainly in the semi-automatic stage, with a small number of machine types and a low degree of automation. They are mainly used for two-row planting and are only suitable for transplanting operations with large plant spacing and large row spacing. There is still a lack of automatic transplanters for densely planted vegetables with multiple rows, small plant spacing, and small row spacing.

The planting mechanism is the core component of the automatic transplanter for densely planted vegetables. Its design is the key to realizing the automatic transplantation of densely planted vegetables and directly determines the planting quality [20–22]. Zhou et al. [23] designed a furrowing multi-link seedling planting mechanism with the function of supporting and pushing seedlings. They analyzed the working principle of the mechanism, established a kinematic model, and conducted a virtual simulation analysis. They verified the correctness of the seedling planting mechanism design and improved the planting qualification rate. However, the mechanism had problems, such as large planting spacing and a small number of planting rows. Wang et al. [24] proposed a three-arm wheel planting mechanism and conducted an approximate multi-pose motion comprehensive design based on a genetic algorithm. They also conducted a simulation analysis and experiments on the three-arm wheel planting mechanism, which verified the feasibility of the mechanism and improved the planting efficiency. However, the planting spacing of the mechanism was 300 mm. Yu et al. [25,26] designed a planetary gear system for a vegetable seedling planting mechanism that could adapt to a large plant spacing. The transplanting spacing was 450 mm. By constructing a new type of non-circular gear transmission ratio function, the kinematic model of the mechanism was optimized, and the planting efficiency reached 100 plants/(row·min). Hu et al. [27] designed a new automatic transplanting mechanism based on a clamping stem. Through the movement law of a double crank-connecting rod mechanism, the static trajectory of a beak shape and the dynamic trajectory of an N shape were designed to meet the requirements of backward transplanting. The effect of the mechanism parameter changes on the seedling movement trajectory was analyzed. A group of parameter combinations that met the requirements of seedling picking and planting was optimized. However, the transplanting spacing was 260 mm and the number of planting rows was one.

The existing planting mechanisms are mainly multi-link mechanisms, hanging basket types, and non-circular planetary gear mechanisms, and the number of planting rows is mainly two rows. None of them can meet the agronomic requirements of densely planted vegetable transplantations, and the two planting resistances are small. There are few studies on duckbill soil penetration resistance, but for an eight-row densely planted vegetables transplanter, the planting resistance increases significantly, requiring higher structural design requirements. This article describes the design of a multi-row duckbill-type planting mechanism driven by the coupling of a motor and cylinder. Based on the go-stop-go chassis control strategy, eight duckbill-type planting end effectors are planted vertically at the same time. This mechanism can realize multiple planting methods, including transplanting operations in rows with a small plant spacing and small row spacing. This paper first analyzes the working principle of the planting mechanism and establishes its kinematic theoretical model, then determines the structural parameters of the driving device and the opening and closing device in the planting mechanism. Secondly, the planting process is simulated using an EDEM 2022-RecurDyn 2024 coupling simulation. The single-factor analysis method and the interactive factor Box–Behnken response surface analysis method are used to select the duckbill shape with the smallest planting resistance. Finally, the rationality of the mechanism is verified through experiments, providing a reference for the design of multi-row densely planted vegetable planting mechanisms.

2. Device Structure and Working Principle

2.1. Structure and Working Principle of Transplanter

The automatic transplanting machine for densely planted vegetables includes a seedling tray conveying mechanism, a seedling taking and throwing mechanism, a seedling tray recovery mechanism, a variable-distance dispersion mechanism, a planting mechanism, a soil covering mechanism, a chassis, etc., as shown in Figure 1. Its working principle: manually take the seedling tray from the seedling rack and place it on the seedling tray conveying mechanism. Then, press the start button and the seedling tray conveying mechanism transports the seedling tray to the seedling taking position. The seedling taking mechanism moves along the linear guide rail to the seedling taking position to take the seedlings. Then, the seedling taking mechanism moves to the top of the dispersion mechanism and puts the seedling pot into the seedling guide barrel. Then, the dispersion barrel of the variable-distance dispersion mechanism is dispersed to the planting row spacing. After opening the seedling barrel, the seedling pot falls into the 8 duckbill planting end effectors of the planter. The planting motor drives the planting mechanism to vertically enter the soil to plant the vegetable seedlings, and a planting cycle is completed.

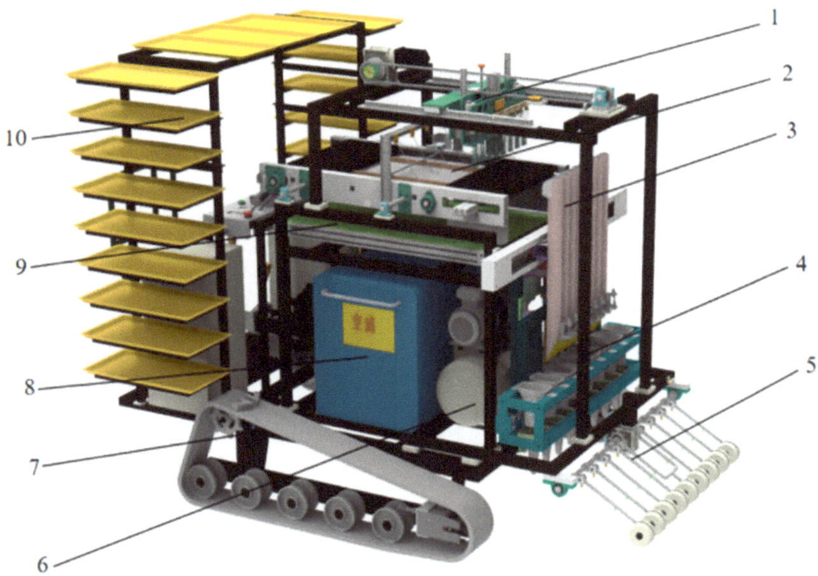

Figure 1. Structural diagram of densely planted vegetables automatic transplanter. 1. seedling taking and dropping mechanism; 2. seedling tray conveying mechanism; 3. variable-distance dispersion mechanism; 4. planting mechanism; 5. soil covering mechanism; 6. air compressor; 7. chassis; 8. power generator; 9. seedling tray recovery mechanism; 10. seedling tray storage rack.

2.2. Structure and Working Principle of Planting Mechanism

The planting mechanism mainly includes a driving motor, a driving rod, a planting frame, a linkage rod, opening and closing cylinders, duckbill-type end effectors, and other components, as shown in Figure 2, and its working principle is shown in Figure 3. The driving motor drives the screw to rotate and the screw rotation drives the screw nut to move up and down. The screw nut is connected to the driving rod, so that the duckbill type planting end effector moves vertically up and down through the guide rail slider mechanism. The duckbill planting end effector is initially located at the initial position above the planting mechanism, as shown in Figure 3a. When the duckbill planting end effector moves vertically downward to the planting position, the opening and closing cylinder works. Opening and closing cylinder 1 pushes out the opening and closing rod opening and closing cylinder 2 contracts the opening and closing rod, and the connecting linkage rod performs a short lateral displacement. The right half of the 8 duckbill planting end effectors connected to the linkage rod works, and the left half of the duckbill planting end effectors rotates at a relative angle through engagement with the right half of the duckbill planting end effectors, thereby opening the duckbill planting end effector, as shown in Figure 3b. The seedling pot falls into the hole formed by the opening of the duckbill planting end effector, and then the driving motor rotates in the opposite direction and the planting mechanism rises vertically to the initial position, as shown in Figure 3c Then, opening and closing cylinder 1 contracts the opening and closing rod, opening and closing cylinder 2 pushes out the opening and closing rod, and the duckbill planting end effector closes, as shown in Figure 3d.

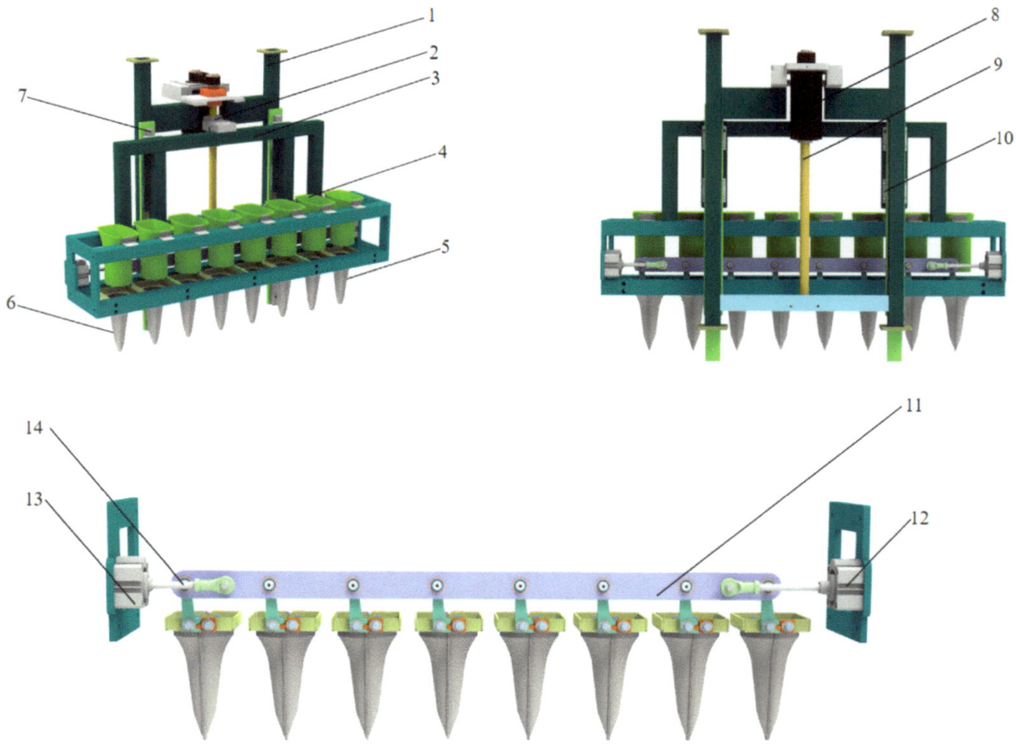

Figure 2. Structural diagram of planting mechanism. 1. planting frame; 2. screw nut; 3. driving rod; 4. seedling guide bucket; 5. right half of the duckbill planting end effectors; 6. left half of the duckbill planting end effectors; 7. guide rail; 8. driving motor; 9. screw; 10. slider; 11. linkage rod; 12. opening and closing cylinder 1; 13. opening and closing cylinder 2; 14. opening and closing rod.

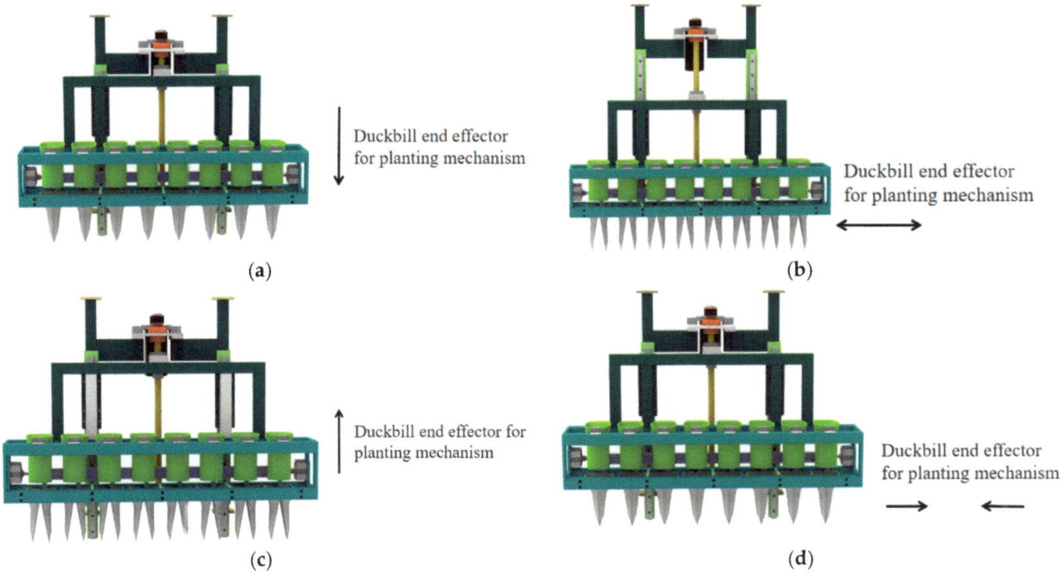

Figure 3. Working principle diagram of planting mechanism: (**a**) duckbill planting end effector descending stage; (**b**) duckbill planting end effector opening stage; (**c**) duckbill planting end effector rising stage; and (**d**) duckbill planting end effector closing stage.

3. Design of Key Components

3.1. Establishment of Kinematic Model of Planting Mechanism

The planting mechanism included a power transmission mechanism and a planting end effector. The design of the power transmission mechanism was the premise for the normal operation of the planting mechanism. In order to meet the requirements of densely planted vegetables with multiple rows, small plant spacing, and small row spacing, a scheme for a motor-driven planting mechanism with a double cylinder driving the planting mechanism was proposed. For the convenience of analysis, each part and the corresponding joint point were simplified in the plane. Among them, the screw driving the driving rod to drive the linkage rod to move up and down was simplified to the screw driving the linkage rod directly to move up and down, and the cylinder was simplified to a single-cylinder drive. The cylinder was used as a driving part to control the telescopic movement of the opening and closing pull rod to drive the linkage rod to move left and right. The top of the linkage rod was connected to the roller bearing in an articulated manner, and the roller bearing produced a relative rolling displacement, thereby driving the duckbill planting end effector to open and close. The simplified diagram of the planting mechanism is shown in Figure 4. L is the effective stroke of the opening and closing cylinder, and D, E, and F are hinge points.

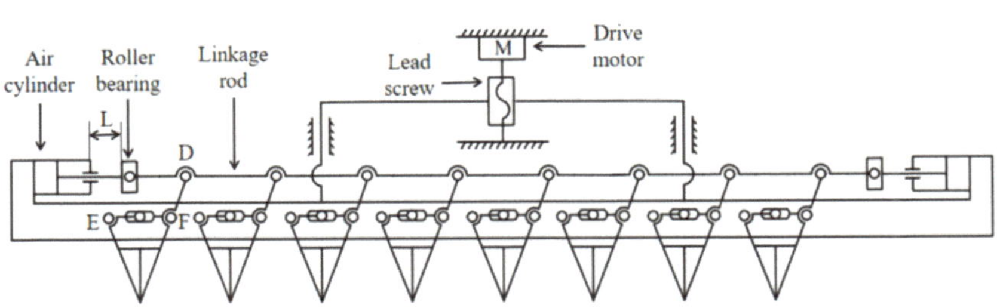

Figure 4. Simplified diagram of planting mechanism.

3.2. *Design of Power Transmission Mechanism*

3.2.1. Design of Motor and Screw

The duckbill planting end effector first accelerates, then moves at a constant speed, and then decelerates. The total weight m of the linkage rod and the 8 duckbill planting end effectors was 10.4 kg. According to the literature [28], the vertical planting resistance of duckbill planting end effectors is between 55–65 N. This paper takes the planting resistance of a single duckbill as 60 N, regards the multi-row planting mechanism as a rigid body, and ignores the deformation factor. Therefore, the planting resistance F_f of the 8 duckbill planting end effectors of the multi-row planting mechanism was 480 N. The force analysis of the screw rod was carried out, and the self-weight of the screw rod was ignored. It can be regarded as a two-force rod, as shown in Figure 5a. The force analysis of the duckbill planting end effector was carried out as shown in Figure 5b.

(**a**) (**b**)

Figure 5. Diagram of force analysis: (**a**) diagram of screw force analysis; and (**b**) diagram of duckbill planting end effector force analysis.

After force analysis, according to the principle of force balance, the force F_m exerted by the drive motor on the screw was solved as shown in Formulas (1)–(3); the lead P_h of the screw was solved by Formula (4); the maximum stroke of the duckbill planting end effector's up and down movement was solved by Formula (5); the maximum torque and power required by the drive motor were solved according to Formulas (6) and (7), and the drive motor was selected.

$$F_{d'} + G - F_f = ma \quad (1)$$

$$G = mg \quad (2)$$

$$F_{d'} = F_d = F_m \quad (3)$$

where G is the weight of the linkage rod and the 8 duckbill planting end actuators, N; m is the mass of the linkage rod and the 8 duckbill planting end actuators, kg, take m = 10.4 kg; a is the acceleration of the duckbill planting end actuator during movement, m/s^2, take a = 0.5 g m/s^2; F_f is the planting resistance of the 8 duckbill planting end actuators, take F_f = 480 N; $F_{d'}$ is the force of the screw on the duckbill planting end actuator, N; F_d is the force of the duckbill planting end actuator on the screw, N; and F_m is the force of the drive motor on the screw, N. From Formulas (1)–(3), it can be seen that F_m is 383.28 N.

$$n \cdot P_h = v \cdot 60 \quad (4)$$

where n is the motor speed, rpm, take n = 3000 rmp; and V is the speed of the duckbill planting end effector during movement, m/s, take v = 0.5 m/s. The screw lead P_h is 10 mm. From the table, the screw diameter d is 15 mm.

$$d \cdot 60 > h \quad (5)$$

where d is the screw diameter, mm; and h is the movement stroke of the duckbill planting end effector, mm, take h = 400 mm. Substituting the data into Formula (5), we get 15 × 6 > 400, and the designed screw parameters are a lead P_h of 10 mm and screw diameter d of 15 mm, which meets the requirements.

$$T_{max} \cdot 2\pi = F_m \cdot P_h \tag{6}$$

where T_{max} is the maximum driving torque required for motor, N·m. The calculation shows that the maximum driving torque T_{max} required for the motor is 0.61 N·m.

$$P_{max} = \frac{T_{max} \cdot n}{2a} \tag{7}$$

where P_{max} is the maximum power required by the motor, kW.

The calculation shows that the maximum power required for the motor is 0.20 kW. Considering the complex soil working environment, the drive motor should have a larger safety factor, so the Mitsubishi HG-KR43J motor was selected, with a rated power of 0.4 kW and a rated torque of 1.3 N·m, which meets the design requirements.

3.2.2. Design of Mechanical-Pneumatic Coupling Duckbill Opening and Closing Mechanism

The opening and closing mode of a single duckbill planting end effector is as follows: point D moves in a circle around point E under the drive of the linkage rod, and the right half duckbill planting end effector opens and closes. The right half duckbill planting end effector and the left half duckbill planting end effector mesh with each other to drive the left half duckbill planting end effector to move. The opening and closing angle of the duckbill planting end effector can be controlled via the telescopic distance of the cylinder rod. By analyzing the kinematic characteristics of the opening and closing control motion joint of the duckbill planting end effector, it can be known that improving the transmission efficiency of the joint and maintaining the stability of the duckbill planting end effector when the displacement constraint condition is the largest can effectively solve the problem of seedling leakage and unstable planting hole diameter during planting, and improve the control ability of the mouth angle and hole diameter. Simplifying the plane, as shown in Figure 6, which is the principle diagram of the duckbill planting end effector, shows that the opening and closing angle of the duckbill planting end effector is controlled by the effective displacement of the D joint [29,30].

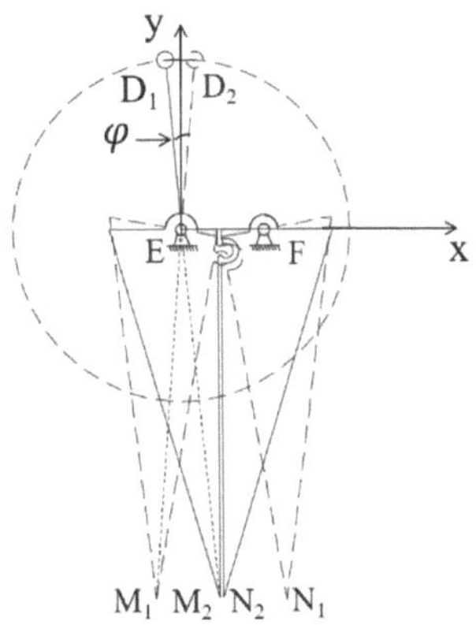

Figure 6. Simplified diagram of duckbill planting end effector.

When the planting mechanism is performing planting operations, the effective stroke distance of the cylinder was set to L. Since the horizontal displacement distance of the D joint is consistent with L, the relationship between the two is shown in Formula (8); since the angle φ is relatively small in actual motion, the distances of D_1E, D_2E, EM_1, EM_2,

FN_1, and FN_2 are approximately regarded as their projections on the XY plane, and the relationship between them is shown in Formula (9). In the triangle OM_1N_1, $\angle M_1ON_1 = \theta$, the half-angle formula can be used to determine the relationship between the side length and θ, as shown in Formula (10). After simplification, the relationship between θ and the side length is shown in Formula (11).

$$L = D_2 - D_1 \tag{8}$$

$$\begin{cases} \dfrac{D_1D_2}{M_1M_2} = \dfrac{D_1E}{EM_2} \\ \dfrac{L}{M_1M_2} = \dfrac{D_1E}{EM_2} \\ M_1M_2 = \dfrac{L \times EM_2}{D_1E} \\ M_1N_1 = 2M_1M_2 = \dfrac{2L \times EM_2}{D_1E} \end{cases} \tag{9}$$

$$\begin{cases} sin\dfrac{\theta}{2} = \dfrac{M_1N_1}{2OM_1} \\ cos\theta = 1 - 2\left(sin\dfrac{\theta}{2}\right)^2 \end{cases} \tag{10}$$

$$cos\theta = 1 - \frac{2L \times EM_2}{D_1E \times OM_1} \tag{11}$$

Therefore, the opening and closing angle of the duckbill unit is controlled by the movement of the linkage rod. When the duckbill planting end effector was opened by 40°, the relevant parameters of the planting module were substituted into the above formula group, and the movement distance of the linkage rod was 10 mm. Since the linkage rod was connected to the cylinder through the cylinder connector, the stroke of the cylinder only needed to be greater than 10 mm. The cylinder controlled the duckbill end effector to open in the soil. Due to the high recirculation of the soil, the opening and closing action needed to be completed in a shorter time. The larger the inner diameter of the cylinder, the faster the cylinder expansion and contraction speed under the same pressure. The duckbill planting end effector needed to overcome a large reaction force when opening and closing in the soil. As shown in Formula (12), when the working pressure is constant, the larger the inner diameter of the cylinder, the greater the force it generates.

$$F = \frac{10Pd^2\pi}{4} \tag{12}$$

where F is the tension at the end of the cylinder, N; P is the working pressure, bar; and d is the diameter of the piston rod in the cylinder, cm.

In summary, the AirTac SDA model 40 × 10 specification cylinder was selected. The inner diameter was 40 mm, the stroke was 10 mm, and it could provide a thrust of 500 N.

3.3. Structural Design of Duckbill Planting End Effector

3.3.1. Strength Analysis of Linkage Rod of Planting Mechanism

Considering the universality of densely planted vegetables, pak choi seedlings with a seedling age of 30–40 days, 4–6 leaves, and a seedling height of 100–130 mm suitable for mechanized transplanting were selected. A single pak choi seedling is shown in Figure 7a. This study selected Shanghai green seedlings cultivated in 128-hole plug trays as the research object. The overall specifications of the plug tray were 540 mm × 280 mm × 40.5 mm, and the size of a single hole was 15 mm × 15 mm at the bottom and 32 mm × 32 mm at the top. The schematic diagram of the seedling pot substrate is shown in Figure 7b.

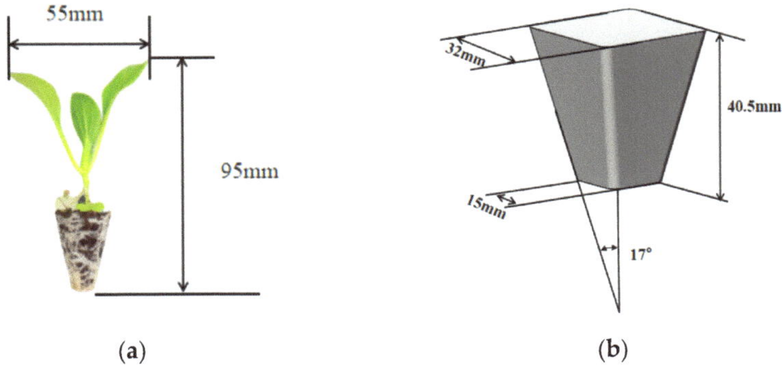

Figure 7. Schematic diagram of seedlings in pot: (**a**) diagram of pak choi seedlings in pot; and (**b**) schematic diagram of matrix in three dimensions.

During the planting operation, the initial posture of the seedling pot in the duckbill planting end effector will affect the posture of the seedling pot falling into the hole, and then determine whether the uprightness of the seedling pot is qualified. Therefore, the seedling pot in the duckbill planting end effector must fit the inner wall of the duckbill planting end effector. According to the shape of the seedling pot matrix, the overall shape of the duckbill planting end effector structure was determined to be conical. As shown in Figure 7a above, the hole tray seedlings cultivated with 128-hole tray seedlings have a space diameter of 50~60 mm and a height of 90~100 mm. Therefore, the diameter of the seedling port of the duckbill unit is determined to be 70 mm. The height of the vegetable seedlings is usually about 90~100 mm, so the height of the duckbill unit was set to 125 mm.

During planting, the duckbill end effector directly interacts with the soil, and the seedling placement area in the upper part of the duckbill end effector is related to the shape of the seedling pot and the volume occupied by the leaves, so there is not much room for optimization design. However, the cone angle part of the duckbill end effector first contacts the soil and is most affected by the resistance of the soil. At the same time, the soil has a certain fluidity. If the cone angle part is optimized to a better level, the planting resistance it encounters can be reduced. Therefore, this study mainly focuses on the cone angle part of the duckbill end effector, as shown in Figure 8.

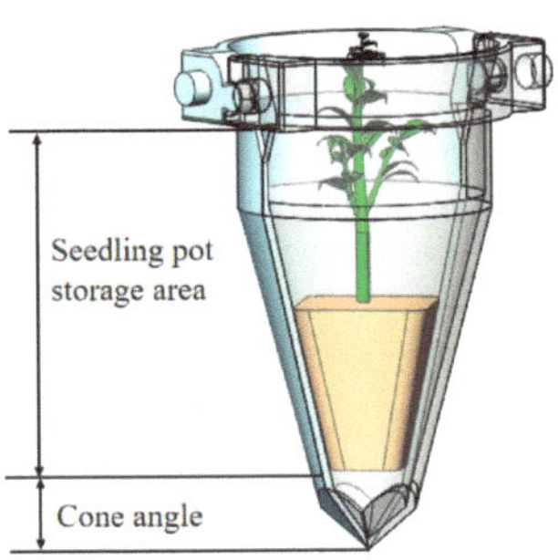

Figure 8. Two-stage duck-billed transplanting end effector.

Based on the external structural design of duckbill implant end effectors in the domestic and foreign literature [31], the following three types are summarized: double incisions, four incisions, and conical types, as shown in Figure 9.

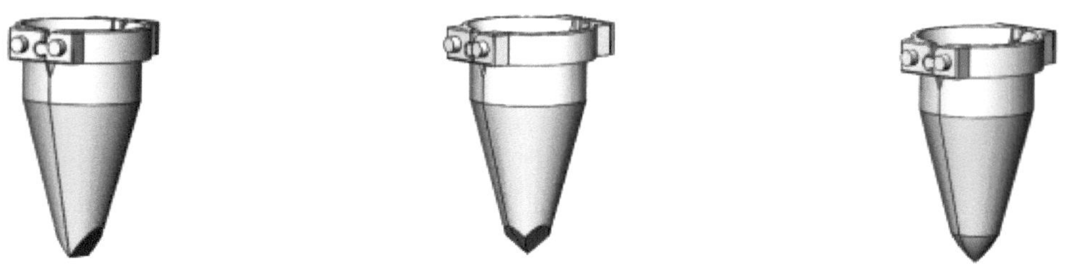

Figure 9. Different types of two-stage duck-billed transplanting end effector shapes: (**a**) double incisions type; (**b**) four incisions type; and (**c**) conical type.

3.3.2. Coupling Simulation Test of Duckbill Planting End Effector

(1) Determination of test parameters

Since the planting mechanism adopts a vertical planting method, the planting speed and planting depth of the duckbill planting end effector were selected as the test parameters of the simulation test. Its movement stroke before contacting the soil was 300 mm, and the duckbill planting depth was set to S_0, so the overall movement stroke of the duckbill was $(300 + S_0)$ mm. The expected planting frequency of the dense planting transplanter was 6000–8000 plants/h, so the average single planting cycle was about 4 s. A single planting cycle can be divided into four processes: the whole machine moves forward 12 cm, the planting mechanism vertically inserts into the planting soil, the duckbill end effector opens and drops the seedlings, and the planting mechanism vertically leaves the soil. The time allocation for each process was about 1 s. Therefore, in order to achieve the expected design goal, the overall movement and falling time of the duckbill end effector was about 1 s, and the total stroke was $(300 + S_0)$ mm. In summary, the planting speed parameters of the duckbill planting end effector were 300 mm/s, 350 mm/s, 400 mm/s, 450 mm/s, and 500 mm/s. Considering the planting requirements of vegetables in dryland transplanting, the planting depth is usually between 60–80 mm. Therefore, the planting depth parameters were 60 mm, 65 mm, 70 mm, 75 mm, and 80 mm. The data Fmax at the maximum resistance during the whole process were selected as the output values of the simulation test.

(2) Test methods

I. Single-factor test method: For the three external shapes, based on the two factors of planting speed and planting resistance, the influence of single factors on the result indicators was analyzed, and each factor adopted five levels. Three external shapes and two influencing factors were divided into 6 groups of tests; each group of tests was divided 5 times, and a total of 30 coupled simulation tests were conducted. The experimental factor level tables are shown in Tables 1 and 2. By changing the same parameters under the three external shapes, a comprehensive comparison is made as to which external shape has the smallest planting resistance.

Table 1. Level table of planting speed test parameters.

Shape	Planting Speed (mm/s)				
Double incisions	300	350	400	450	500
Four incisions	300	350	400	450	500
Conical	350	350	400	450	500

Table 2. Level table of planting depth test parameters.

Shape	Planting Depth (mm)				
Double incisions	300	350	400	450	500
Four incisions	300	350	400	450	500
Conical	350	350	400	450	500

II. Interactive Factor Response Surface Experiment Method: This study uses Design-Expert 13 software to perform interactive factor response surface experimental design. Based on multiple groups of interactive test data, the design compared and analyzed which shape structure has the lowest planting resistance. Based on three shape structure classifications and two influencing factors, three levels were set for each test. Two test parameters were established, namely planting speed and planting depth, and one determination parameter, the shape structure. The indicator response value is the planting resistance. Using the 3-Level Factorial Design method, 39 groups of experiments were conducted.

(3) Coupling simulation test based on EDEM 2022-RecurDyn 2024 software

This simulation test used the multi-body dynamic simulation software RecurDyn 2024 to simulate the motion state of the planting mechanism driven by the motor. It was coupled with the soil model simulated by EDEM 2022 in the discrete element analysis software to analyze the maximum value F_{max} of the dynamic resistance of the duckbill planting end effector under the dynamic reaction force of the soil. The simulation is shown in Figure 10.

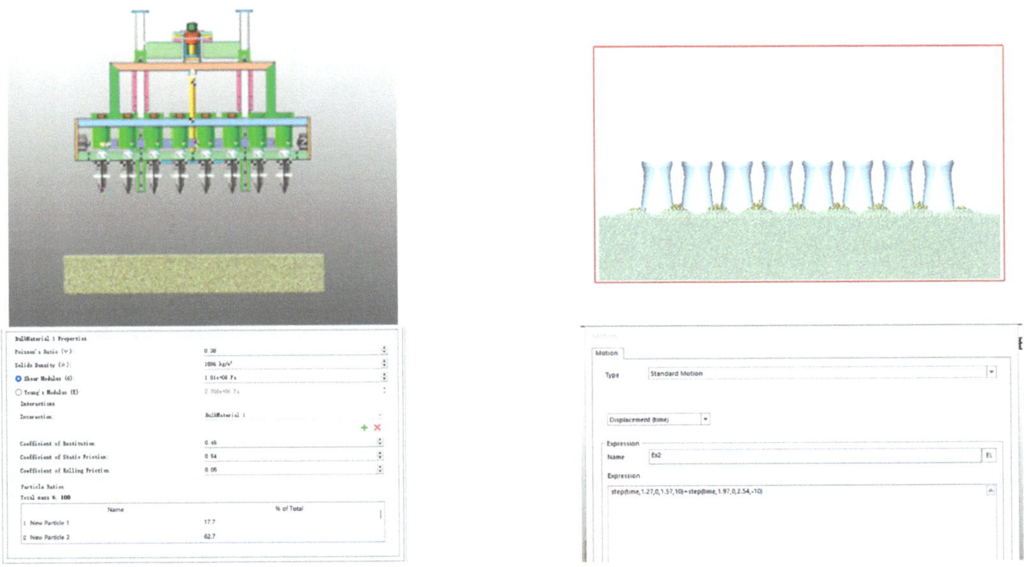

Figure 10. Coupling simulation of duckbill planting end effector.

(4) Analysis of single-factor effect

First, the single-factor test method was used to define the planting duckbill as two test factors, planting speed and planting degree, and conduct coupling simulation tests on three duckbill shapes. At the same time, other test factors remained unchanged. The maximum value of the dynamic reaction force of the soil on the planting duckbill was obtained as shown in Tables 3 and 4; the overall trend of F_{max} is shown in Figure 11.

Table 3. Test data for different transplanting speeds.

Test Group	Shape	Planting Speed (mm/s)				
1	Double incisions	300	350	400	450	500
	Planting resistance (N)	259.5	261.4	275.9	285.7	303.6
2	Four incisions	300	350	400	450	500
	Planting resistance (N)	358.1	361.8	381.2	405.2	415.7
3	Conical	300	350	400	450	500
	Planting resistance (N)	315.7	321.8	348.5	360.8	381.6

Table 4. Test data for different transplanting depths.

Test Group	Shape	Planting Depth (mm)				
1	Double incisions	60	65	70	75	80
	Planting resistance (N)	225.3	246.8	278.9	339.3	368.4
2	Four incisions	60	65	70	75	80
	Planting resistance (N)	307.9	328.6	345.7	380.5	401.5
3	Conical	60	65	70	75	80
	Planting resistance (N)	275.4	304.7	348.8	389.5	433.7

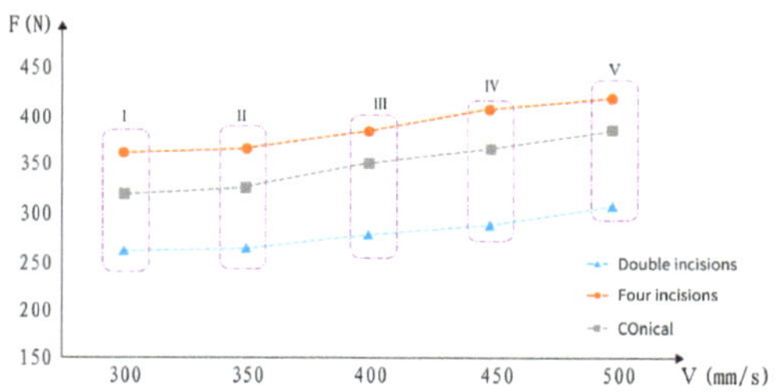

(a)

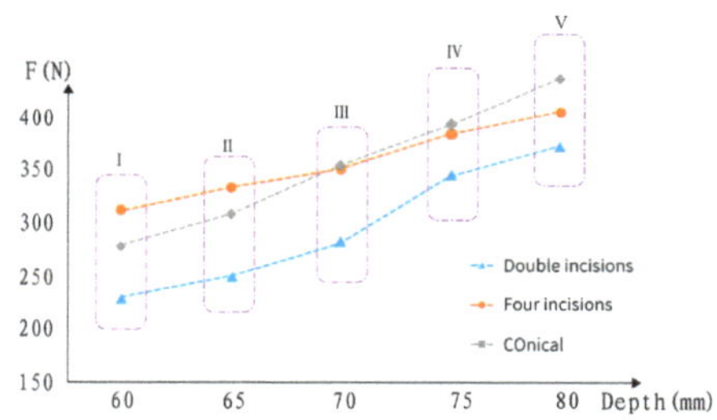

(b)

Figure 11. Effects of different factors on the planting resistance of duckbill planting end effector: (**a**) effect of single-factor planting speed on the resistance of duckbill planting end effector; and (**b**) effect of single-factor planting depth on the resistance of duckbill planting end effector.

For three different duckbill shapes, at different levels of soil penetration speed, the changes in soil penetration resistance of eight duckbills at the same time are shown in Figure 11. The soil penetration resistance increases slightly with the increase in soil penetration speed. The reason for that is that deep soil has consistent soil fluidity. When the speed of the duckbill entering the soil is greater, the time it stays in the soil is shorter. The shorter the soil flow time, the greater the interference effect of the duckbill on the soil. Therefore, the reaction of the soil on the duckbill is more powerful. For the three types of duckbill shapes, under the same experimental factors, five comparative experiments on soil penetration resistance were conducted. The results showed that in the range of 300–500 mm/s, the double-cut duckbill had smaller soil penetration resistance. However, in order to verify whether the soil penetration speed factor significantly affects the soil penetration resistance and whether the conclusion that the double-notch soil penetration resistance is smaller is correct, a double-interaction-factor response surface test must be conducted for verification.

For three different duckbill shapes and at different levels of soil penetration depth, the simultaneous soil penetration resistance of eight duckbill-type planting end effectors is shown in Figure 11b. The soil penetration resistance increases with the soil penetration depth, and the trend is a slight increase. The reason for that is that when the soil penetration depth is greater, the contact area between the duckbill surface and the soil is larger, and the friction force formed is greater, which hinders the duckbill penetration into the soil. For the three duckbill shapes, under the same experimental factors, five comparative experiments on soil penetration resistance were conducted. The results showed that in the range of 60–80 mm, the double-cut duckbill-type planting end effector had smaller soil penetration resistance. In order to verify whether the soil penetration depth test factors significantly affect the soil penetration resistance and whether the conclusion that the double-notch soil penetration resistance is smaller is correct, a double-interaction-factor response surface test must be conducted for verification.

(5) Interactive Factor Response Surface Analysis

After setting up the Design-Expert software, 39 groups of tests were required. RecurDyn-EDEM coupling tests were performed based on the test parameters, and the statistics are shown in Table 5.

Table 5. Effects of speed and depth on different duckbill planting end effectors.

Test Group	Depth (mm)	Speed (mm/s)	Shape	Resistance (N)
1	60	400	Double incisions	229.5
2	60	500	Double incisions	268.3
3	70	400	Double incisions	302.8
4	60	300	Conical	275.6
5	60	400	Four incisions	320.5
6	70	400	Double incisions	303.5
7	70	400	Four incisions	356.8
8	70	400	Double incisions	309.5
9	70	400	Conical	342.5
10	70	400	Four incisions	325.6
11	80	400	Double incisions	416.5
12	70	400	Conical	339.8
13	80	500	Conical	469.8
14	70	300	Double incisions	290.6
15	60	500	Four incisions	344.3
16	70	300	Four incisions	319.8
17	70	400	Double incisions	312.5
18	60	300	Four incisions	308.6
19	80	500	Double incisions	453.8
20	80	400	Conical	353.6
21	80	500	Four incisions	423.3
22	60	500	Conical	338.6
23	70	500	Four incisions	379.8
24	60	400	Conical	298.6
25	70	300	Conical	320.4
26	80	300	Four incisions	424.1
27	70	500	Double incisions	318.6
28	80	400	Four incisions	454.1
29	70	400	Conical	341.8
30	70	400	Four incisions	340.4
31	70	500	Conical	382.7
32	70	400	Double incisions	330.5
33	70	400	Four incisions	356.8
34	60	300	Double incisions	205.6
35	70	400	Conical	345.6
36	70	400	Four incisions	338.7
37	70	400	Conical	342.6
38	80	300	Conical	421.3
39	80	300	Double incisions	376.8

The Design-Expert software was used to perform multiple regression fitting on the above data to observe the effectiveness of the penetration depth and penetration speed parameters. It can be seen in Table 6 that the p value of the multiple regression fitting variance in the duckbill soil penetration resistance is less than 0.0001, which proves that the effect of the model is significant. The p values of duckbill soil penetration depth, duckbill planting velocity, and duckbill shape are all less than 0.0001, which proves that two experimental factors and one clear factor can have a greater impact on soil penetration resistance. It can be verified that the two factors have a significant effect in the single-factor test.

Table 6. Variance analysis of soil penetration resistance of different duckbill planting end effectors.

Parameter	Sum of Squares	Degrees of Freedom	Mean Square	F	p	
Model	1.19×10^5	17	6984.75	21.93	<0.0001	significant
A-Depth	80,423.22	1	80,423.22	252.46	<0.0001	
B-Velocity	10,582.7	1	10,582.7	33.22	<0.0001	
C-shape	14,145.9	2	7072.95	22.2	<0.0001	
AB	112.12	1	112.12	0.35	0.5593	
AC	5052.37	2	2526.18	7.93	0.0027	
BC	642.23	2	321.12	1.01	0.3819	
A2	2718.67	1	2718.67	8.53	0.0082	
B2	537.02	1	537.02	1.69	0.2082	
ABC	324.26	2	162.13	0.51	0.6084	
A2C	1367.6	2	683.8	2.15	0.1418	
B2C	1807.85	2	903.93	2.84	0.0811	
Residual	6689.83	21	318.56			
Lost fitting term	5460.42	9	606.71	5.92	0.0029	significant
Pure deviation	1229.4	12	102.45			
Sum	1.25×10^5	38				

From the response surface diagrams of the planting resistance value changes in duckbill planting end effectors with different shapes under the action of interactive factors in Figure 12a–c, it can be seen that the planting resistance of the duckbill planting end effector with a double-cut shape is significantly smaller than that of the duckbill planting end effector with a four-cut and conical shape. As shown in Figure 13, a comparative analysis shows that the overall distribution point of the duckbill planting end effector with a double-cut shape is smaller than that of the duckbill planting end effector with a four-cut and conical shape. Through single-factor experiments and two-factor interactive response surface experiments, it can be seen that the planting resistance of the duckbill planting end effector with a double-cut shape is the smallest, so the duckbill planting end effector with a double-cut shape is used.

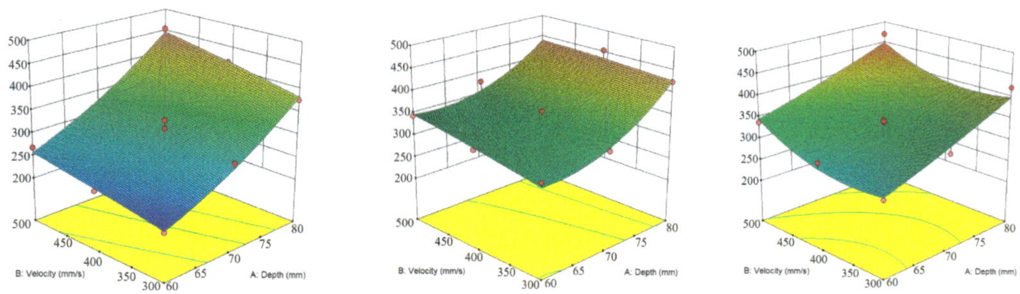

Figure 12. Response surface diagram of the planting resistance changes in duckbill planting end effectors with different shapes: (**a**) response surface diagram of planting resistance changes in duckbill planting end effector with double incisions; (**b**) response surface diagram of planting resistance changes in duckbill planting end effector with four incisions; and (**c**) response surface diagram of planting resistance changes in duckbill planting end effector with conical shape.

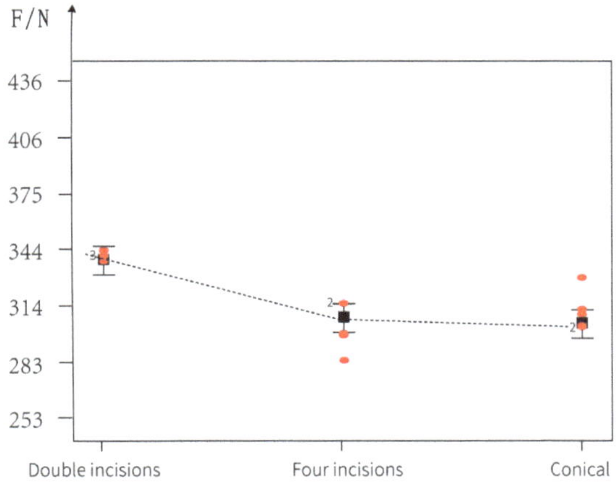

Figure 13. Response surface diagram of the planting resistance changes in duckbill planting end effectors with different shapes.

4. Resistance Verification Test of Duckbill Planting End Effector

4.1. Test Design

The test device consisted of a soil trough, a YZC131 weighing sensor module, an HX711 module, and an STM32 microcontroller. The weighing sensor module was installed at the bottom of the soil trough. The sensor formed a full-bridge circuit through multiple strain gauges. The change in mass in the soil trough caused the strain gauge to deform, resulting in a change in its resistance value to form an analog signal. The HX711 module converted the analog signal into a digital signal and sent it to the STM32 microcontroller. Through the serial port debugging assistant, the value of the soil resistance could be sent to a computer. Through the KEIL 5 programming software in the laptop used, the program was written in C language, and the program was burned into the single-chip computer to detect the soil resistance. For the duckbill planting end effector with a double-cut structure, experiments were carried out based on different planting speeds and different planting depths, and the actual measured values were compared and analyzed to see if they were consistent with the simulation data. In order to simulate a more realistic working environment, the test bench was built on a wheeled frame to complete the soil trough verification test, as shown in Figure 14.

Figure 14. Resistance test of duckbill planting end effector.

4.2. Analysis of Test Results

The test was repeated three times for different planting speeds and planting depths to obtain the maximum value of the planting resistance in the dynamic process, as shown in Tables 7 and 8.

Table 7. Test parameters for different planting speeds.

Test Group		Planting Speed (mm/s)				
		300	350	400	450	500
Resistance (N)	1	265.9	265.8	286.7	301.6	318.9
	2	272.2	263.9	283.6	286.9	320.8
	3	270.9	253.6	290.8	291.6	334.8
Average value		259.5	261.4	275.9	285.7	303.6

Table 8. Test parameters for different transplanting depths.

Test Group		Planting Depth (mm)				
		60	65	70	75	80
Resistance (N)	1	238.6	253.1	272.4	308.9	356.8
	2	240.8	258.9	269.8	315.4	349.1
	3	236.1	253.4	271.8	305.6	360.2
Average value		259.5	238.5	255.1	271.3	309.9

Tables 7 and 8 are plotted and compared with the simulation data, as shown in Figures 15 and 16. By comparison, it is found that the data obtained from the actual test and the data obtained from the simulation test have the same basic trend. The maximum resistance difference at a planting speed of 500 mm/s is 21.2 N, which is 6.98% different from the simulation value, and the other error values are all below 5%. When the planting depth is 75 mm, the actual planting resistance value differs from the simulation value by 29.4 N, which is 8.67% different from the simulation value, and the other errors are all below 5%. The reason for the error is that during the soil transportation process, due to problems such as transportation, the physical properties of the soil after rotary tillage change, resulting in errors between the actual planting resistance and the simulation value. However, the planting resistance trends in the simulation test and the bench test are the same, and the data are basically within the allowable error range, thus proving that the duckbill planting model is correct.

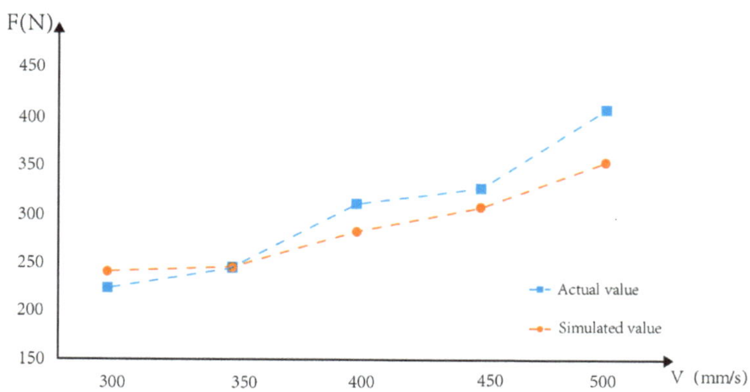

Figure 15. The comparison of planting resistance at different planting speeds.

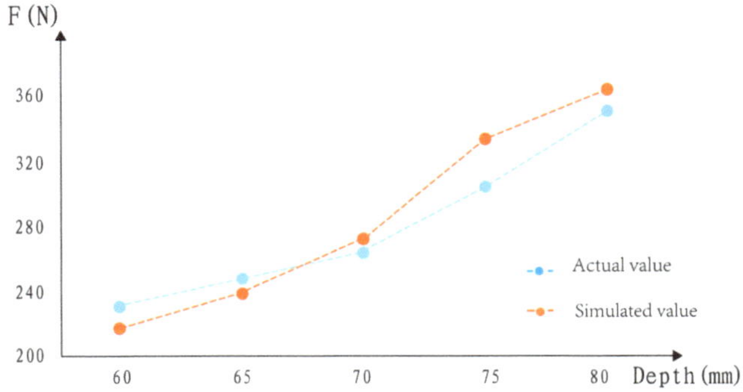

Figure 16. The comparison of planting resistance at different planting depths.

5. Planting Performance Test

5.1. Test Conditions

The test site was the Jiangxinzhou Agricultural Demonstration Base in Zhenjiang City. Before transplanting, a rotary tillage and ridging machine was used to loosen and ridge the soil. The ridge width was set to 1100 mm and the ridge height was 120 mm. The test machine is shown in Figure 17. This test used 128-hole pak choi seedlings cultivated in a greenhouse as the test object. The seedling age was about 40 days, the substrate compaction degree was 1.0, and the pot body moisture content ranged from 55% to 60%, as shown in Figure 18.

Figure 17. Automatic transplanter for densely planted vegetables.

Figure 18. 128 holes for pak choi seedlings.

5.2. Test Design and Evaluation Index

(1) Test Design

The parameters, such as the plant spacing of the densely planted vegetable automatic transplanter, were set. The row spacing of the multi-row densely planted transplanter was 120 mm, and the plant spacing was adjusted between 100 and 250 mm. The smaller the plant spacing, the higher the control difficulty and action execution accuracy of the transplanter. Therefore, a plant spacing of 100 mm was selected for the test. In order to verify that the densely planted transplanter could achieve stable and efficient results for vegetables with different planting depth requirements, three common planting depths of 60 mm, 70 mm, and 80 mm were selected. The machine operated at a forward speed of 1.4 km/h. After completing the continuous planting operation, the results were statistically analyzed according to the planting indices.

(2) Evaluation Index

This study is about an automatic transplanter for densely planted vegetables. The goal is to achieve efficient planting with small plant spacing and small row spacing. Therefore, the planting success rate, plant spacing variation coefficient, and planting efficiency were selected as evaluation indicators.

I. Planting success rate

The planting success rate is the ratio of the number of qualified plantings to the total number of plants tested. The formula is as follows:

$$R = \frac{Z - M - E - O - R}{Z} \tag{13}$$

where Z is the total number of test plants, plant; M is the number of exposed seedlings, plant; E is the number of missed seedlings, plant; O is the number of fallen seedlings, plant; and R is the number of replanted seedlings, plant.

II. Plant spacing variation coefficient

The plant spacing of the automatic transplanter for densely planted vegetables was set according to the specific vegetable variety. However, due to the complexity of field work or the influence of the working state of the transplanter, there may be plant spacing errors during planting, which is called the plant spacing variation coefficient. The calculation formula is as follows:

$$X = \frac{\sum_{i=1}^{n} X_i}{n} \tag{14}$$

$$S_x = \sqrt{\frac{1}{n-1}\sum_{i=1}^{n}(X_i - \overline{X})^2} \tag{15}$$

$$C_v = \frac{S_x}{X} \cdot 100\% \tag{16}$$

where $\overline{X}$ is the average plant spacing, mm; S_X is the standard deviation of plant spacing, mm; and C_V is the coefficient of variation of plant spacing, %.

III. Planting efficiency

Densely planted vegetables have a high planting density and must be planted efficiently. In order to verify whether the multi-row dense planting transplanter can achieve a planting efficiency of 6000~8000 plants/h, a field test was required. The calculation formula is as follows:

$$F = \frac{Z}{t} \times 3600 \tag{17}$$

where F is the planting frequency, plant/hour; Z is the number of seedlings planted in a single test, grain; t is the time of a single test, second.

5.3. Test Results and Analysis

The test planting effect is shown in Figure 19. The test results are shown in Tables 9–11.

Figure 19. Planting mechanism performance test: (**a**) test effect; and (**b**) measurement of plant spacing.

Table 9. Planting success rate.

Planting Depth (mm)	Group	Number of Plantings (Plants)	Successful Plantings (Plants)	Missed Plantings (Plants)	Exposed Seedlings (Plants)	Lodging (Plants)	Replanting (Plants)	Planting Success Rate (%)	Average Planting Success Rate (%)
60 mm	1	640	621	5	6	5	3	97.03%	96.09%
	2	640	615	6	9	3	7	96.09%	
	3	640	609	9	5	10	7	95.15%	
70 mm	4	640	628	3	4	2	3	98.13%	96.81%
	5	640	614	8	10	5	3	95.90%	
	6	640	617	3	8	7	5	96.41%	
80 mm	7	640	619	6	7	3	3	96.72%	96.97%
	8	640	627	2	4	3	4	97.97%	
	9	640	610	7	6	8	9	95.31%	

Table 10. Results of plant spacing variation coefficient.

Planting Depth (mm)	Group	Number of Plantings (Plants)	Average Plant Spacing (mm)	Standard Deviation of Plant Spacing (mm)	Coefficient of Variation of Plant Spacing (%)
60 mm	1	640	101.4	3.42	3.46
	2	640	102.4	4.68	
	3	640	96.4	2.35	
70 mm	4	640	102.5	4.21	3.61
	5	640	97.3	4.11	
	6	640	96.1	2.40	
80 mm	7	640	98.4	3.53	3.59
	8	640	102.1	4.01	
	9	640	100.9	3.27	

Table 11. Field time of test.

Group	1	2	3	4	5	6	7	8	9	Average
Time (s)	321	346	332	314	308	332	310	328	315	322.9

Aiming to optimize the planting success rate and plant spacing variation coefficient, planting performance tests were conducted at planting depths of 60 mm, 70 mm, and 80 mm.

The results showed that when the planting depth was 60 mm, the planting qualification rate was 96.09% and the plant spacing variation coefficient was 3.46%; when the planting depth was 70 mm, the planting qualification rate was 96.81% and the plant spacing variation coefficient was 3.61%; when the planting depth was 80 mm, the planting qualification rate was 96.97% and the plant spacing variation coefficient was 3.59%; the average planting success rate was 96.62% and the average plant spacing variation coefficient was 3.55%, both of which meet the relevant standards. Regarding the planting efficiency, nine field tests were conducted, each of which used five trays of 128-hole seedling pots. The test started from the movement of the planting mechanism and ended when the last row of seedling pots was planted. On average, 640 seedlings were planted in each test, and the time taken was 322.9 s. The planting efficiency was 7135 plants/h, which meets the dense planting and transplanting design standard of 6000–8000 plants per hour.

During the test, it was found that some seedlings had fewer roots, and the matrix did not form a soil pot shape, resulting in poor rooting of the seedlings. Some seedlings could not be completely removed by the seedling removal mechanism, and part of the matrix had been scattered before the seedlings fell into the duckbill planting end effector, resulting in missing seedlings or lodging of the seedlings during planting, leading to planting failure Because the height and expansion of the seedlings were not exactly the same, the duckbill planting end effector often took the seedlings out when closing, resulting in planting failure at the same duckbill planting end effector closing height. After the duckbill planting end effector was opened, the seedling pot fell into the pit. Due to the poor soil covering effect of the soil covering and suppression device, the seedlings tilted or even fell, resulting in planting failure. The above problems will be solved and avoided in subsequent research.

Our proposed solution is to strictly and accurately control the moisture content, substrate ratio, compaction degree, and other factors of the seedlings in the initial seedling raising stage to ensure that the seedling substrate is formed and has good packing properties, thereby improving the qualified planting rate. By modifying the closing height of the duckbill planting end effector, extending the opening time of the duckbill planting end effector, and reducing the phenomenon of the seedlings being taken out when the duckbill planting end effector is closed, the qualified planting rate can be improved. We should also aim to optimize and improve the soil covering and suppression device to improve the reflowability of the soil after planting, reduce the phenomenon of seedlings lodging in the pot, and improve the qualified planting rate.

6. Conclusions

(1) Based on the agronomic requirements of densely planted vegetables with multiple rows, small plant spacing, and small row spacing, an eight-row planting mechanism driven by a motor and a cylinder coupling was designed to overcome the problem of large plant spacing and row spacing in existing mechanized planting structures. The feasibility of the planting mechanism was verified by establishing a kinematic model.

(2) The RecurDyn 2024-EDEM 2022 coupling simulation method was used to compare and analyze the planting resistance of three duckbill planting end effectors with different external structures based on the two experimental factors of planting speed and planting depth. The external structure with the smallest planting resistance was selected. Design-Expert software was used to analyze the results. Single-factor analysis and two-factor interactive analysis were performed. It was concluded that the duckbill planting end effector with a double-cut structure had the smallest planting resistance.

(3) The planting resistance verification system of the duckbill planting end effector was used to analyze the changes in planting resistance at different planting speeds and planting depths. The correctness of the simulation model was verified by comparing the actual planting resistance value with the simulation test value. A performance test of the planting mechanism was used to verify the coordination of the key components and the planting effect. The test results showed that the planting qualification rate of the prototype reached 96.62%, the various components had good coordination, and

the planting spacing variation coefficient was only 3.55%. The planting efficiency reached about 7135 plants/h, which meets the agronomic requirements of dense vegetable transplanting.

(4) Next, the kinematic model needs to be further refined and optimized, considering more variables in practical applications such as soil moisture, terrain changes, etc., to improve the adaptability and stability of the planting mechanism; developing an intelligent control system to automatically adjust the planting parameters will improve work efficiency and adaptability. This study solves the problem that existing planting mechanisms cannot meet the agronomic requirements of densely planted vegetables, reduces labor costs, improves production efficiency, and promotes the development of agricultural mechanization in the direction of efficiency, precision, and intelligence.

Author Contributions: Conceptualization, J.S. and J.H.; methodology, J.S., J.H. and J.L.; software, J.S. and J.L.; validation, J.S. and J.H.; formal analysis, J.S., J.H. and M.Y.; investigation, J.S. and T.Z.; resources, J.H.; data curation, R.Y. and W.L.; writing—original draft preparation, J.S.; writing—review and editing, J.S., J.H. and J.L. All authors have read and agreed to the published version of the manuscript.

Funding: This work was supported by the Shanghai Green Leafy Vegetable Industry Technology System Construction—Development and Application of Key Technologies for high-density green Leafy Vegetable transplanting [Shanghai Nongke (2023) No. 2], the Jiangsu Province Agricultural Science and Technology Independent Innovation Fund Project (CX (22)2022), and the Jiangsu Province Key Research and Development Plan—Modern Agriculture Project (BE2021342). A Project Funded by the Priority Academic Program Development of Jiangsu Higher Education Institutions (No. PAPD-2023-87).

Institutional Review Board Statement: Not applicable.

Data Availability Statement: Data are contained within the article.

Conflicts of Interest: The authors declare no conflicts of interest.

References

1. Periasamy, V.; Gouncer, D.V.M.; Ramasamy, K. Factors influencing the performance of mechanical end effector during automatic transplanting of tomato seedlings. *J. Appl. Nat. Sci.* **2022**, *14*, 227–231. [CrossRef]
2. Wu, G.; Wang, S.; Zhang, A.; Xiao, Y.; Li, L.; Yin, Y.; Li, H.; Wen, C.; Yan, B. Optimized design and experiment of a self-covering furrow opener for an automatic sweet potato seedling transplanting machine. *Sustainability* **2023**, *15*, 13091. [CrossRef]
3. Sharma, A.; Khar, S. Design and development of a vegetable plug seedling transplanting mechanism for a semi-automatic transplanter. *Sci. Hortic.* **2024**, *326*, 112773. [CrossRef]
4. Ji, D.; Liu, L.; Zeng, E.; Zhang, G.; Liu, Y.; Diao, H.; Tian, S.; Zhao, Z. Design and Experimental Study of a Traction Double-Row Automatic Transplanter for *Solanum lycopersicum* Seedlings. *Horticulturae* **2024**, *10*, 692. [CrossRef]
5. Ji, D.; Tian, S.; Wu, H.; Zhao, B.; Gong, Y.; Ma, J.; Zhou, M.; Liu, W. Design and experimental verification of an automatic transplant device for a self-propelled flower transplanter. *J. Braz. Soc. Mech. Sci. Eng.* **2023**, *45*, 420. [CrossRef]
6. Markumningsih, S.; Hwang, S.-J.; Kim, J.-H.; Jang, M.-K.; Nam, J.-S. Stress Simulation on Cam-Type Transplanting Device of Semiautomatic Vegetable Transplanter. *Agriculture* **2023**, *13*, 2230. [CrossRef]
7. Miah, M.S.; Rahman, M.M.; Hoque, M.A.; Ibrahim, S.M.; Sultan, M.; Shamshiri, R.R.; Ucgul, M.; Hasan, M.; Barna, T.N. Design and evaluation of a power tiller vegetable seedling transplanter with dibbler and furrow type. *Heliyon* **2023**, *9*, e17827. [CrossRef] [PubMed]
8. Liu, W.; Tian, S.; Wang, Q.; Jiang, H. Key technologies of plug tray seedling transplanters in protected agriculture: A review. *Agriculture* **2023**, *13*, 1488. [CrossRef]
9. Paudel, B.; Basak, J.K.; Jeon, S.W.; Lee, G.H.; Deb, N.C.; Karki, S.; Kim, H.T. Working speed optimisation of the fully automated vegetable seedling transplanter. *J. Agric. Eng.* **2024**, *55*, 1569. [CrossRef]
10. Sharma, A.; Khar, S.; Chaudhary, D.; Thakur, P. Study of biometric attributes of plug type tomato seedlings pertinent to transplanter design. *Indian J. Ecol.* **2023**, *50*, 503–507.
11. Habineza, E.; Ali, M.; Reza, M.N.; Woo, J.-K.; Chung, S.-O.; Hou, Y. Vegetable transplanters and kinematic analysis of major mechanisms: A review. *Korean J. Agric. Sci.* **2023**, *50*, 113–129. [CrossRef]

12. Elwakeel, A.E.; Mazrou, Y.S.; Eissa, A.S.; Okasha, A.M.; Elmetwalli, A.H.; Makhlouf, A.H.; Metwally, K.A.; Mahmoud, W.A.; Elsayed, S. Design and Validation of a Variable-Rate Control Metering Mechanism and Smart Monitoring System for a High-Precision Sugarcane Transplanter. *Agriculture* **2023**, *13*, 2218. [CrossRef]
13. Li, M.; Xiao, L.; Ma, X.; Yang, F.; Jin, X.; Ji, J. Vision-based a seedling selective planting control system for vegetable transplanter. *Agriculture* **2022**, *12*, 2064. [CrossRef]
14. Stubbs, S.; Colton, J. The Design of a Mechanized Onion Transplanter for Bangladesh with Functional Testing. *Agriculture* **2022**, *12*, 1790. [CrossRef]
15. Chen, K.; Yuan, Y.; Zhao, B.; Jin, X.; Lin, Y.; Zheng, Y. Finite element modal analysis and experiment of rice transplanter chassis. *Int. J. Agric. Biol. Eng.* **2022**, *15*, 91–100. [CrossRef]
16. Paradkar, V.; Raheman, H.; Rahul, K. Development of a metering mechanism with serial robotic arm for handling paper pot seedlings in a vegetable transplanter. *Artif. Intell. Agric.* **2021**, *5*, 52–63. [CrossRef]
17. Li, P.; Yun, Z.; Gao, K.; Si, L.; Du, X. Design and Test of a Force Feedback Seedling Pick-Up Gripper for an Automatic Transplanter. *Agriculture* **2022**, *12*, 1889. [CrossRef]
18. Yang, Q.; Zhang, R.; Jia, C.; Li, Z.; Zhu, M.; Addy, M. Study of dynamic hole-forming performance of a cup-hanging planter on a high-speed seedling transplanter. *Front. Mech. Eng.* **2022**, *8*, 896881. [CrossRef]
19. Zhang, M.; Jiang, Z.; Jiang, L.; Wu, C.; Yang, Y. Design and Test of the Seedling Cavitation and Lodging Monitoring System for the Rape Blanket Seedling Transplanter. *Agriculture* **2022**, *12*, 1397. [CrossRef]
20. Khadatkar, A.; Mathur, S. Design and development of an automatic vegetable transplanter using a novel rotating finger device with push-type mechanism for plug seedlings. *Int. J. Veg. Sci.* **2022**, *28*, 121–131. [CrossRef]
21. Yang, J.; Zhou, M.; Yin, D.; Yin, J. Design and Development of Rice Pot-Seedling Transplanting Machinery Based on a Non-Circular Gear Mechanism. *Appl. Sci.* **2024**, *14*, 1027. [CrossRef]
22. Zhao, X.; Hou, Z.; Zhang, J.; Yu, H.; Hao, J.; Liu, Y. Study on the Hole-Forming Performance and Opening of Mulching Film for a Dibble-Type Transplanting Device. *Agriculture* **2024**, *14*, 494. [CrossRef]
23. Haili, Z.; Wei, Y.; Gaohong, Y.; Bin, W.; Bingliang, Y. Optimization Design and Experiments of Ditching Multi-bar Seedling Planting Mechanism. *Trans. Chin. Soc. Agric. Mach.* **2023**, *54*, 79–86.
24. Lei, W.; Liang, S.; Yadan, X.; Gaohong, Y.; Gervais, N.L.; Jiahui, H. Multi-pose Motion Synthesis of Three-arm Gear Train Planting Mechanism Based on Genetic Algorithm. *Trans. Chin. Soc. Agric. Mach.* **2022**, *53*, 70–77.
25. Gaohong, Y.; Zhenpiao, L.; Lehui, X.; Peng, Z.; Chuanyu, W. Optimization design and test of large spacing planetary gear train for vegetable pot-seedling planting mechanism. *Trans. Chin. Soc. Agric. Mach.* **2015**, *46*, 38–44.
26. Gaohong, Y.; Ye, J.; Shushu, C.; Bingliang, Y.; Jinbo, G.; Xiong, Z. Design and test of clipping-plug type transplanting mechanism of rice plug-seedling. *Trans. Chin. Soc. Agric. Mach.* **2019**, *50*, 100–108.
27. Hu, S.; Hu, M.; Yan, W.; Zhang, W. Design and experiment of an integrated automatic transplanting mechanism for picking and planting pepper hole tray seedlings. *Agriculture* **2022**, *12*, 557. [CrossRef]
28. Shi, L.; Zhao, W.; Sun, W. Parameter calibration of soil particles contact model of farmland soil in northwest arid region based on discrete element method. *Trans. Chin. Soc. Agric. Eng.* **2017**, *33*, 181–187.
29. Daqing, Y.; Nuoyi, Z.; Maile, Z.; Yuchao, Y.; Size, Y.; Jinwu, W. Optimal Design and Experiment of High Speed Duckbill Planting Mechanism with Variable Catch-seedling Attitude. *Trans. Chin. Soc. Agric. Mach.* **2020**, *51*, 66–72.
30. Ji, J.; Yang, L.; Jin, X.; Gao, S.; Pang, J.; Wang, J. Design and parameter optimization of planetary gear-train slip type pot seedling planting mechanism. *Trans. CSAE* **2018**, *34*, 83–92.
31. Long, Z.; Bo, Y.; Xin, J. Structure Design of Duck Mouth Transplanting End Actuator Based on TRIZ Theory. *Agric. Eng.* **2020**, *10*, 76–80.

agriculture

Article

Design and Preliminary Experiment of Track Width Adjustment System for Sprayer Based on Integral Separated Fuzzy Proportional Integral Derivative Control Strategy

Lu Tan [1,2], Mei Jia [1,2], Jin He [1,2,*], Xintong Su [1,2], Quanyu Wang [1,2], Hanyu Yang [1,2] and Hang Li [1,2]

[1] College of Engineering, China Agricultural University, Beijing 100083, China; tanluctrc@cau.edu.cn (L.T.); jiamei@cau.edu.cn (M.J.); s20233071554@cau.edu.cn (X.S.); wangquanyu@cau.edu.cn (Q.W.); yanghanyu1102@foxmail.com (H.Y.); hli@cau.edu.cn (H.L.)

[2] Key Laboratory of Agricultural Equipment for Conservation Tillage, Ministry of Agricultural and Rural Affairs, Beijing 100083, China

* Correspondence: hejin@cau.edu.cn; Tel.: +86-010-6273-7300

Abstract: Different agronomic requirements, production conditions, and crop species result in varying row spacings. To address the issue of seedling damage caused by pressure when a fixed track width sprayer operates in different row spacings and enhance the accuracy of track width adjustment, this study designed a track width adjustment system for a sprayer based on the agronomic requirements for field management during the early and mid-stages of corn growth and the entire growth period of wheat in Henan Province, China. The designed track width adjustment system for the sprayer comprised transmission mechanisms, telescopic track width adjustment mechanisms, and an electro-hydraulic control system. The control system achieved a precise track width adjustment by controlling the movement of the hydraulic cylinders through electrical signals, forming a closed-loop adjustment system with the aid of sensors. Four control schemes are proposed: classical PID, integral separated PID, fuzzy adaptive PID, and integral separated fuzzy PID. Simulation experiments were conducted using MATLAB to compare these schemes. The results indicated that the integral separated fuzzy PID exhibited the fastest response and highest steady-state accuracy. The performance of the track width adjustment system was validated through field experiments. The results demonstrate that the stability coefficient of variation for the track width adjustment was 3.04%, which is below the 10% threshold required by agricultural machinery standards. Additionally, the average error of the track width adjustment was 13.42 mm, indicating high precision and effectively reducing seedling compression damage during plant protection operations.

Keywords: electro-hydraulic control; fuzzy adaptive rule; integral separated PID; sprayer; track width adjustment

Citation: Tan, L.; Jia, M.; He, J.; Su, X.; Wang, Q.; Yang, H.; Li, H. Design and Preliminary Experiment of Track Width Adjustment System for Sprayer Based on Integral Separated Fuzzy Proportional Integral Derivative Control Strategy. *Agriculture* **2024**, *14*, 1247. https://doi.org/10.3390/agriculture14081247

Academic Editor: Mustafa Ucgul

Received: 22 June 2024
Revised: 24 July 2024
Accepted: 26 July 2024
Published: 28 July 2024

1. Introduction

As China's major grain crops, the national total production of wheat and corn in 2023 reached 136.59 million tons and 288.84 million tons, respectively, accounting for 61.18% of the country's total grain production. Consequently, enhancing the production capacity of wheat and corn is crucial for ensuring China's food security [1,2]. Plant protection technology, which mitigates the threats posed by diseases, pests, and weeds to crops, creates favorable growing and developing conditions for crops, thus helping to ensure crop quality and increase yields [3]. Therefore, the adoption of efficient and comprehensive plant protection technologies during the growth of wheat and corn is an inevitable choice for modern agricultural production.

Boom sprayers are widely used in field plant protection operations with their advantages of high spraying quality and operational efficiency. In recent years, China has placed increasing emphasis on the research and promotion of plant protection machinery, conducting extensive studies on boom sprayers. By incorporating electronic control technology and hydraulic technology and equipping them with GPS or Beidou Navigation Satellite Systems, these sprayers are continually evolving toward being more intelligent, efficient and environmentally friendly [4–6].

China's diverse ecological environment and complex terrain result in varying agronomic requirements and field operation needs for crop planting methods across different regions [7]. This diversity leads to fixed track width sprayers crushing crops during operations in fields with different row spacings, causing crop damage and reducing yields Therefore, designing an intelligent track width adjustment system for the chassis of the sprayer, which accommodates various planting row spacings, is of great significance for enhancing the sprayer's adaptability to different environments, promoting increased crop yields, and reducing production costs.

Track width adjustment devices are widely used in agricultural machinery such as agricultural chassis, rear wheels of wheeled tractors, and high-clearance vehicles with four-wheel diamond arrangements. They can be categorized into stepwise adjustment and stepless adjustment, depending on whether the track width can be adjusted to any position within the adjustable range [8]. Currently, the stepwise adjustment method for track width mainly relies on manual adjustments, typically implemented through mechanisms such as wheel hub cap clamping, rim flipping, and hub spacers. For instance, the Dongfanghong LX750H, manufactured by China YTO Group Corporation in Luoyang, China, achieves track width adjustment by replacing the front and rear axles. The Case New Holland (China) Management Co., Ltd. located in Shanghai produces the TS6.120 high-clearance tractor, which adjusts its track width by flipping the tires. The Case IH Patriot™ 3230 sprayer, also manufactured by the company, adapts to different row spacings through optional configurations, with an adjustment range of 3050 to 3990 mm and an adjustment increment of 25 mm. Its track width position stop must be manually determined, and the sliding bar is marked with raised text indicating the track width [9]. While stepwise adjustment is reliable and widely used, it is time-consuming and labor-intensive. At the same time, it increases the cost of some sprayers and lacks precision.

Currently, both domestically and internationally, electro-hydraulic control technology or fully electronic control methods are predominantly used for stepless adjustment of the track width in agricultural machinery. This is achieved by driving the corresponding mechanical structures through hydraulic cylinders or electric actuators to implement track width adjustment functions. The primary mechanical structures used include linkage mechanisms, telescopic adjustment sleeves, sliding bearings, and gear racks. Among these, linkage mechanisms and telescopic adjustment sliding sleeve mechanisms are widely used. Zhang, Q. [10] employed a hydraulic mechanism to drive a parallel four-bar linkage, thereby adjusting the track width and ground clearance. This design ensures high stability for the sprayer under hilly and mountainous working conditions. Liu et al. [11] designed a self-propelled adjustable sprayer that uses eight connecting rods to hinge the walking main beam and the frame together, with each hydraulic cylinder independently controlling a single walking main beam, allowing for single-side track width adjustments. Although linkage adjustment of the track width is compact and enables stepless adjustment, it is structurally complex and difficult to manufacture.

Compared to linkage mechanisms, a telescopic adjustment sliding sleeve structure is simpler and it is easier to achieve any desired track width adjustment [12]. Li, J. [13] utilized the telescopic motion of hydraulic cylinders to directly drive a sprayer wheel with leg mounting brackets to move laterally along guide rails. A Watt balance linkage ensures that the movement distance of the left and right wheel leg mounting brackets remains equal. The structure is compact, but it lacks a design for the software control component. Xia et al. [14] designed a track width adjustment device for the rear wheels of wheeled tractors, which connects the rim and hydraulic adjustment system through the inner and outer rings of a sliding bearing. The hydraulic cylinder piston is moved by a directional valve, and the sliding bearing drives the rim to move left and right. Chipriana et al. [15] designed and implemented a high-clearance sprayer where the movable support wheel frames on both sides are mounted on the framework through U-shaped arms, and the track width is adjusted by moving the two support wheel frames along sliding rails with hydraulic cylinders. The high-clearance sprayer designed by Italy's SPAPPERI for tobacco is equipped with real-time monitoring technology, allowing for real-time track width adjustments from the cab [16]. Zhang et al. [17] studied a hydraulically adjustable chassis with machine vision, where a microcontroller plans the optimal driving route based on images of the drive wheel's front view captured by a camera. It then controls digital hydraulic cylinders to move the fixed frame along guide rods, with the drive wheel and motor moving along with the fixed frame. This chassis enables independent left and right track width adjustments and real-time automatic track width adjustments based on ground conditions.

In the aforementioned study, although the use of a fully electronic control for track width adjustment offers higher precision and eliminates hydraulic impact, its load-bearing capacity is relatively small, making it unsuitable for large agricultural machinery. Therefore, the current practical application of stepless track width adjustment mainly relies on electro-hydraulic control technology. The electro-hydraulic control technology for track width adjustment in sprayers typically uses open-loop control, which has the issue of excessive hydraulic cylinder impact, leading to significant wear on tires, frames, and guide rails, and also lacks precision. Adding sensors to the electro-hydraulic control technology for closed-loop adjustment can improve the accuracy of track width adjustment.

Therefore, to avoid seedling damage caused by fixed track width sprayers when operating between different row spacings, and to improve the stability and precision of the sprayer's track width adjustment, this paper designed a track width adjustment system for the sprayer based on electro-hydraulic control and an integral separated fuzzy PID control strategy. The performance of the system was validated through field experiments.

2. Materials and Methods

2.1. Overall Structure of the Track Width Adjustment System

The 3WP-280 self-propelled boom sprayer is utilized as the platform for the track width adjustment system. As illustrated in Figure 1, the track width adjustment system is primarily comprised of a transmission mechanism, telescopic track width adjustment mechanism, and electro-hydraulic control system.

Figure 1. Overall structure of the track width adjustment system. 1—telescopic track width adjustment mechanism; 2—electro-hydraulic control system; 3—transmission mechanism.

The transmission mechanism transfers the engine's power to the gearbox, generator, and hydraulic pump, thereby providing a power source for the electro-hydraulic control system and the wheels. The telescopic track width adjustment mechanism connects the left and right support wheel frames to the central body chassis. Under the driving force of the hydraulic cylinder, the support wheel frames can slide along the sprayer's crossbeam. The electro-hydraulic control system detects the real-time adjustment distance of the track width and adjusts the lateral movement speed and distance of the movable tire support bracket on both sides in real-time based on the detected track width values.

2.2. *Design of Key Components of the Track Width Adjustment System*

2.2.1. Transmission Mechanism

The transmission mechanism includes the drive shaft assembly and the wheel-side drive components. As shown in Figure 2, in the drive shaft assembly, two double-row chain couplings sequentially connect the differential half-shaft to the splined driving shaft and the splined driven shaft sleeve to the sprocket box input shaft. The splined driven shaft sleeve rotates with the splined driving shaft while being able to slide left and right along the driving shaft, facilitating track width adjustment during the sprayer's operation. The power transmission path is sequentially as follows: differential left and right half-shafts, double-row chain coupling, splined driving shaft, splined driven shaft sleeve, double-row chain coupling, and sprocket box input shaft.

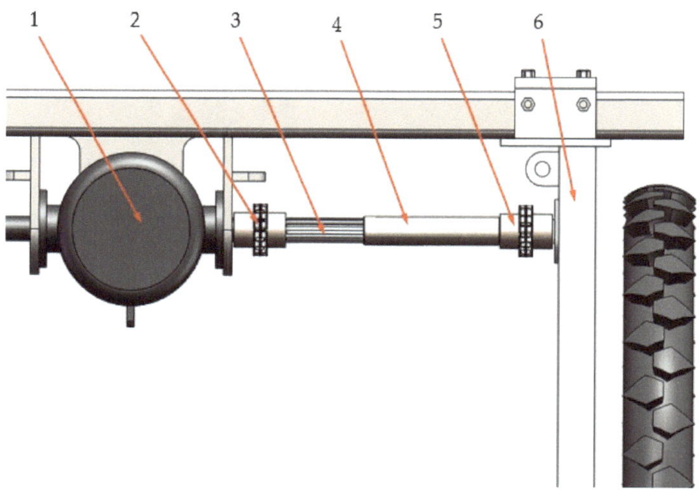

Figure 2. Schematic diagram of the drive shaft assembly. 1—differential; 2, 5—double-row chain coupling; 3—splined driving shaft; 4—splined driven shaft sleeve; 6—sprocket box input shaft.

As shown in Figure 3, the wheel side drive component mainly consists of a sprocket box, which provides power to the tires. Inside the sprocket box, the sprocket chain connects the power input shaft to the power output shaft. The power output shaft is then connected to the tire through a spline. The power transmission path for the wheel-side drive components is sequentially as follows: sprocket box input shaft, sprocket chain, sprocket box output shaft, and wheel.

Figure 3. Schematic diagram of the wheel-side drive components. 1—sprocket box output shaft; 2—sprocket chain; 3—sprocket box input shaft.

2.2.2. Telescopic Track Width Adjustment Mechanism

Table S1 of the supplementary material presents the row spacing sizes for corn and wheat under different agronomic requirements in Henan Province. Based on the values in the table and considering the need to enhance the maneuverability of the sprayer, this study designed a telescopic track width adjustment mechanism to meet the field management requirements for spanning 7 rows of wheat throughout its growth cycle and 2 rows of corn in the early to mid-growth stages in Henan Province. The adjustable track width parameters range from 1200 to 1600 mm, with a ground clearance of 900 mm.

To enhance the stability of the track width adjustment process and meet road travel requirements, the telescopic adjustment sleeve is employed for track width adjustment. The telescopic track width adjustment mechanism consists of a sprayer crossbeam, adjustment sleeve, sliding bearings, lock nuts, and gap adjustment bolts. Its structure is shown in Figure 4. The movable support wheel frames on both sides are fixedly connected below the adjustment sleeve. The adjustment sleeve and the sprayer crossbeam are connected via sliding bearings. When the adjustment sleeve slides along the crossbeam, the movable support wheel frames move laterally. Due to machining errors and the necessity of assembly clearance, the crossbeam and sliding bearings cannot be in perfect contact. To ensure uniform contact between the sliding bearings and the crossbeam and to reduce tire wobble during track width adjustment, the gap adjustment bolt and lock nut mechanism is used to eliminate the clearance.

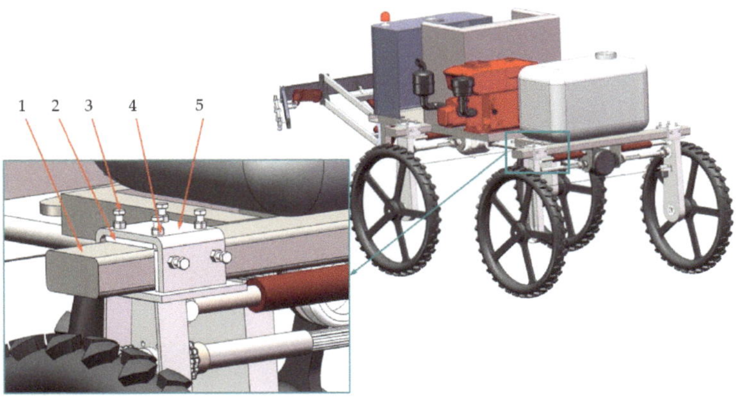

Figure 4. Telescopic track width adjustment mechanism. 1—sprayer crossbeam; 2—sliding bearings; 3—gap adjustment bolts; 4—lock nuts; 5—adjustment sleeve.

As shown in Figure 5, in the hydraulic system, one end of the hydraulic cylinder body is fixedly connected to the sprayer crossbeam, while the piston rod is connected to the movable support wheel frames. During track width adjustment, the electro-hydraulic control system controls the extension and retraction of the hydraulic cylinder, driving the movable support wheel frames on both sides to slide along the direction of the crossbeam, thereby achieving the adjustment of the track width.

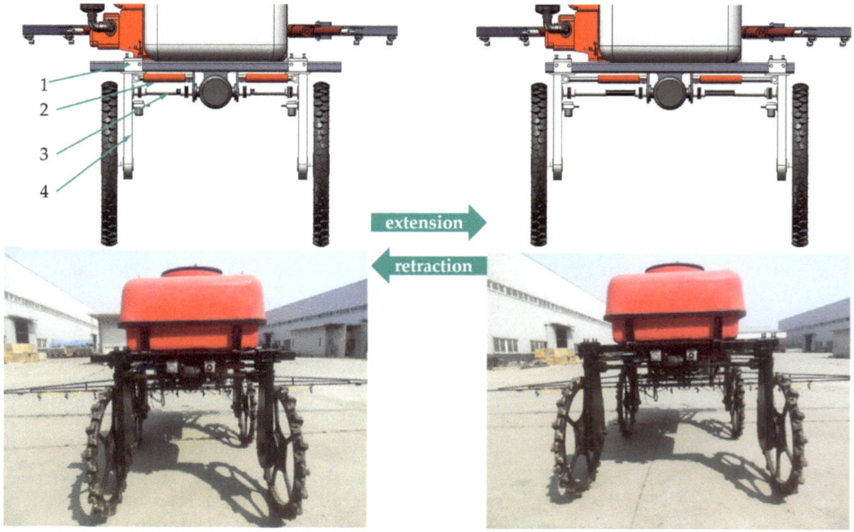

Figure 5. Track width adjustment process. 1—telescopic track width adjustment mechanism; 2—electro-hydraulic control system; 3—transmission mechanism; 4—the movable support wheel frame.

2.3. Electro-Hydraulic Control System

2.3.1. Overall Framework of the Electro-Hydraulic Control System

As shown in Figures 6 and 7, the electro-hydraulic control system comprises the human–machine interface, central processing module, detection devices, and hydraulic system. The human–machine interface, which consists of a display screen and matrix keyboard, is used to input the target track width adjustment value and transmit this signal to the central processing module. The central processing module reads signals from each component, performs calculations, and outputs control signals based on the PWM signal duty cycle to control the actions of the hydraulic system.

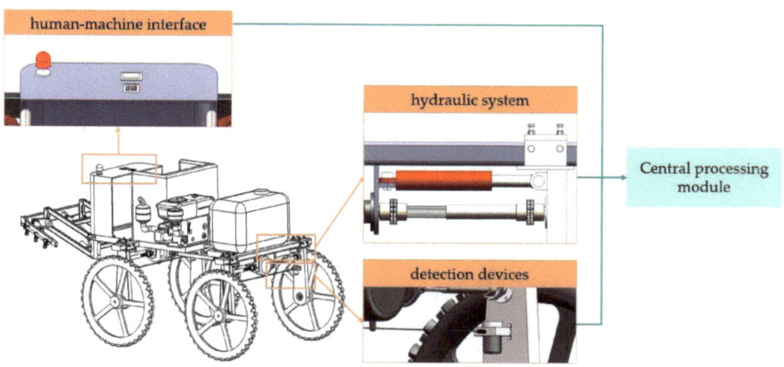

Figure 6. Overall structure of electro-hydraulic control system. Note: The central processing module is an Arduino Mega 2560 development board.

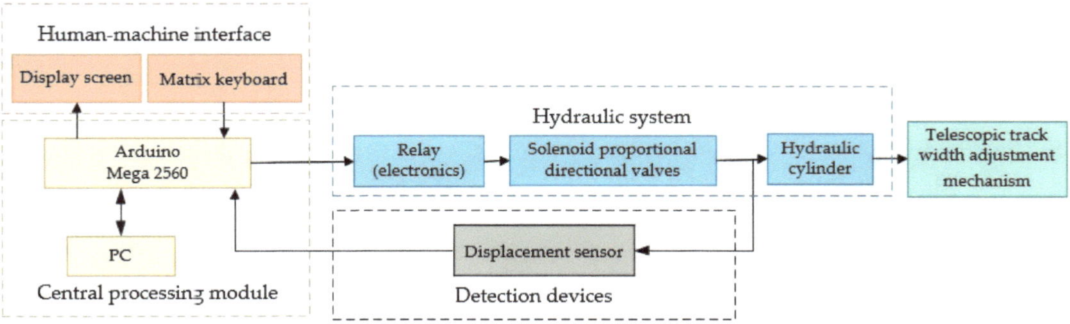

Figure 7. Overall framework of the electro-hydraulic control system.

The detection devices are used to detect real-time track width information and transmit signals to the central processing module. If the detection devices only measure the overall left and right track width, the wheels on both sides adjust simultaneously during the spacing adjustment process. In this condition, when the overall track width meets the requirements, discrepancies in the hydraulic fluid distribution can lead to inconsistent lateral adjustment distances between the left and right wheels, affecting the stability and operational effectiveness of the sprayer. Therefore, to reduce the error in the track width adjustment between the left and right sides, each wheel is controlled individually, with the detection devices measuring the vertical distance from each wheel to the central axis of the chassis.

The central processing module is the core of the entire electro-hydraulic control system and must be capable of flexible, real-time, and accurate data processing, serial communication, and high reliability. To fulfill the system's requirement for independent control of four wheels, the MEGA2560 R3, designed by the Italian-origin Arduino (Ivrea, Italy), is utilized. This board features an ATMEGA2560-16AU chip manufactured by Microchip Technology Incorporated in the Chandler, AZ, USA. The detection devices use draw-wire sensors installed inside each movable support wheel frame, with the draw-wire attached to the center of the sprayer crossbeam, and with each draw-wire's linear motion direction parallel to the crossbeam. Since the track width parameters are adjustable from 1200 mm to 1600 mm, the displacement sensor range must be at least 800 mm. The sensor needs to provide stable, accurate, and timely feedback on the hydraulic cylinder position with millimeter-level precision. Considering these factors and cost, an MPS pulse-type draw-wire displacement sensor was selected.

2.3.2. Control System Main Program

The main control program of the system was developed in the Arduino environment using C language. Figure 8 shows the main program flowchart. First, the system powers on and initializes the devices. Initialization involves configuring the input and output pins, setting the initial pin levels, configuring the serial port, and setting up external interrupts for the pins. After initialization, the system enters the main loop, where it first retrieves data from the matrix keypad to set the target track width. The microcontroller uses external interrupts to obtain real-time distance measurements from the draw-wire sensors and compares the received displacement values with the input set values. Through calculations, it generates error signals to select the output channel, essentially determining the stroke and direction of the proportional solenoid valve, thereby altering the flow and direction of hydraulic fluid entering the hydraulic cylinder. An alarm is triggered if the input track width set value exceeds the maximum allowable distance.

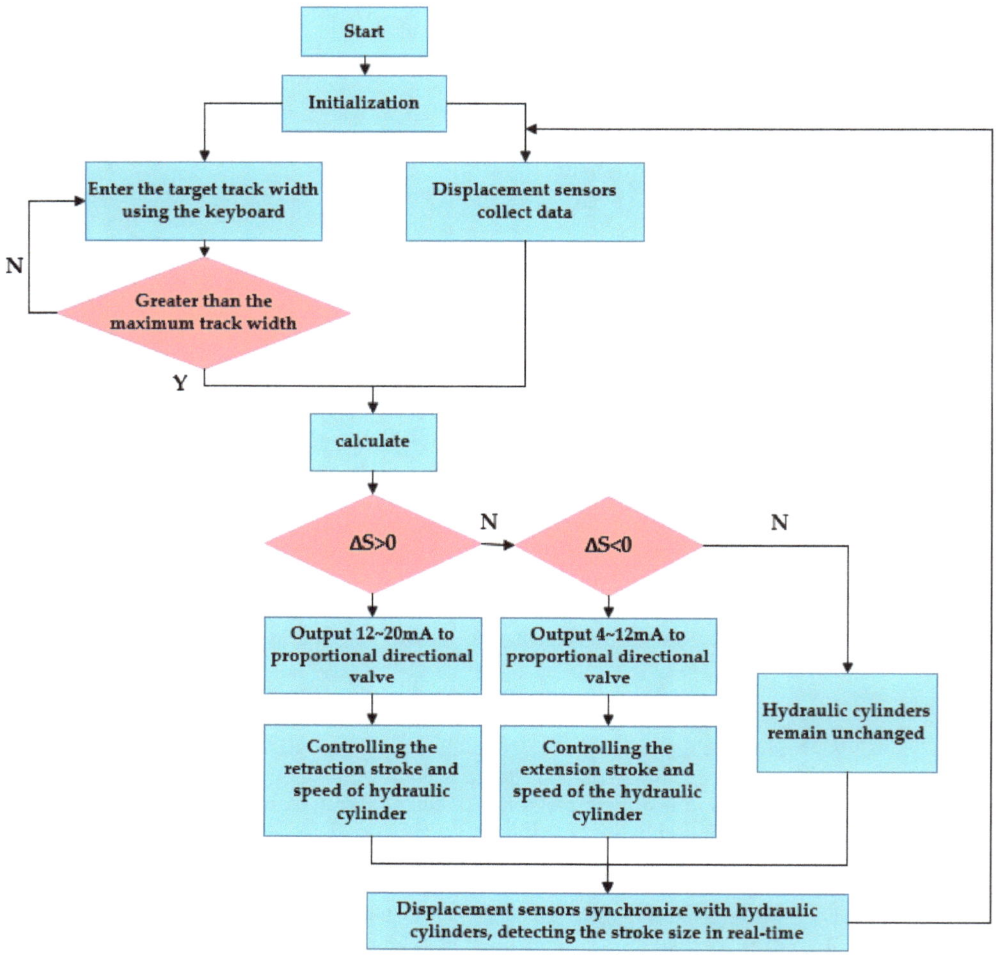

Figure 8. The main program flowchart.

2.3.3. Hydraulic System

1. Principle of the hydraulic system

The hydraulic system primarily consists of an oil tank, solenoid proportional directional valve, hydraulic pump, hydraulic cylinder, engine, bidirectional hydraulic lock, and relief valve, as illustrated in Figure 9. The hydraulic pump, acting as the power source for the hydraulic system, draws in oil, converting low-pressure oil into high-pressure oil. The high-pressure oil exits the hydraulic pump and enters the solenoid proportional directional valve's inlet. It then passes through the bidirectional hydraulic lock and ball valve before reaching the piston rod side of the hydraulic cylinder, while the oil from the other side of the cylinder flows back to the oil tank [18–22]. To accommodate wide–narrow row seeding, a bidirectional hydraulic lock is incorporated. When the machine is in operation and the solenoid proportional directional valve is in the neutral position, the bidirectional hydraulic lock secures the hydraulic fluid at both ends of the double-acting hydraulic cylinder, maintaining the position of the piston rod and, consequently, the track width.

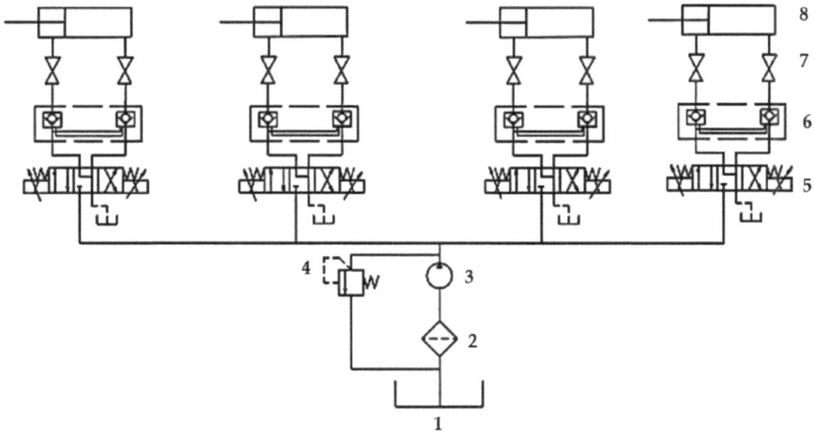

Figure 9. Schematic diagram of the hydraulic system. 1—oil tank; 2—filtering apparatus; 3—hydraulic pump; 4—relief valve; 5—solenoid proportional directional valve; 6—bidirectional hydraulic lock; 7—ball valves; 8—hydraulic cylinder.

When the right-side wheel needs to reduce its spacing, the operator inputs the target track width value through the human–machine interface. The controller receives the current track width value detected by the sensor and the target value input from the keyboard. It then calculates the error signal, optimizes the signal using the PID control, and determines the appropriate output channel to control the corresponding solenoid proportional directional valve for the right-side wheel. The solenoid proportional directional valve directs the hydraulic fluid to the right side of the hydraulic cylinder, causing the piston rod to retract to the left, thereby moving the movable support wheel frame leftward and reducing the track width. Once the target distance is reached, the solenoid proportional directional valve resets to the neutral position, allowing the sprayer to operate in the field at the predetermined track width. Conversely, if the right-side track width needs to be increased, the piston rod extends to the right, moving the movable support wheel frame rightward and increasing the track width. Upon reaching the target distance, the solenoid proportional directional valve resets to the neutral position, enabling the sprayer to operate in the field at the specified track width. The same process applies to the left-side wheel.

2. Design of the hydraulic cylinder

When adjusting the track width, the movable support wheel frame is mainly subjected to the driving force of the hydraulic cylinder piston rod and the friction between the ground and the tire. According to the force–balance relationship, the driving force is equal to the friction force. Therefore, the load on a single hydraulic cylinder satisfies the following equation:

$$F = \frac{1}{4} fG \tag{1}$$

where f is the friction coefficient between the rubber material of the tire and the ground, with values ranging from 0.5 to 0.7; G is the total weight of the sprayer, N; and F is the load on the hydraulic cylinder, N.

When the sprayer is fully loaded with liquid and the wheelbase is being adjusted, its total weight is approximately 6000 N. Under these conditions, the hydraulic cylinder experiences its maximum load, with $f = 0.7$. Substituting these values into Equation (1), the maximum load on the hydraulic cylinder is obtained as $F = 1050$ N.

According to the hydraulic design manual, the initial system pressure P is selected to be 10 MPa.

The hydraulic cylinder bore diameter D is:

$$D = \sqrt{\frac{4F\lambda}{\pi P \eta_m}} \tag{2}$$

where η_m is the mechanical efficiency of hydraulic cylinders and λ is the hydraulic cylinder round trip speed ratio.

The hydraulic cylinder piston rod diameter d is:

$$d = D\sqrt{\frac{\lambda - 1}{\lambda}} \tag{3}$$

Taking η_m as 1.33 and λ as 0.9, and substituting the values of F and P into Equations (2) and (3), the calculations yield the hydraulic cylinder bore diameter D and piston rod diameter d. After consulting the hydraulic manual and selecting the appropriate sizes from the GB/T 2348-2018 standard [23], the hydraulic cylinder bore diameter D is determined to be 25 mm, with the piston rod diameter d being 16 mm.

The wheel track adjustment range of the sprayer is from 1200 mm to 1600 mm, so the hydraulic cylinder displacement must exceed 200 mm. Therefore, a hydraulic cylinder stroke of 240 mm is selected.

2.4. Control Scheme Simulation Experiment

In the process of adjusting the track width, the hydraulic system control faces issues such as a significant impact force from the hydraulic cylinders, slow response time, and low control precision. To address these problems, four control strategies are proposed: classical PID, integral separated PID, fuzzy adaptive PID, and integral separated fuzzy PID. Simulations of these four control strategies are conducted using the Simulink tool in MATLAB. Based on the simulation results, the optimal control strategy will be selected.

2.4.1. Design of Control Strategies

Considering that the track width is independently controlled by adjusting the four wheels separately, we can focus on simulating the track width variable of a single wheel under different control schemes. The interference between the wheels can be simplified, based on specific working conditions, into a disturbance load torque applied to a single wheel. By designing appropriate controller parameters, this disturbance load torque can be suppressed, enabling the track width control system to meet the desired performance requirements.

Compared to the classical PID control, the integral separated PID control disables the integral action when the deviation between the system output and the target value is large. This prevents excessive overshoot caused by integral accumulation. When the deviation is small, the integral control is introduced to eliminate the steady-state error and improve the control precision [24–28].

The fuzzy adaptive PID control adjusts the parameters of the PID controller in real-time based on fuzzy rule algorithms, adapting to varying working conditions during system operation. This enhances the system's ability to handle complex scenarios [29,30]. The principles for tuning the fuzzy PID algorithm parameters are as follows: when the absolute value of the deviation e is large and the system is in the response phase, a larger proportional gain K_p is required to speed up the response and prevent a sudden increase in e. To prevent integral saturation, a smaller integral time constant T_i should be used. Additionally, to avoid derivative saturation and prevent significant overshoot in the system response, the influence of the derivative term should be reduced. Figure 10 illustrates the system structure of the fuzzy adaptive PID control.

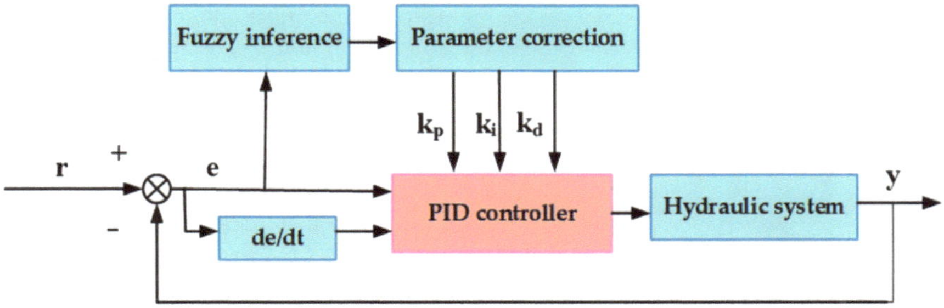

Figure 10. System structure of the fuzzy adaptive PID control.

By combining fuzzy adaptive rules with the integral separated PID control, the integral separated fuzzy PID control scheme is obtained and its system structure is illustrated in Figure 11.

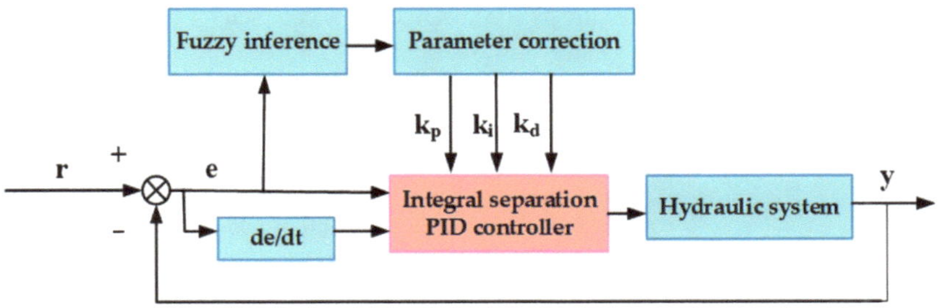

Figure 11. System structure of the integral separated fuzzy PID control.

2.4.2. Mathematical Model of Hydraulic System

Before simulating the four control schemes, the mathematical model of the hydraulic system is first constructed using local linearization of the system [31].

1. Mathematical model of hydraulic cylinders

Make the following assumptions:

- The connecting pipelines between the valve and the hydraulic cylinder are symmetrical, short, and thick, allowing the pressure loss and dynamics within the pipelines to be neglected.
- The pressure within each working chamber of the hydraulic cylinder is equal.
- The initial volumes of the oil inlet and return chambers are equal.
- The oil temperature and bulk modulus are constant.
- Both internal and external leakage in the hydraulic cylinder are laminar flows.

The simplified flow continuity equation can be obtained as follows:

$$q_L = A_p \frac{dx_p}{dt} + C_{ip} p_L + \frac{V_t}{4\beta_e} \frac{dp_L}{dt} \tag{4}$$

where A_p is the piston area, m^2; x_p is the piston displacement, m; C_{ip} is the hydraulic cylinder internal leakage coefficient, $m^3/(Pa{\cdot}s)$; β_e is the effective bulk modulus of elasticity (including the mechanical flexibility of fluids, connecting pipes and cylinders), Pa; $V_t = 2V_0$ is the total compressed volume, m^3; and p_L is the pressure in the low-pressure chamber (return chamber) of the hydraulic cylinder, Pa.

2. Mathematical model of solenoid proportional directional valve

The model of an electro-hydraulic proportional directional valve primarily involves the flow-pressure equations of the valve ports. Here, q_1 and q_2 represent the flow rates entering the high-pressure chamber and low-pressure chamber of the motor through the servo valve, respectively. p_1 and p_2 are, respectively, the pressures in the high-pressure and low-pressure chambers of the motor. p_s and p_e are, respectively, the supply pressure and return pressure. x_{sp} represents the displacement of the spool of the electro-hydraulic proportional valve.

$$\begin{cases} q_1 = C_d w x_{sp}\sqrt{\frac{2}{\rho}(p_s - p_1)} \\ q_2 = C_d w x_{sp}\sqrt{\frac{2}{\rho}(p_2 - p_e)} \end{cases} \quad x_{sp} > 0 \tag{5}$$

$$\begin{cases} q_1 = C_d w x_{sp}\sqrt{\frac{2}{\rho}(p_1 - p_e)} \\ q_2 = C_d w x_{sp}\sqrt{\frac{2}{\rho}(p_s - p_2)} \end{cases} \quad x_{sp} < 0 \tag{6}$$

To improve simulation efficiency, the linearization method is used to further simplify Equations (5) and (6). Assuming the valve is a zero-lapped four-way spool valve, with four matched and symmetrical throttle windows, constant supply pressure p_s, and zero return pressure p_0, the linearized flow equation for the zero-lapped four-way spool valve is given by:

$$q_L = K_q x_v - K_c p_L \tag{7}$$

The dynamic analysis of position servo systems is often conducted under zero-displacement operating conditions, where the increments and variables are equal. For a matched and symmetric zero-lapped four-way sliding valve, the flow q_1 and q_2 through both control channels is equal to the load flow q_L. During dynamic analysis, it is essential to consider the impact of the external leakage and compressibility of the hydraulic cylinder, which results in the inflow and outflow rates of the hydraulic cylinder being unequal. To simplify the analysis, the load flow is defined as:

$$q_L = \frac{q_1 + q_2}{2} \tag{8}$$

Consider an electro-hydraulic proportional valve with the current Δi as the input parameter and the no-load flow rate $q_0 = K_q x_v$ as the output parameter. In most electro-hydraulic proportional systems, the dynamic response of the proportional solenoid valve is often higher than that of the power components. In this context, the transfer function of the proportional solenoid valve can be represented by a second-order oscillatory system:

$$W_{sv}(s) = \frac{Q_0}{\Delta I} = \frac{K_{sv}}{\frac{s^2}{\omega_{sv}^2} + \frac{2\zeta_{sv}}{\omega_{sv}}s + 1} \tag{9}$$

where K_{sv} is the flow gain of the solenoid proportional valve, m^2/s; ω_{sv} is the inherent frequency of the solenoid proportional valve, rad/s; and ζ_{sv} is the damping ratio of the solenoid proportional valve.

3. Mathematical model of solenoid proportional directional valve controlled hydraulic cylinder system

Based on the force balance equation, the linearized equations of the hydraulic cylinder and the electro-hydraulic proportional valve are combined to form the mathematical model of the valve-controlled hydraulic cylinder system:

$$A_p p_L = m_t \frac{d^2 x_p}{dt^2} + B_p \frac{dx_p}{dt} + K_p x_p + F_L \tag{10}$$

where m_t is the total mass of the piston and load converted to the piston, kg; B_p is the viscous damping coefficient of the piston and load, N·s/m; K_p is the load spring stiffness, N/m; and F_L is the external load force, N.

Performing the Laplace transform of Equations (4), (7) and (10):

$$Q_L = K_q X_v - K_c P_L \tag{11}$$

$$Q_L = A_p s X_p + C_{ip} P_L + \frac{V_t}{4\beta_e} s P_L \tag{12}$$

$$A_p p_L = m_t s^2 X_p + B_p s X_p + K_p X_p + F_L \tag{13}$$

The transfer function block diagram is obtained as shown in Figure 12.

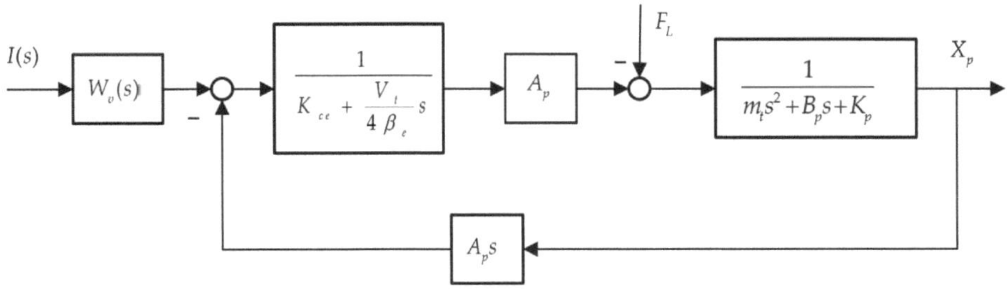

Figure 12. The transfer function block diagram of valve-controlled hydraulic cylinder system.

As shown in Table S2 of the supplementary material, the parameters for each hydraulic component were configured, and the values were incorporated into the aforementioned control block diagram. This process yielded the mathematical model presented in Equation (14).

$$W_{sv}(s) = \frac{0.00013}{\frac{s^2}{200^2} + \frac{1.414s}{200} + 1} \tag{14}$$

2.4.3. Construction of Simulation Models for Control Schemes

Based on the transfer function models of the main components in the hydraulic cylinder system controlled by the proportional valve, four control scheme models were constructed in Simulink. The same PID parameters (K_p = 1.5, K_i = 2, K_d = 0) were set for the simulation experiments. The construction of the classical PID and the integral separated PID models is relatively straightforward and is not described in detail in this paper. Figures 13 and 14 show, respectively, the simulation models of the classical PID and the integral separated PID.

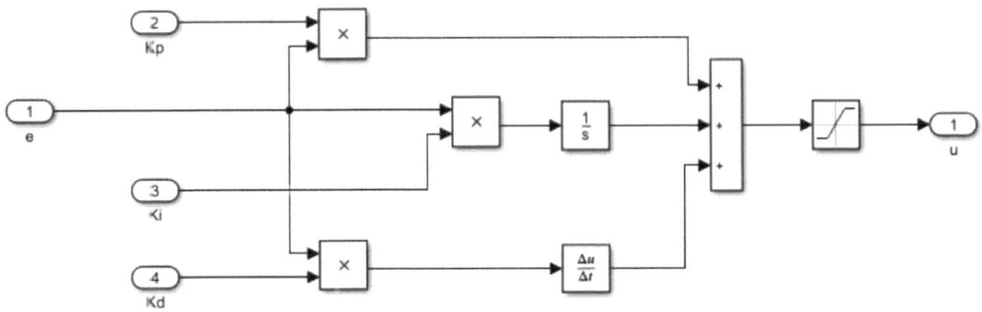

Figure 13. Simulink model for classical PID controller.

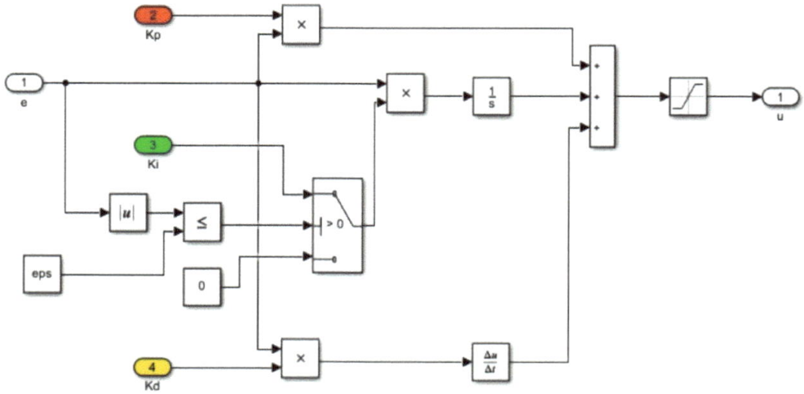

Figure 14. Simulink model for integral separated PID controller.

The design diagram of the fuzzy controller is shown in Figure 15a. Firstly, the range of the input variable e is set to [−6, 6], and the range of the input variable ec is set to [−1, 1]. Both variables e and ec include the NB, NS, ZE, PS, and PB. Secondly, the output variable K_p is set with a range of [0, 1]. The output variables K_i and K_d are both set with a range of [−1, 1]. These three output variables also include ZE, PO, PS, PM, and PB.

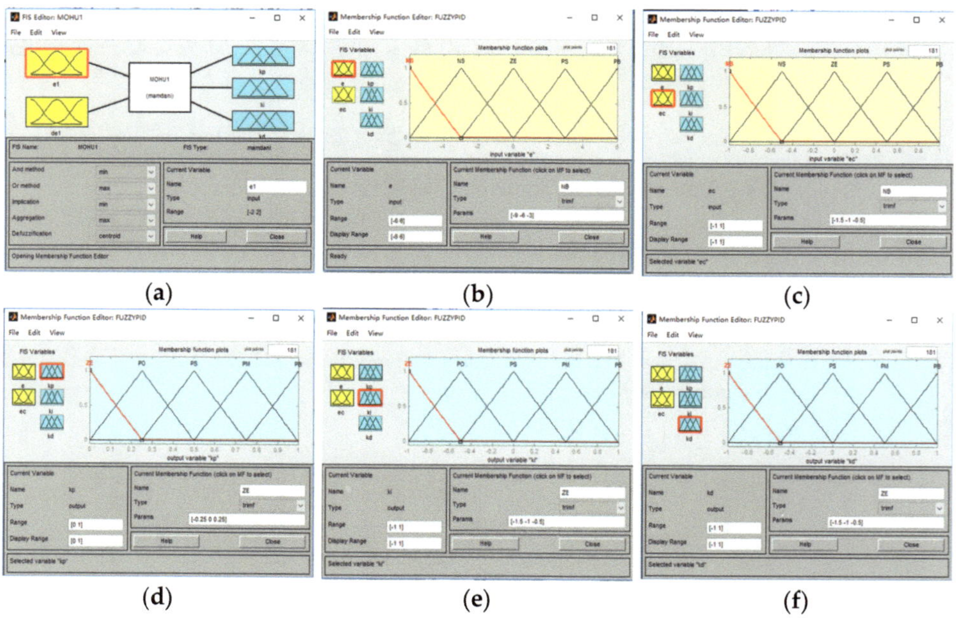

Figure 15. Fuzzy controller design diagram. (**a**) Fuzzy controller design diagram; (**b**) Input variable e; (**c**) Input variable ec; (**d**) Output variable K_p; (**e**) Output variables K_i; (**f**) Output variables K_d.

Based on the fuzzy PID rules, the output variable values are calculated. They are created and stored in real-time on the computer in the form of a lookup table. By directly querying the output table values with the fuzzy linguistic variables of error e and error change ec, the PID increment values are obtained. The controller model constructed in Simulink is shown in Figure 16.

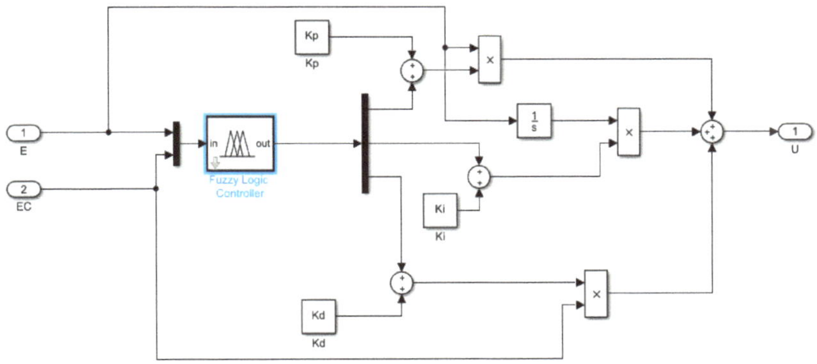

Figure 16. Simulink model for fuzzy adaptive PID controller.

Figure 17 illustrates the real-time adjustment of the three controller data. It can be observed that the fuzzy rules effectively adjust the proportional and integral coefficients during the initial stage, while the adjustment of the differential coefficient is relatively minor. This indicates that the track width control system primarily achieves an input signal tracking response through the proportional integral controller. Figure 18 shows the input–output surface view of the fuzzy inference system.

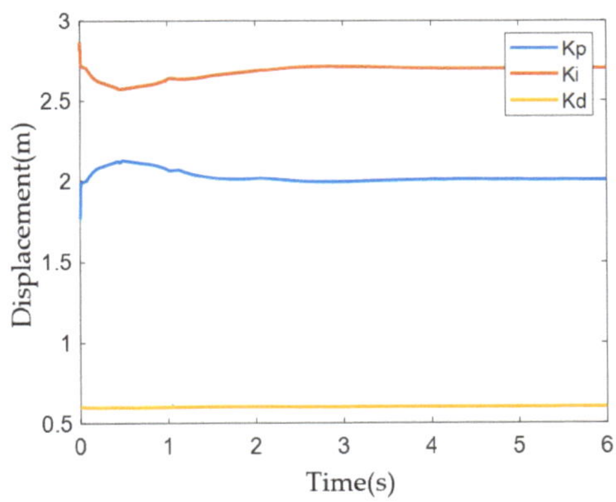

Figure 17. The real-time adjustment of the controller data.

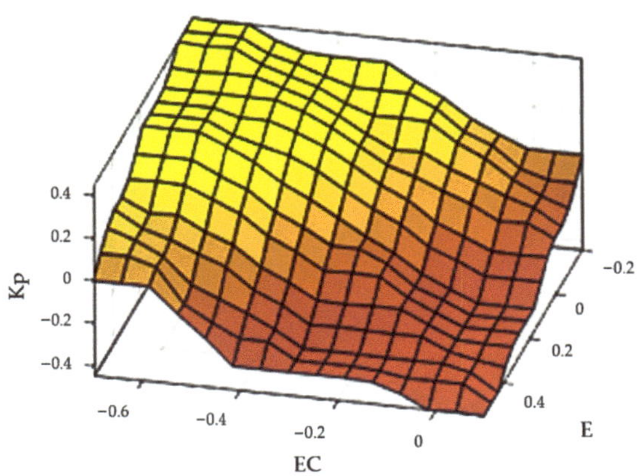

Figure 18. The input–output surface view of the fuzzy inference system.

Combining the fuzzy adaptive rule with the integral separated PID controller, the integral separated fuzzy PID control scheme model was constructed in Simulink, as shown in Figure 19. In this model, the part combining the integral separated PID and fuzzy control is encapsulated within the Subsystem module.

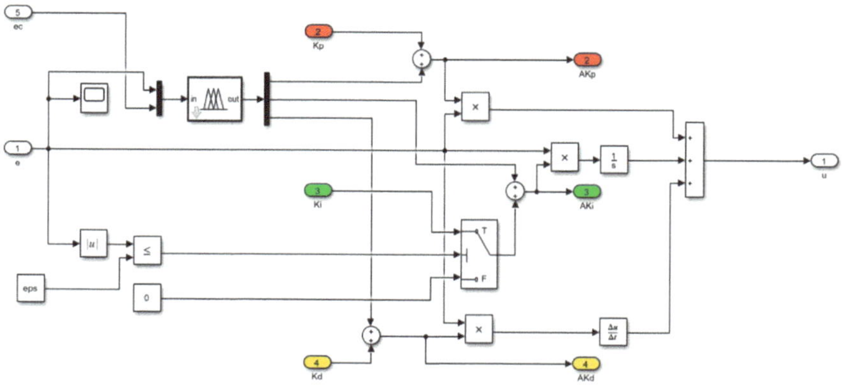

Figure 19. Simulink models for integral separated fuzzy PID controller.

Taking the scenario of increased track width as an example, the simulation was conducted where the displacement of the single-side track width was adjusted from 600 mm to 800 mm, corresponding to a hydraulic rod stroke adjustment from 0 to 200 mm. A step signal was input to simulate the track width adjustment process, with the step signal trigger time set to 0 s.

2.5. Field Performance Experiment

2.5.1. Description of the Field Experiment Area

A stability and precision test for row spacing adjustment was conducted in a wheat field in Wuzhi County, Henan Province. The wheat plants were approximately 20 to 40 cm tall. As shown in Figure 20, the experimental wheat field utilized wide–narrow row seeding, with a wide row spacing of 20 cm and a narrow row spacing of 15 cm. The field surface was level, and the soil was relatively loose.

(**a**) (**b**)

Figure 20. Measurement of spacing in wide–narrow row seeding. (**a**) Wide row spacing; (**b**) Narrow row spacing.

2.5.2. Experimental Treatments

1. Track Width Adjustment Stability Experiment Treatment

Prior to the experiment, the sprayer was adjusted to 1400 mm (spanning 8 rows of wheat). During the experiment, the operator controlled the sprayer speed at 0.5 m/s, maintaining a track width of 1400 mm while navigating the field. A measuring tape was used to measure the track width at 1 m intervals along the sprayer's operating path, resulting in 20 sets of actual track width data.

2. Track Width Adjustment Precision Experiment Treatment

To verify the precision of row spacing adjustment, an error analysis was conducted for the process of adjusting the sprayer track width from 1400 mm to 1550 mm (from 8 rows to 9 rows) and from 1550 mm to 1400 mm (from 9 rows to 8 rows). During the experiment, the operator set the target track width via the human–machine interface and controlled the sprayer speed at 0.5 m/s. A measuring tape was used to measure the track width at 1 m intervals along the sprayer's operating path.

When the relative errors between the measured track widths for five consecutive times and the target track width were less than the stability coefficient obtained from the stability test, the sprayer was considered to have achieved the predetermined track width. At this point, the average of these five track width measurements was taken as the final track width adjustment value of the sprayer. The adjustment process from 8 rows to 9 rows and from 9 rows to 8 rows was each repeated six times in the same field plot. Figure 21 illustrates the experimental process.

Figure 21. Field experiment process.

3. Results and Discussion

3.1. Simulation Experiment Data Collection and Analysis

The hydraulic cylinder displacement response curves for four different control schemes were obtained through simulation, as shown in Figure 22.

From the figure, it can be observed that under a given input of 800 mm, the classic PID control strategy exhibits the highest overshoot. This occurs because when the system starts or undergoes significant changes in the target value, the system experiences a substantial deviation in a short period. This deviation causes an accumulation in the integral calculation of the PID operation, resulting in a particularly large output and significant system oscillations.

In comparison to the classic PID control strategy, the integral-separated PID control strategy exhibits a smaller overshoot. This improvement is attributed to the addition of a gating switch in the integral-separated PID control, which turns off the integrator when the deviation exceeds a set threshold.

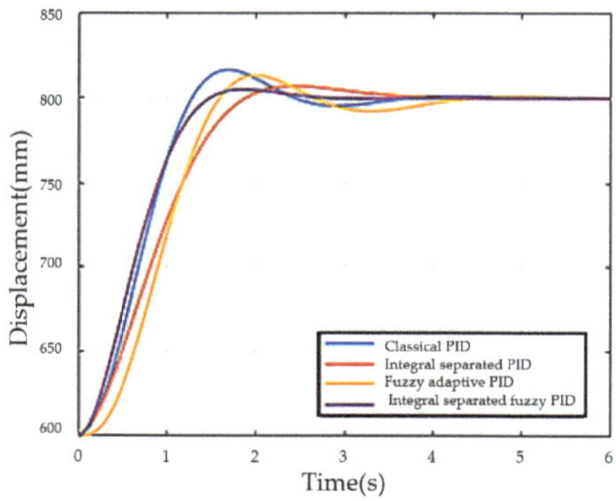

Figure 22. Comparison of displacement response curves for four different control schemes.

The fuzzy adaptive PID, relative to the classic PID control strategy, shows improvements in both overshoot and settling time due to its capability to adjust the PID controller parameters in real-time based on system conditions.

The integral separated fuzzy PID control strategy achieves a rapid response speed and the smallest overshoot. This control strategy can achieve the desired track width within 2.08 s, maintaining an overshoot of around 2.37%, effectively reducing the number of oscillations and achieving high steady-state accuracy. The integral separated fuzzy PID control strategy reduces overshoot by weakening the integral term in the presence of large system deviations. Additionally, the inclusion of fuzzy control allows for further fine-tuning of control parameters, optimizing performance beyond the basic configuration. Thus, it demonstrates superior control performance compared to the other three strategies.

Therefore, the integral separated fuzzy PID control strategy is confirmed as the optimal control scheme for the track width adjustment system.

3.2. *Track Width Adjustment Stability Experiment Data Collections and Analysis*

Processing the data from the 20 groups in Figure 23 yields an average actual track width of 1398.05 mm with a standard deviation of 42.50 mm. According to Equation (15), the calculated track width stability coefficient is 3.04%, which is below the threshold of 10%. This indicates that the track width adjustment system is relatively stable and meets the standards for agricultural machinery and agronomic requirements.

$$V = \frac{S}{a} \times 100\% \quad (15)$$

where S is the standard deviation of track width, mm; a is the average track width, mm; and V is the track width stability coefficient.

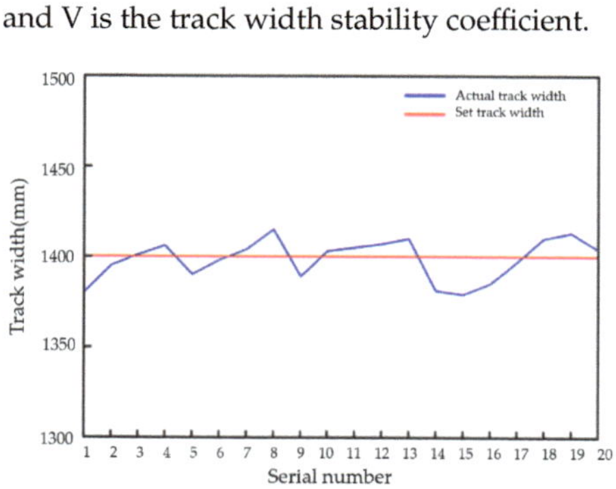

Figure 23. The actual track width measurement value and set value. Note: The actual track width measurement is represented by the blue line, while the red line indicates the set track width value.

3.3. Track Width Adjustment Precision Experiment Data Collections and Analysis

Table 1 presents the experimental data on track width adjustment errors. A comparative analysis between the final track width adjustment values of the sprayer and the target track width values reveals that the track width adjustment errors are all within 21 mm or less, with an average error of 13.42 mm. The accuracy of the track width adjustment is relatively high.

Table 1. Track width adjustment error.

Experiment Number	Initial Track Width (mm)	Target Track Width (mm)	Final Track Width Adjustment Value (mm)	Error (mm)
1	1400	1550	1563	13
2	1400	1550	1571	21
3	1400	1550	1562	12
3	1400	1550	1558	8
4	1400	1550	1564	14
5	1400	1550	1555	5
6	1550	1400	1380	20
7	1550	1400	1415	15
8	1550	1400	1410	10
9	1550	1400	1393	7
10	1550	1400	1421	21
11	1550	1400	1415	15
12	1400	1550	1563	13
Average	/	/	/	13.42

4. Conclusions

To improve the adaptability of the sprayer to different operating conditions, reduce crop damage, and enhance the stability and accuracy of the track width adjustment, this paper employed electro-hydraulic control technology to design a track width adjustment system for sprayers based on an integral separated fuzzy PID control strategy.

The designed track width adjustment system consists of a transmission mechanism, a telescopic track width adjustment mechanism, and an electro-hydraulic control system. By analyzing the agronomic requirements for field management during the early growth stages of corn and throughout the growth stages of wheat in Henan Province, China, the key parameters for the track width adjustment mechanism were determined. The focus was on designing the electro-hydraulic control system, which comprises a human–machine interface, a central processing module, detection devices, and a hydraulic system. Four control strategies were proposed: classic PID, fuzzy adaptive PID, integral separated PID, and integral separated fuzzy PID. Simulink was used to simulate these control strategies, and the results indicated that the integral separated fuzzy PID control strategy was the most effective, significantly reducing system oscillations, improving the response speed, and enhancing the track width adjustment accuracy. Field test results showed that the stability difference coefficient of the track width adjustment system was 3.04%, which is below the 10% agricultural machinery standard. The average error during track width adjustment was 13.42 mm, demonstrating high accuracy.

The findings of this study provide a research foundation for the development of a more intelligent, precise, and stable track width variable chassis for sprayers and other agricultural machinery. Future efforts should focus on further field testing and system optimization to enhance the stability and reliability of track width adjustments. Additionally, integrating image sensors to collect real-time data on crop row spacing for comparisons with the track width adjustments would enable automatic adjustment to the optimal track width.

Supplementary Materials: The following supporting information can be downloaded at: https://www.mdpi.com/article/10.3390/agriculture14081247/s1, Table S1. Agronomic requirements for corn and wheat; Table S2. Parameter setting for each hydraulic component.

Author Contributions: Conceptualization, L.T., M.J. and J.H.; Funding acquisition, J.H.; Methodology, L.T. and M.J.; Software, M.J.; Validation, L.T., X.S. and H.L.; Writing—original draft, L.T.; Writing—Review and editing, J.H., Q.W. and H.Y. All authors have read and agreed to the published version of the manuscript.

Funding: This research was funded by the National Key R&D Program of China no. 2022YFD1500704-1.

Institutional Review Board Statement: Not applicable.

Data Availability Statement: The data that support the findings of this study are available from the corresponding author upon reasonable request.

Conflicts of Interest: The authors declare no conflicts of interest.

References

1. National Bureau of Statistics of China. *China Statistical Yearbook-2023*; China Statistics Press: Beijing, China, 2023.
2. Shen, Z.; Han, T.F. Temporal and spatial variability of relative yield differences between wheat and maize and their response to nitrogen fertilizer. *Chin. J. Agric. Sci.* **2023**, *56*, 2724–2737.
3. Zheng, R.J.; Luan, L. Application of boom sprayers in the control of diseases, pests, and weeds in Rice-Wheat rotation Areas. *Mod. Agric. Sci. Technol.* **2023**, *15*, 99–103.
4. Sun, W.F.; Liu, H.Y. Design and experiment of PID control variable spraying system based on neural network tuning. *Trans. Chin. Soc. Agric. Mach.* **2020**, *51*, 55–64+94.
5. Zhou, Z.Y.; Zhou, M.J. Research on automatic leveling control system of boom sprayer with rotor suspension. *Trans. Chin. Soc. Agric.* **2022**, *53*, 70–79.
6. Xu, Y.D.; Liu, G.H. Four-wheel synchronous steering sprayer tracking control based on fuzzy MPC. *Control Eng.* **2024**, 1–9. [CrossRef]
7. Wan, T.; Wang, W.J. Development research and analysis of variable chassis technology for agricultural tractors. *Tractor Farm Transp.* **2020**, *47*, 4–6.
8. Zuo, Z.J. Research and Design of Variable Ground Clearance and Track Width Tractor Chassis. Master's Thesis, Shandong University, Jinan, China, 2016.
9. Case New Holland. Case "Patriot" Patriot~(TM) 3230 sprayer, powerful. *Agric. Mach.* **2020**, *4*, 29.
10. Zhang, Q. Stability Study of Variable Track Power Platform in Hilly and Mountainous Areas. Master's Thesis, Xihua University, Chengdu, China, 2023.
11. Liu, D.X.; Qi, J.H. Multifunctional Self-Propelled Adjustable Boom Spraying Detasseling Machine. Siping, China, 2015. Available online: http://search.cnipr.com/search!doDetailSearch.action (accessed on 25 July 2024).
12. Wang, M.; Xie, F. Current development status of plant protection spraying devices in China. *Agric. Eng.* **2019**, *9*, 4–7.
13. Li, J. Design of an Agricultural Variable Wheelbase Electric Mobile Platform. Master's Thesis, Chongqing Three Gorges University, Chongqing, China, 2021.
14. Xia, J.L.; Guan, C.S. A Device for Adjusting the Rear Wheel Spacing of a Wheeled Tractor. Yancheng, China, 2014. Available online: http://search.cnipr.com/search!doDetailSearch.action (accessed on 25 July 2024).
15. Chipriana, A.A.; Puig Mauri, O.C. Equips para la mechanization del cultivate explanations frustules de Alta densipan. *Investig. Agrar. Prod. Prot.* **2000**, *15*, 125–142.
16. Blanco, M.N.; Michael, R.A. Real-time monitoring of spray drift from three different orchard sprayers. *Chemosphere* **2019**, *222*, 46–55. [CrossRef] [PubMed]
17. Zhang, X.; Xie, Y. Machine Vision-Based Obstacle Avoidance Hydraulic Chassis with Variable Wheel Track and Its Usage Method. 2018. Available online: http://search.cnipr.com/search!doDetailSearch.action (accessed on 25 July 2024).
18. Wang, X.C.; Shang, Z.G. Current research status and development trends of plant protection machinery design and manufacturing technology. *Mod. Agric. Sci. Technol.* **2017**, *2*, 282–288.
19. Huang, M.; Zheng, D.K. Synchronization study of sugarcane cultivator with rhombic self-propelled chassis based on AMESim. *J. Agric. Mech. Res.* **2017**, *39*, 7–12.
20. Dou, L.J. Design and Research of Adjustable Track Steering System for High-Clearance Self-Propelled Sprayer. Master's Thesis, Chinese Academy of Agricultural Mechanization Sciences, Beijing, China, 2012.
21. Fan, G.Q.; Meng, X.G. Agricultural High-Clearance Operation Machine. 2015. Available online: http://search.cnipr.com/search!doDetailSearch.action (accessed on 25 July 2024).
22. Cui, Y.J.; Zhang, Z.L. Kinematic simulation analysis of fruit picking manipulator based on ADAMS. *J. Agric. Mech. Res.* **2008**, *4*, 59–61.

23. *GB/T 2348-2018*; Fluid Power Systems and Components—Cylinder Bores and Piston Rod Diameters. Standardization Administration of China: Beijing, China, 2018.
24. Kwok, K.E.; Chong, P.M. A model based augmented PID algorithm. *J. Process Control* **2000**, *10*, 9–18. [CrossRef]
25. Xue, D.Y. *Control System Computer-Aided Design—MATLAB Language and Application*; Tsinghua University Press: Beijing, China, 1996.
26. Liu, J.K. *Advanced PID Control and Simulation*; Electronics Industry Press: Beijing, China, 2000; pp. 1–8.
27. Qiu, L.; Zeng, G.E. Comparative study on several PID controller parameter tuning methods. *Autom. Technol. Appl.* **2005**, *24*, 28–31.
28. Han, P.P. Research and Design of Fuzzy Self-Tuning PID Controller. Master's Thesis, Hebei University of Technology, Tianjin, China, 2010.
29. Sun, C.Y. *Principles and Design of Measurement and Control Systems*; Beijing University of Aeronautics and Astronautics Press: Beijing, China, 2003; pp. 160–166, 174.
30. Zhao, H.Q. Rapid empirical tuning method for PID control parameters. *Chem. Ind. Autom. Instrum.* **1997**, *24*, 43–45.
31. Liu, J.Q.; Ding, P. Research on remote steering system of tractor based on fuzzy PID control. *J. Agric. Mech.* **2024**, 1–8. [CrossRef]

agriculture

MDPI

Article

Optimizing Contact Force on an Apple Picking Robot End-Effector

Hongwei Zhang, Wei Ji *, Bo Xu and Xiaowei Yu

School of Electrical and Information Engineering, Jiangsu University, Zhenjiang 212013, China; zhanghongwei991016@163.com (H.Z.); xubo@ujs.edu.cn (B.X.); 18879209223@163.com (X.Y.)
* Correspondence: jiwei@ujs.edu.cn

Abstract: The quality of apple picking affects the sales of apples, and the grasping force of the end effector of an apple picking robot is very important for apple picking. It is easy to cause apple damage due to excessive contact force, or when the contact force is too small to grasp the apple. However, the current research lacks an analysis of the minimum stable grasping force of apples. Therefore, in order to realize the stable grasping of apples by the end-effector of a picking robot and reduce fruit damage, this study first analyzes the grasping stability of the end-effector based on the force closure theory, and comprehensively considers the force closure constraints, nonlinear friction cone constraints and the introduction of torque constraints. Next, the constraint conditions are processed using an obstacle function, and a penalty factor is introduced to construct an optimization model of the contact force distribution of the end-effector. Then, the improved Newton method is used to grasp and solve the contact force distribution model. Under the premise of selecting the penalty factor, the optimal contact force of grasping an apple is determined using a method of numerical example simulation analysis, and the validity of the solution is verified. In order to verify the reliability of the contact force distribution optimization model, the practical significance of the method for apple grasping is verified in an actual grasping experiment. The actual experiment shows that the method can provide the minimum stable grasping force to the end-effector to achieve stable grasping.

Keywords: apple picking robot; end-effector; grasping stability; optimization of contact force; constraints

Citation: Zhang, H.; Ji, W.; Xu, B.; Yu, X. Optimizing Contact Force on an Apple Picking Robot End-Effector. *Agriculture* **2024**, *14*, 996. https://doi.org/10.3390/agriculture14070996

Academic Editors: Mustafa Ucgul and Massimo Cecchini

Received: 14 May 2024
Revised: 18 June 2024
Accepted: 24 June 2024
Published: 25 June 2024

1. Introduction

The apple picking robot hardware consists mostly of a chassis, manipulator, end-effector, and camera [1,2]. The end-effector, in particular, has a crucial function as it is responsible for the separation and picking of apples [3]. Hence, it is crucial to investigate the optimization of contact force distribution throughout the apple picking process.

Researchers have conducted numerous studies on the ways to optimize grasping stability and the contact force of items. Chen et al. [4] constructed regularization optimization problems using obstacle functions, gave a comprehensive representation of objective functions with different dimensions, and introduced a penalty factor to form an augmented optimization objective function. For a specific operation task, a more compact, stable or relaxed flexible grasping scheme can be obtained by adjusting the penalty factor. James et al. [5] employed sliding detection to guarantee the stability of the gripping procedure. The reliability of the detection method in unstructured situations was confirmed by employing the support vector product to identify sliding and test various sliding scenes. In their study, Mavrakis et al. [6] introduced a novel approach for assessing the stability of robot grasping. This method involves determining the intrinsic stiffness matrix of a two-finger grasp and using it to establish a stability metric. The metric takes into account many factors such as local contact curvature, material properties, contact force, and target mass. Xu et al. [7] proposed a fast force-closure grasp synthesis (FFCGS) method for an anthropomorphic hand to efficiently grasp unknown objects. The outcomes of this study have great significance in promoting the motion planning of robot hand-arm systems and upper-limb prostheses. Li et al. [8] expanded upon the existing grasping stability theory for rigid items to examine the stability of grabbing semi-ripened tomatoes. They determined that curved fingers are better suited for securely holding tomatoes compared to disc-shaped fingers.

While numerous scholars have examined the impact of different factors on grasping stability, it is important to optimize the contact force applied to the fruit tissue to ensure stability without causing damage. Buss et al. [9] introduced a recursive optimization technique that is well-suited for dynamic situations. The purpose of this method is to determine an appropriate set of internal grasping forces that can accomplish both grasping stability and minimize grasping energy. In their study, Li et al. [10] aimed to identify the contact point location during dynamic operations that fulfills the force closure criterion based on the equivalency between the friction point constraint and the positive definite symmetric matrix found by Buss, they employed an iterative optimization algorithm to maximize the contact force of the jaw and determine the appropriate clamping force. In order to facilitate a systematic and precise examination of the force closure of MCDRMs with a cable routing design, Liang et al. [11] presented a novel force-closure analysis approach for MCDRMs that takes into account both cable coupling and friction effects. Nonlinear contact force distribution issues are defined by Liu et al. [12] as smooth manifold optimization problems that relate to a positive definite matrix that is linearly limited. The contact force is optimized using a quadratic index gradient flow method that is based on a low-dimensional description matrix. Li et al. [13] introduced a technique for evaluating the stability of envelope grasping using geometric analysis. They also developed an optimal method for planning envelope grasps, which involves determining the necessary deployment length of the DRH and ensuring a secure grip.

The above research has basically considered various constraints of force closure and optimized the grasping force optimization algorithm to a certain extent, but the applications for agricultural picking robots is relatively few. This study will conduct further research on the basis of these studies and introduce torque constraints on the premise of synthesizing the above constraints. At the same time, this study will consider optimizing the grasping force of the grasping model under new constraints, and apply the idea of the penalty function [14] to the apple picking robot in order to achieve higher computational efficiency and finally achieve acquisition of the minimum stable grasping force of the apple picking robot, which is also the focus of this research. The main contributions of this research are as follows:

(1) The grasping stability of the end-effector will be analyzed based on the force closure theory for an apple picking robot, and the optimal model of the end-effector contact force distribution will be constructed by introducing a penalty factor;
(2) The improved Newton method will be applied to the contact force distribution model, and the high computational efficiency of this method will be analyzed using numerical simulation;
(3) The actual grasping experiment will further verify that this method can achieve the minimum stable grasping force of an apple picking robot and realize the stable grasping of the apple.

The rest of the paper is organized as follows. Section 2 introduces the stability analysis of the end-effector based on the force closure theory, establishes the optimization model of force distribution, and introduces how to solve it. Section 3 details the numerical example calculation, simulation and experimental analysis and discussion. Section 4 summarizes the full paper.

2. Materials and Methods

The contact force and joint velocity vector in the operating space represent the translation of the generalized contact force and velocity screw of each finger in the operating space. This can be manifested as the coordination among numerous fingers [15]. The friction point contact model is used as the grasping model for force closure analysis as the end-effector used in this paper is a rigid device [16], which has three fingers and will be introduced in detail in the later experimental equipment introduction.

2.1. Friction Cone Constraint and Analysis

The contact force of the end-effector grasping the apple f_i can be separated into a unidirectional parameter f_{iz} and two intersecting parameters f_{ix} and f_{iy}, as seen in Figure 1.

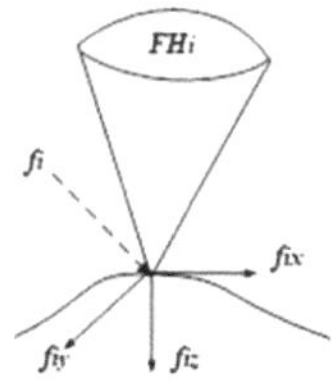

Figure 1. Schematic diagram of contact force components.

To avoid the apple from sliding, the contact force components must satisfy the constraint condition [17]. The constraint condition is shown in Equation (1).

$$\|(f_{ix}, f_{iy})\|_2 = \sqrt{\left(f_{ix}^2 + f_{iy}^2\right)} \leq \mu_i f_{iz}, \ i = 1, 2 \ldots n \tag{1}$$

Here, μ_i represents the friction coefficient.

The friction constraint condition at the contact point i of the apple picking robot end-effector can be mathematically represented as a second-order cone, as illustrated in Equation (2).

$$FH_i = \left\{ f_i \in R^3 \middle| \frac{1}{\mu_i} \sqrt{\left(f_{ix}^2 + f_{iy}^2\right)} \leq f_{iz} \right\}, \ i = 1, 2 \ldots n \tag{2}$$

Here, FH_i represents the friction constraint at contact point i of the end-effector.

The above friction cone constraint conditions can be made equivalent using a linear matrix inequality (LMI), that is, $f_i \in FH_i$ can be expressed as the form shown in Equation (3).

$$\begin{bmatrix} \mu_i f_{iz} & 0 & f_{ix} \\ 0 & \mu_i f_{iz} & f_{iy} \\ f_{ix} & f_{iy} & \mu_i f_{iz} \end{bmatrix} \geq 0 \tag{3}$$

2.2. Force Closure Constraint and Analysis

To accurately represent the overall impact of the contact force, it is essential to convert the contact force f_i of each fingertip of the end-effector into the coordinate system where the apple is positioned. The contact coordinate system at each contact point is denoted as b_i, while the coordinate system for the apple is denoted as a_i. The position and orientation of b_i relative to a_i are defined as $w_{abi} = (P_{abi}, R_{abi})$. Consequently, the contact force f_i at each contact point can be represented in the apple coordinate system [18].

$$\begin{bmatrix} R_{abi} & 0 \\ \hat{P}_{abi} R_{abi} & R_{abi} \end{bmatrix} \times B_{bi} \times f_i = G_i f_i \tag{4}$$

Here, $\hat{P}_{abi}$ is expressed as a cross product operator, G_i is the clamping matrix obtained at the contact point i, and B_{bi} is the force spiral base of the contact type with friction point contact.

To stably grasp the apple, the grasping should satisfy the following force balance relationship:

$$GF = \sum_{i=1}^{n} G_i f_i = -W_{ext} \tag{5}$$

where G represents the grasping matrix at each contact point in the apple coordinate system, F represents the contact force at each contact point in the apple coordinate system, and $-W_{ext}$ represents the external force spiral suffered by the apple. Only by ensuring that the force balance equation has a solution, and the solution is inside the friction cone, can it be explained that the apple grasp is a force-closed grasp.

The judgment condition of the contact force sealing of the end=effector is shown in Figure 2.

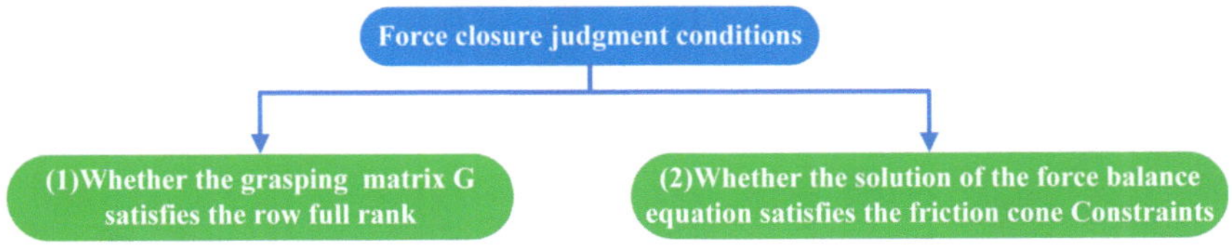

Figure 2. Force closure judgment conditions.

Therefore, combined with the analysis of Sections 2.1 and 2.2, when the end-effector grabs the apple, the three fingers must meet the force closure condition when they are in a balanced state, that is, the contact force spiral and the external force spiral satisfy the force balance relationship shown in Equation (5), and the grasping matrix G row is full rank.

2.3. Torque Constraint and Analysis

The contact force exerted by the fingertip is a result of the joint torque produced by the end-effector. Therefore, a constraint is imposed on the joint torque of the end-effector. This work utilizes the Jacobian matrix to represent the mathematical correlation between the joint torque and the fingertip contact force.

$$J^T f_i = \tau, i = 1, 2 \ldots n \tag{6}$$

Here, J is the Jacobian matrix of the end finger, and τ is the joint torque of the end finger.

In summary, in order to ensure that the damage caused by apple picking is as small as possible, the grasping force of the end-effector needs to be optimized under the premise of satisfying the constraints.

To minimize or prevent damage to the fragile and vulnerable apple, it is essential to optimize the contact force applied to it [19].

2.4. Modeling Force Distribution Optimization for End-Effector

As shown in Figure 3, for optimization of the contact force of the end-effector, it is necessary to comprehensively consider the three constraints to ensure the stability of the entire grasping process.

Therefore, the optimization objective can be described as:

$$\begin{gathered}
\text{minmize } F_{\max} = \max\left\| f_{ix}, f_{iy}, f_{iz} \right\|, i = 1, 2 \ldots n \\
\text{subject to } \frac{1}{\mu_i}\sqrt{\left(f_{ix}^2 + f_{iy}^2\right)} \le f_{iz}, i = 1, 2 \ldots n \\
\sum_{i=1}^{n} G_i f_i = -w_{ext} \\
\tau_l \le J_H^T f_i \le \tau_r
\end{gathered} \tag{7}$$

The barrier function is a constraint function. The characteristic of this function is that as the point approaches the feasible region boundary of the optimization problem, its value tends to infinity. The main function deals with the inequality constraint by adding a penalty term to the objective function, so that the optimization problem is easier to deal with. Equation (8) is selected to deal with the inequality constraints in the optimization model from the obstacle function [20] and is used as the penalty term for the optimization goal.

$$k(u) = \begin{cases} 0 & u > 0 \\ \infty & u \leq 0 \end{cases} \tag{8}$$

For the torque constraint, this is processed by a logarithmic barrier function, which can be defined as Equation (9). The domain of the barrier function is $dorm\,\gamma = \{x \in R^n | g_j(x) \geq 0, j = 1, 2 \cdots n\}$. At the same time, $g_j(x) = c_j - d_j^T x$ defines the torque constraint form.

$$\gamma(x) = -\sum_{j=1}^{n} \ln(g_j(x)) \tag{9}$$

The expression of the torque constraint processed by the obstacle function can be expressed as Equations (10) and (11).

$$\gamma_1(f_i) = -\ln\left(\tau_r - J_H^T f_i\right) \tag{10}$$

$$\gamma_2(f_i) = -\ln\left(-\tau_l + J_H^T f_i\right) \tag{11}$$

For the friction cone constraint, the constraint is a second-order inequality constraint. A second-order cone is defined in the real range, which satisfies the following conditions:

$$\beta = \left\{x \in R^{i+1} | \|x_1, x_2 \ldots x_i\|_2 \leq \alpha x_{i+1}\right\}, i = 1, 2 \ldots n \tag{12}$$

Finally, after processing, the friction cone constraint processing results required in this paper are as follows:

$$\delta(x) = -\ln\left((\mu_i f_{iz})^2 - \|f_{ix}, f_{iy}\|_2^2\right), i = 1, 2 \ldots n \tag{13}$$

The optimization challenge for the distribution of contact forces in the end-effector primarily focuses on minimizing the contact force component while satisfying the given constraints. Thus, due to the penalty imposed on the constraint border of the friction cone, the contact force is prevented from approaching the boundary of the friction cone, resulting in a reduction in the magnitude of the joint torque. The construction of the optimization model for the distribution of contact force on the end-effector is demonstrated in Equation (14) [21].

$$\begin{gathered} \omega(f_i, r_k) = \max\|f_{iz}, f_{ix}, f_{iy}\| - r_k \sum_{u=1}^{N} \frac{1}{\gamma_u(f_i)} + \frac{1}{\sqrt{r_k}} \sum_{v=1}^{M} \left[\delta_v(f_i)^2\right] \\ \text{minimize } \omega(f_i, r_k) \\ \text{subject to } \sum_{i=1}^{M} G_i f_i = -w_{ext} \end{gathered} \tag{14}$$

Here, r_k represents the penalty factor, N represents the number of end-effector joints, the second part represents the obstacle term, the third is the penalty term, and M represents the number of friction cones in contact. At the same time, the values of r_k and $\frac{1}{\sqrt{r_k}}$ are selected according to the interior point method and the exterior point method, respectively.

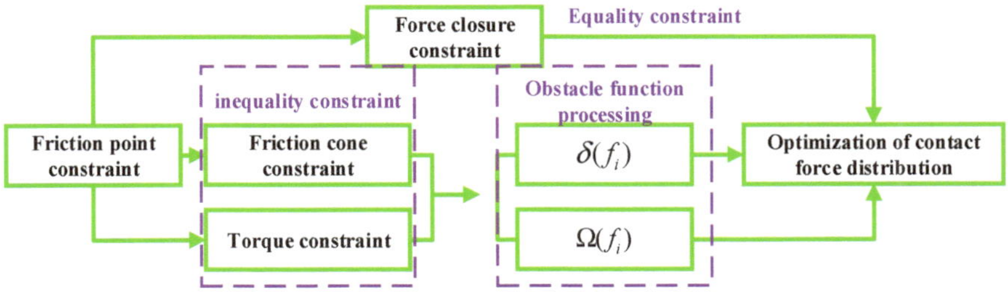

Figure 3. Contact force distribution optimization flow chart.

2.5. Solution to the Contact Force Distribution Optimization Model of End-Effector

The contact force distribution optimization of the three-finger end-effector has been modeled above, and the optimization model will be solved below. The goal of this part is to find the optimal contact force in multiple sets of feasible contact force solutions to minimize the grasping damage on apples. Firstly, the solution set of the force closure constraint condition is decomposed into two parts: the orthogonal special solution f_i^p and the homogeneous solution f_i^h, as shown in Equation (15).

$$f_i = f_i^p + f_i^h = G^+ w_{ext} + M\sigma \tag{15}$$

Here, G^+ represents the pseudo-inverse of the grasping matrix G , the matrix M represents the null space of the external force w_{ext}, and σ represents the vector weight of the external force null space w_{ext}. The special solution f_i^p represents the grasping force exerted by the end-effector to balance the external force on the target apple. The homogeneous solution f_i^h represents the grasping internal force exerted by the end-effector under the zero space of the grasping matrix G , and its role is mainly reflected in improving the grasping stability on the apple.

Therefore, the core objective of dynamic force optimization is transformed as shown in Equation (16).

$$\text{minimize } \vartheta\left(G^+ w_{ext} + M\sigma\right) \tag{16}$$

The quadratic norm defined by the Hessian matrix is used as the step search direction of the Newton iteration algorithm, and the stop criterion of the search is determined by the defined Newton reduction $\sigma + \beta\Delta\sigma_{nt}$ will be iterated as a new predictor for each round:

$$\sigma* = \sigma + \beta\Delta\sigma_{nt} = \sigma - \beta\nabla^2 C(\sigma)^{-1}\nabla C(\sigma) \tag{17}$$

Subsequently, the suitable magnitude of the step β is ascertained using the backtracking line search technique inside the range of (0, 1) to guarantee that each iteration remains within the interval of the viable region [22]. Prior to meeting the search stop requirement, the algorithm will continue searching for a target value that surpasses the current forecast value.

Typically, when the initially chosen feasible solution is far from the desired solution we need to find, the Newton iteration method will require more iterations, and this will significantly decrease computational efficiency. In this part, the solution set of the convex optimization problem exhibits continuity. This means that while the input of $\sigma*$ varies smoothly, the output value similarly changes smoothly. Additionally, an estimated solution for the ideal starting value can be obtained [23].

If a feasible initial iteration value cannot be obtained, the specific solution method and steps for the contact force distribution optimization model are shown in Figure 4.

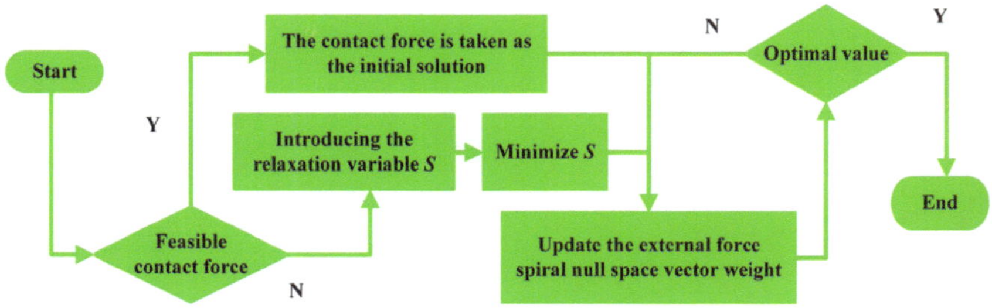

Figure 4. Contact force distribution optimization solution flow chart.

Firstly, a strictly feasible initial solution satisfying the constraint condition of contact force optimization is sought. If there is such a solution, the iterative solution is directly solved. If it does not exist, the relaxation variable S is introduced, and the following optimization description is constructed:

$$\begin{gathered} \text{minimize s} \\ \text{subject to } f_i + se \in FH_i, i = 1,2\ldots n \\ \sum_{i=1}^{n} G_i f_i = -w_{ext} \\ \tau_l \le J_H^T f_i \le \tau_r \end{gathered} \tag{18}$$

where, the relaxation variable $S \in R$, $S \geq -1$ guarantees that the solution cannot reach negative infinity, $e = (0,0,1)$.

In the above second-order optimization problem, the relaxation variable S can be interpreted as a hypothetical force connected to the perpendicular component of the contact force f_i to satisfy the optimization requirements of the grasping issue. The ultimate objective of the optimization problem outlined in Equation (18) is to minimize the virtual force, aiming for a value of 0 or below [21], while ensuring that S is greater than or equal to -1. If the optimization problem can find a suitable optimal solution, f_i can serve as the initial feasible answer for the original optimization problem related to apple grasping contact force. If the value of the relaxation variable S is consistently greater than 0 in the final solution, it signifies that optimization of the original problem is not possible.

The initial feasible solution in the optimization problem of the relaxation variable S described in Equation (18) can be selected in the following ways. A set of minimum norm solutions satisfying the force closure constraint is taken as the initial contact force f_i:

$$f_i = -G_i^T \left(G_i G_i^T \right)^{-1} w_{ext} \tag{19}$$

For the initial relaxation variable S, simply select the value that satisfies the following conditions as its initial value:

$$s > \max\left\{ \frac{1}{\mu_i} \sqrt{\left(f_{ix}^2 + f_{iy}^2 \right)} \le f_{iz}, i = 1,2\ldots n \right\} \tag{20}$$

At this point, the obtained (f_i, s) can be used as the initial feasible solution for the optimization problem of the relaxation variable S constructed using Equation (18).

3. Result and Discussion

3.1. Optimization Example and Simulation of End-Effector Contact Force Distribution

3.1.1. Numerical Example

This section uses MATLAB R2023A to implement the solution for the contact force distribution optimization problem from the previous section. Since only the optimization solution is verified, and the influence of the penalty factor r_k on the optimization model is not considered, a fixed penalty factor is selected for calculation to verify the feasibility of the optimization solution. Finally, this is compared with the semi-definite programming algorithm in order to obtain better computational efficiency, and optimization of the minimum stable apple grasping force is verified through simulation.

It is assumed that the end-effector is a three-fingered hand, and the apple is grasped according to the position shown in Figure 5. The apple has a uniform centroid in its orthocenter, a mass of 0.3 kg, and a radius of 50 mm. The three-fingered fingertips of the end-effector are in contact with the apple. The positions are c_1, c_2, and c_3. Zhang et al. [24] conducted a friction measurement experiment, and obtained the friction coefficient. In this paper, the appropriate friction coefficient is selected according to this friction experiment. The friction coefficient is $\mu_i = 0.5$, and the object coordinate system and the contact coordinate system are established at the apple's center and the contact point, respectively.

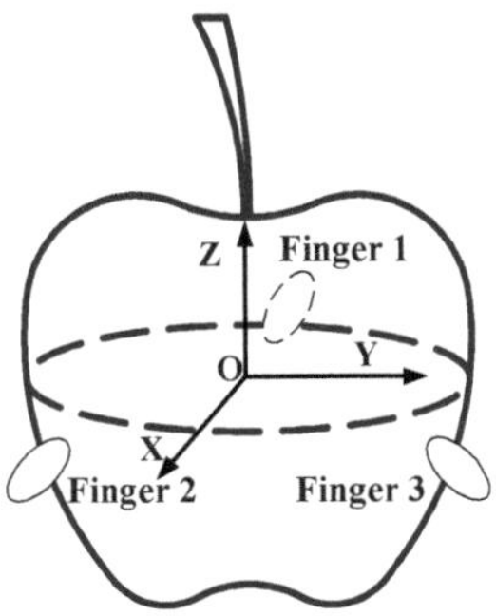

Figure 5. Apple grasping diagram.

The simulation involves using a three-finger end-effector to grip the apple. The specific parameters for this numerical example can be found in Table 1. The contact point coordinates specify the location of the three fingers in relation to the coordinate system of the object.

Table 1. Contact point parameters of the end-effector.

Finger Number	X	Y	Z
1	−48.51	0	−12.26
2	22.45	−41.25	−12.26
3	22.45	41.25	−12.26

Since the coordinates of the contact points are the representation of the contact points in the object coordinate system, the pose of the contact coordinate system relative to the object coordinate system can be calculated according to the coordinates of the contact points, and the grasping matrix G_i of the three fingers can be obtained. The complete grasping matrix G can be obtained by combining them:

$$G = \begin{bmatrix} 0 & 1 & 0 & 1 & 0 & 0 & 1 & 0 & 0 \\ -0.5929 & 0 & 0.7914 & 0 & -0.8137 & -0.5812 & 0 & 0.8137 & -0.5812 \\ -0.7914 & 0 & 0.5929 & 0 & 0.5812 & 0.8137 & 0 & 0.5812 & 0.8137 \\ -7.2689 & 0 & 9.7026 & 0 & -33.9504 & -40.6906 & 0 & 33.9504 & 26.4396 \\ -33.6424 & -12.26 & 25.2042 & -12.26 & -13.0749 & -18.2672 & -12.26 & -13.0749 & -18.2672 \\ 25.2042 & 0 & -33.6424 & 41.25 & -18.2672 & -13.0749 & -41.25 & 18.2672 & -13.0749 \end{bmatrix}$$

The contact force vector is:

$$f = \left[f_{1,x}, f_{1,y}, f_{1,z}, f_{2,x}, f_{2,y}, f_{2,z}, f_{3,x}, f_{3,y}, f_{3,z}\right] \tag{21}$$

The friction cone constraint is:

$$FH = FH_1 \times FH_2 \times FH_3 \tag{22}$$

$$FH_i = \left\{ f_i \in R^3 \middle| \frac{1}{\mu_i} \sqrt{\left(f_{ix}^2 + f_{iy}^2\right)} \le f_{iz}, i = 1,2,3 \right\} \tag{23}$$

The Jacobian matrix of the end-effector fingertips is:

$$J_H = \begin{bmatrix} 0 & 54.276 & 102.244 & 0 & 0 & 0 & 0 & 0 & 0 \\ 0 & -6.283 & 83.584 & 0 & 0 & 0 & 0 & 0 & 0 \\ 0 & 0 & 0 & -61.576 & 0 & 114.012 & 0 & 0 & 0 \\ 0 & 0 & 0 & 13.014 & 0 & -96.376 & 0 & 0 & 0 \\ 0 & 0 & 0 & 0 & 0 & 0 & -61.576 & 0 & 114.012 \\ 0 & 0 & 0 & 0 & 0 & 0 & -13.014 & 0 & 96.376 \end{bmatrix}$$

Each joint torque of the end-effector is $\tau_l = -3$ N·m, $\tau_r = 3$ N·m. According to the selected penalty factor $r_k = 4$, a set of optimal contact forces is obtained:

$$f = [2.483, 1.679, 4.073, 1.779, 1.613, 2.048, 2.569, 1.463, 3.164]^T$$

Applying the semi-definite programming approach to determine the ideal contact force in the given gripping scenario will need additional floating-point operations, resulting in a significant decrease in computational efficiency. The precise calculation efficiency is displayed in Table 2, and these are compared in the same context of a computer with the Windows 11 operating system and MATLAB R2023A (9.14.0.2206163) 64-bit (win64) software.

Table 2. Computational efficiency comparison index.

Contrast Ratio	Method of this Article	Semidefinite Programming Algorithm
Computing time (ms)	0.346	0.675
Number of convergences	5	9
Number of floating-point operations (kflops)	7	19

3.1.2. Grasping Simulation

The previous section verified the efficiency of the improved Newton iterative algorithm in solving the optimization problem of contact force distribution under fixed penalty factors. This section will continue to study the selection of penalty factors in MATLAB R2023A by selecting appropriate penalty factors for the entire optimization model of contact force distribution, simulating the three-finger grasp and observing the law of contact force change.

In the simulation process, the apple object coordinate system is always kept at the position of $p = (0, -4, 2.5)$ relative to the end-effector coordinate system. At this time, the contact position and vector of each end-effector fingertip are:

$$R_{oc1} = [0.87, 0, 0.5] \quad R_{oc2} = [-0.8, 0.5, 0] \quad R_{oc3} = [0.8, 0.5, 0]$$
$$p_{oc1} = [0.87, 15, -2.5] \quad P_{oc2} = [13, 7.5, 4.35] \quad P_{oc3} = [-14, 7.5, 4.35]$$

The penalty factor r_k is gradually increased from 0.1 to 100 with a step size of 0.1, and the optimal contact force normal component f_{iz} of each finger is recorded in turn. The variation of the normal contact force f_{iz} with the penalty factor r_k is shown in Figure 6. By observing the change rule of the following figure, it can be expected that when the penalty factor r_k gradually increases, the contact force of the end-effector's three fingertips will be significantly reduced, and this will also increase the possibility of grasping failure.

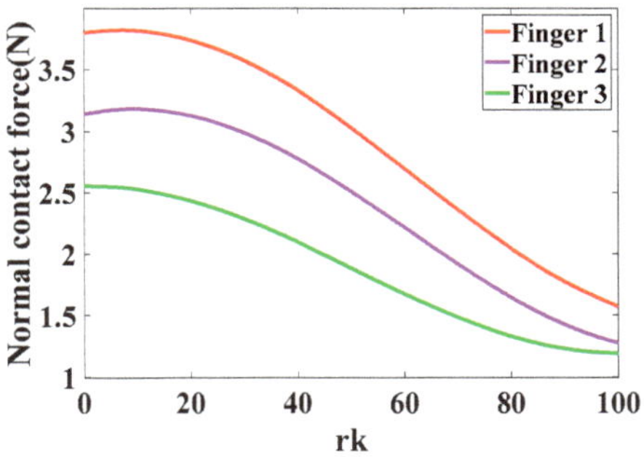

Figure 6. Contact force variation diagram.

To establish a well-defined boundary for the penalty factor r_k, the friction residual and the torque residual are introduced. These residuals serve as approximations to assess the proximity of the constraint boundary of the friction cone and the difference between the joint torque and the limit torque of each finger, respectively. The torque residual and friction residual, which are normalized, are depicted in Figure 7, showcasing their variation with the penalty factor r_k as indicated in Equations (24) and (25).

$$\varphi_\tau = \text{minimize}\left\{|\tau_i - \tau_{\lim}|\right\}, i = 1,2,3 \tag{24}$$

$$\varphi_F = \text{minimize}\left\{\mu_i f_{iz} - \left\|f_{ix}, f_{iy}\right\|_2\right\}, i = 1,2,3 \tag{25}$$

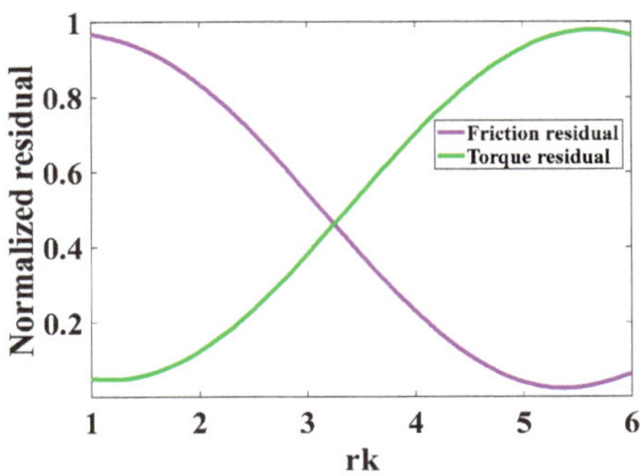

Figure 7. Residual variation diagram.

It can be seen from the Figure 7 that when r_k is large, the penalty for the joint torque is more severe, and the joint torque is small. At this time, there is a risk of grasping failure. When r_k is small, the penalty near the edge of the friction cone is more severe. At this time, the torque is larger and the grasp is more stable. Therefore, when the friction residual φ_F is equal to the torque residual φ_τ, the penalty factor value is the best trade-off between improving grasping stability and reducing apple damage.

The selected penalty factor $r_k = 3.25$ is substituted into the contact force optimization model for grasp simulation. The gravity center of the apple is still maintained at the position relative to the end-effector coordinate system $p = (0, -4, 2.5)$. The movement time of the simulation is 3 s. The law of the contact force generated by the three fingertips over time and the torque residuals are shown in Figures 8 and 9. From the diagram, it can be observed that the contact force component generated by the three fingertips does not exceed the torque limit value. This shows that during the whole simulated grasping process, the three fingers never slide, and the grasping is stable. Finally, the whole simulation proves that the contact force distribution optimization scheme proposed in this study is feasible.

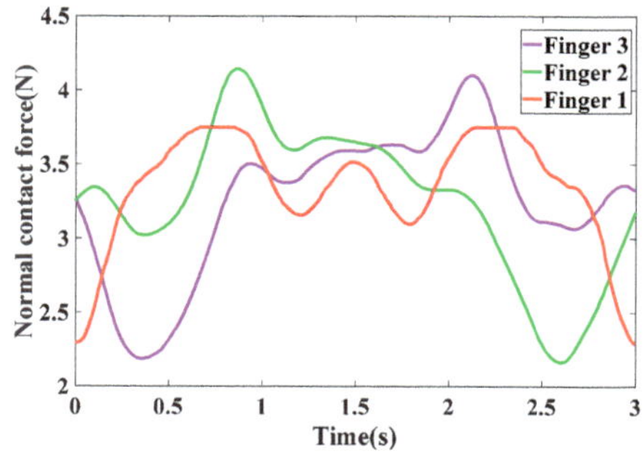

Figure 8. Contact force variation law.

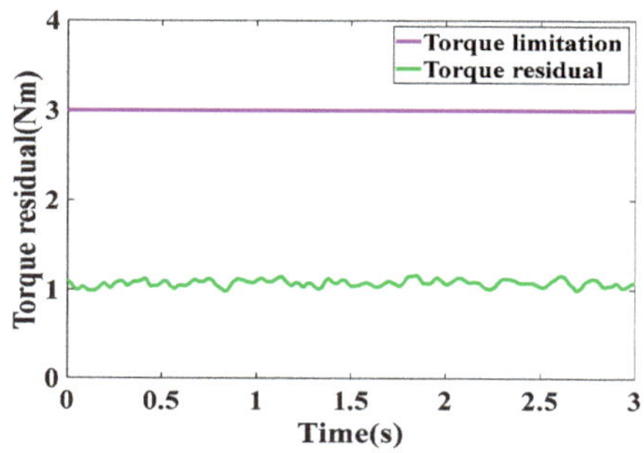

Figure 9. Torque residual.

The optimal contact force of the apple picking robot end-effector developed in the laboratory is solved by the contact force optimization scheme proposed in this section. Finally, a set of solutions are obtained as follows:

$$f = [4.597, 3.648, 5.152, 4.779, 3.793, 5.178]^T$$

3.2. Contact Force Distribution Optimization Experiment of End-Effector

3.2.1. Experimental Equipment

The apple picking robot is shown in Figure 10a, which consists of a chassis, a robot arm, an end-effector, a camera, and a controller. The experimental equipment will be used to verify the optimization of grasping force in practical experiments.

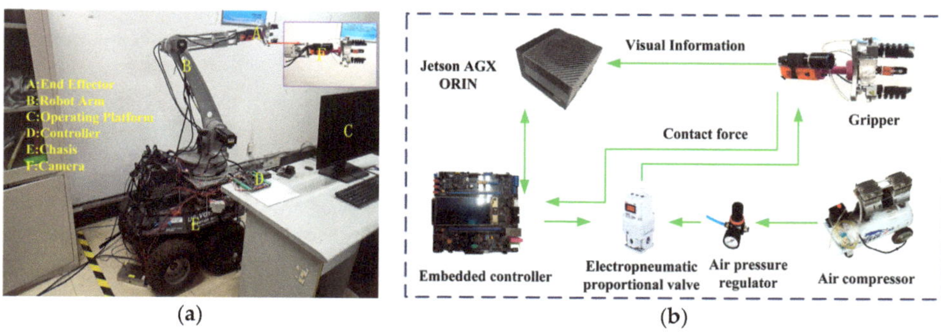

Figure 10. The experimental equipment. (**a**) Apple picking robot; (**b**) the control system diagram of the end-effector.

The control system diagram of the end-effector is shown in Figure 10b, which consists of an end-effector, a Jetson AGX ORIN, an embedded controller, a force sensor, a cylinder, and an electric proportional valve. The end-effector is mounted on a Festo push rod for picking apples. During the grasping task, the grasping force signal between the end-effector and the apple is detected by the force sensor and transmitted to the embedded controller. The embedded controller adjusts the electric proportional valve according to the error to achieve force control. Finally, it determines whether the apple slips or the grasping is successful based on whether there is a sudden change in the grasping force.

3.2.2. Grasping Experiment and Analysis

The numerical samples and simulation verify the practicality of the contact force optimization approach when grasping apples. Nevertheless, throughout grasping, it is possible for the position of contact and the friction coefficient to undergo alterations. These changes can be specified to follow a normal distribution with a change rate of 0.05. The following equation can be employed to define the threshold of grasping stability:

$$s = \underset{i=1,2}{\text{minimize}}\|f_{ix}, f_{iy}, f_{iz}\| - \underset{i=1,2}{\text{optimize}}\|f_{ix}, f_{iy}, f_{iz}\| \tag{26}$$

where, $\underset{i=1,2}{\text{minimize}}\|f_{ix}, f_{iy}, f_{iz}\|$ represents the minimum grasping force required to satisfy the grasping stability condition in the actual grasping process, and $\underset{i=1,2}{\text{optimize}}\|f_{ix}, f_{iy}, f_{iz}\|$ represents the optimal grasping force obtained by the contact force distribution optimization algorithm. When $S < 0$, it indicates that there is an unmet constraint condition for grasping failure, otherwise it means that the grasp is successful.

Figure 11 shows the success rate results obtained by 1000 Monte Carlo simulation probability analysis iterations for each penalty factor under different penalty factors. If the above-mentioned conditions for successful grasping are met, it is recorded as Y, otherwise it is recorded as N. Until the end of the iteration, the number of successful grasps is T_Y, and the number of failed grasps is T_N, so the success rate of grasping can be calculated as:

$$P_Y = T_Y/(T_N + T_Y) \tag{27}$$

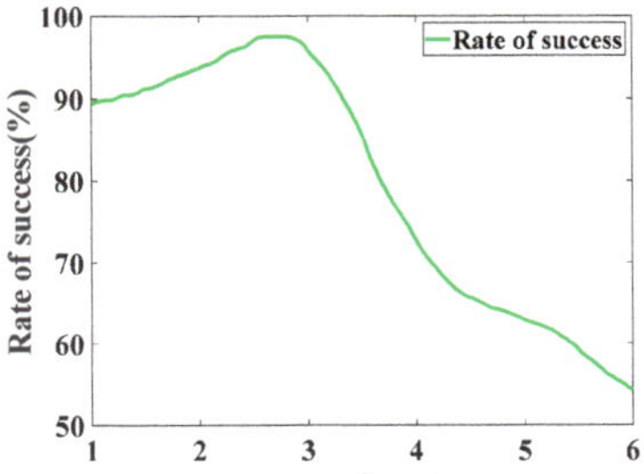

Figure 11. Curve of grasp success rate.

It can be observed from the above figure that when the penalty factor is in the range of (1, 3.3), the grasping success rate is higher, basically above 90%, and the contact force distribution optimization scheme proposed in this paper also shows high stability. When the value of r_k exceeds this interval, the probability of sliding during the actual grasping process will increase. An r_k value in the interval of (1, 3.3) and outside this interval are simulated several times.

In the actual experiment, a red Fuji apple was used as the grasping object for grasping verification. The red Fuji apple peel was smooth, shiny, medium thick, tough, and waxy. The relatively stable apple grasp results and the sliding grasp results can be observed, as shown in Figure 12a,b.

Figure 12. Comparison of grasping phenomena with different penalty factors. (**a**) Stable grasp, (**b**) sliding grasp.

The aforementioned studies demonstrate that selecting the penalty factor within the range of (1, 3.3) results in the best success rate for grasping, hence confirming the reliability of the contact force optimization model. To calculate the lowest stable grasping force input for the end-effector controller, we will monitor the damping control variable [25] to detect any sudden changes in the real grabbing force, which may indicate whether the apple is sliding. Figure 13 displays the experimental results of the grasping force distribution with different penalty factors.

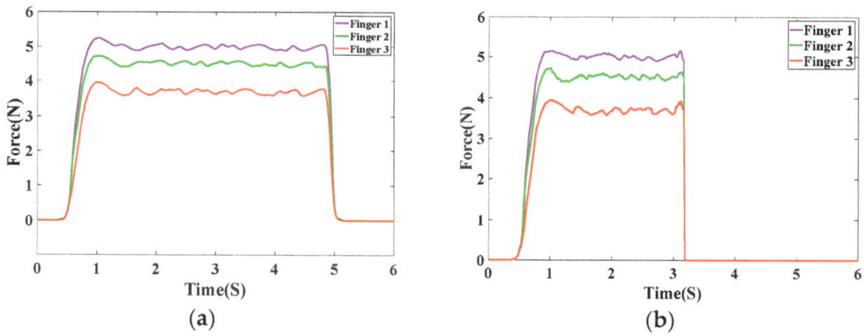

Figure 13. Experimental results of the grasping force distribution with different penalty factors. (**a**) Stable grasp; (**b**) sliding grasp.

The experimental results show that there is no sudden change in contact force during the grasping process, that is, the apple does not slide. When the penalty factor is selected within the range of (1, 3.3), the optimal contact force solved by the contact force optimization model can result in stable grasping.

3.3. Discussion

In this study, the grasp stability of end-effectors was analyzed based on the force closure theory, the constraint conditions were treated using an obstacle function, and an optimal model of end-effector contact force distribution was constructed by introducing penalty factors. Then, the improved Newton method was used to solve the contact force distribution model. The optimal contact force for grasping apples was determined using numerical case simulation analysis on the premise of selecting the penalty factor, and the effectiveness of the solution was verified. Finally, the practical significance of the method for apple grasping is verified in an actual grasping experiment. The experiment showed that the method can provide the minimum stable grasping force for the end-effector to achieve a stable grasp, and avoid sliding caused by a too small grasp force or fruit damage caused by a too large grasp force. Although this method realizes a stable grasp of the apple, the computational cost still exists. In many special weather conditions such as rain, dew or high humidity, apples have less friction, which has not been studied. This would require a lot of programming. The development of soft manipulator claws or slip detection are good research directions, which could also achieve the lossless and stable grasp of apples. In the future, we will study the sliding friction and sliding signal detection of an apple grasp.

4. Conclusions

In this paper, the grasping stability of end-effectors was analyzed based on the force closure theory, considering the force closure constraint, the nonlinear friction cone constraint and the torque. The constraint conditions were processed using an obstacle function, and the contact force distribution optimization model of a three-finger end-effector was constructed by introducing a penalty factor. The improved Newton method was used to solve the contact force distribution model, and through numerical simulation analysis, it was found that the optimal contact force for grasping a target apple can be determined under the premise of selecting the penalty factor. The calculation efficiency was compared with that of the semi-definite programming algorithm. It was found that the calculation time of the method proposed in this paper was 0.346 ms, while the semi-definite cost-effective rule required 0.675 ms, which verified the efficiency of the proposed method. It was found that the optimal contact force solved by the optimization method did not exceed the torque limit by using MATLAB simulation. The iterative method of Monte Carlo simulation probability analysis was used to calculate the optimal contact force required for grasping stability under different penalty factors for multiple iterations. It was found that when the penalty factor was selected within the range [1,3.3], the grasping success rate was the highest, basically reaching more than 90%. The actual grasping experiment further verified that this method can provide the minimum stable grasping force needed for the apple picking robot and achieve the stable grasping of apples. Based on the results of this experiment, we will consider comparing research on the friction force and friction coefficient of an apple under wet and dry conditions in the future, and further study the compliance control strategy based on sliding signal analysis.

Author Contributions: Conceptualization, W.J. and H.Z.; methodology, B.X. and H.Z.; software, H.Z. and X.Y.; validation, X.Y.; formal analysis, H.Z.; investigation, W.J.; data curation, H.Z.; resources, W.J.; writing—original draft preparation, H.Z.; writing—review and editing, W.J. and X.Y.; visualization, B.X.; supervision, W.J.; project administration, W.J.; funding acquisition, B.X. and W.J. All authors have read and agreed to the published version of the manuscript.

Funding: This research was funded by the National Natural Science Foundation of China (No. 61973141) and a project funded by the Priority Academic Program Development of Jiangsu Higher Education Institutions (No. PAPD).

Institutional Review Board Statement: Not applicable.

Data Availability Statement: Data are contained within the article.

Conflicts of Interest: The authors declare no conflicts of interest.

References

1. Zhang, K.; Lammers, K.; Chu, P.; Li, Z.; Lu, R. System design and control of an apple harvesting robot. *Mechatronics* **2021**, *79*, 102644. [CrossRef]
2. Ji, W.; He, G.; Xu, B.; Zhang, H.; Yu, X. A new picking pattern of a flexible three-fingered end-effector for apple harvesting robot. *Agriculture* **2024**, *14*, 102. [CrossRef]
3. Gongal, A.; Amatya, S.; Karkee, M.; Zhang, Q.; Lewis, K. Sensors and systems for fruit detection and localization: A review. *Comput. Electron. Agric.* **2015**, *116*, 8–19. [CrossRef]
4. Chen, Z.; Wu, Q.; Hong, C.; Zhang, X. Multi-fingered grasping force optimization based on generalized penalty-function concepts. Robot. *Auton. Syst.* **2021**, *135*, 103672. [CrossRef]
5. James, J.; Lepora, N. Slip detection for grasp stabilization with a multi-fingered tactile robot hand. *IEEE Trans. Robot.* **2021**, *37*, 506–519. [CrossRef]
6. Mavrakis, N.; Hao, Z.; Gao, Y. On-orbit robotic grasping of a spent rocket stage: Grasp stability analysis and experimental results. *Front. Robot. AI* **2021**, *8*, 652681. [CrossRef]

7. Xu, W.; Guo, W.; Shi, X.; Sheng, X.; Zhu, X. Fast force-closure grasp synthesis with learning-based sampling. *IEEE Robot. Autom. Lett.* **2023**, *8*, 4275–4282. [CrossRef]
8. Li, Z.; Li, P.; Yang, H.; Wang, Y. Stability tests of two-finger tomato grasping for harvesting robots. *Biosyst. Eng.* **2013**, *116*, 163–170. [CrossRef]
9. Buss, M.; Hashimoto, H. Dextrous hand grasping force optimization. *IEEE Trans. Robot. Autom.* **1996**, *12*, 406–418. [CrossRef]
10. Li, Q.; Gao, D.; Deng, H. The sealing of the contact force of the heavy-duty clamping device. *J. Mech. Eng. Sci.* **2010**, *46*, 36–42. [CrossRef]
11. Liang, Z.; Jiang, B.; Quan, P.; Lin, H.; Lou, Y.; Di, S. Force-closure analysis of multilink cable-driven redundant manipulators considering cable coupling and friction effects. *IEEE-ASME Trans. Mechatron.* 2023, *in press*. [CrossRef]
12. Liu, Z.; Jiang, L.; Yang, B. Task-oriented real-time optimization method of dynamic force distribution for multi-fingered grasping. *Int. J. Humanoid Robot.* **2022**, *19*, 2250013. [CrossRef]
13. Li, G.; Xu, P.; Qiao, S.; Li, B. Stability analysis and optimal enveloping grasp planning of a deployable robotic hand. *Mech. Mach. Theory* **2021**, *158*, 104241. [CrossRef]
14. Liu, Q.; Xu, Y.; Zhou, Y. A class of exact penalty functions and penalty algorithms for non-smooth constrained optimization problems. *J. Glob. Optim.* **2020**, *76*, 745–768. [CrossRef]
15. Liu, Y.; Jiang, D.; Tao, B.; Qi, J.; Jiang, G.; Yun, J.; Huang, L.; Tong, X.; Chen, B.; Li, G. Grasping posture of humanoid manipulator based on target shape analysis and force closure. *Alex. Eng. J.* **2022**, *61*, 3959–3969. [CrossRef]
16. Dharbaneshwer, S.; Thondiyath, A. Contact area-based modeling of robotic grasps using deformable solid mechanics. *Int. J. Appl. Mech.* **2021**, *13*, 2150038. [CrossRef]
17. Benkhira, E.; Fakhar, R.; Mandyly, Y. Analysis and numerical approximation of a contact problem involving nonlinear hencky-type materials with nonlocal coulomb's friction law. *Numer. Funct. Anal. Optim.* **2019**, *40*, 1291–1314. [CrossRef]
18. Wang, Y.; Dehio, N.; Kheddar, A. On inverse inertia matrix and contact-force model for robotic manipulators at normal impacts. *IEEE Robot. Autom. Lett.* **2022**, *7*, 3648–3655. [CrossRef]
19. Righetti, L.; Buchli, J.; Mistry, M.; Kalakrishnan, M.; Schaal, S. Optimal distribution of contact forces with inverse-dynamics control. *Int. J. Robot. Res.* **2013**, *32*, 280–298. [CrossRef]
20. Liu, J. Converse barrier functions via Lyapunov functions. *IEEE Trans. Autom. Control* **2022**, *67*, 497–503. [CrossRef]
21. Deng, Z.; Fang, B.; He, B.; Zhang, J. An adaptive planning framework for dexterous robotic grasping with grasp type detection. *Robot. Auton. Syst.* **2021**, *140*, 103727. [CrossRef]
22. Chen, Y.; Sun, Q.; Guo, Q.; Gong, Y. Dynamic modeling and experimental validation of a water hydraulic soft manipulator based on an improved newton-euler iterative method. *Micromachines* **2022**, *13*, 130. [CrossRef] [PubMed]
23. Agrawal, A.; Barratt, S.; Boyd, S. Learning convex optimization models. *IEEE-CAA J. Autom. Sin.* **2021**, *8*, 1355. [CrossRef]
24. Zhang, X.; Yin, L.; Shang, S.; Chung, S. Mechanical properties and microstructure of Fuji apple peel and pulp. *Int. J. Food Prop.* **2022**, *25*, 1773–1791. [CrossRef]
25. Ji, W.; Tang, C.; Xu, B.; He, G. Contact force modeling and variable damping impedance control of apple harvesting robot. *Comput. Electron. Agric.* **2022**, *198*, 107026. [CrossRef]

agriculture

Article

Parametric Design and Genetic Algorithm Optimization of a Natural Light Stereoscopic Cultivation Frame

Dongdong Jia [1,2], Wengang Zheng [2], Xiaoming Wei [2], Wenzhong Guo [2], Qian Zhao [2] and Guohua Gao [1,*]

1 College of Mechanical & Energy Engineering, Beijing University of Technology, Beijing 100124, China; jiadd@nercita.org.cn
2 Research Center of Intelligent Equipment, Beijing Academy of Agriculture and Forestry Sciences, Beijing 100097, China; zhengwg@nercita.org.cn (W.Z.); weixm@nercita.org.cn (X.W.); guowz@nercita.org.cn (W.G.); zhaoq@nercita.org.cn (Q.Z.)
* Correspondence: ggh_bjut@126.com

Abstract: Vertical farming (VF) is an emerging cultivation frame that maximizes total plant production. However, the high energy-consuming artificial light sources for plants growing in the lower and middle layers significantly affect the sustainability of the current VF systems. To address the challenges of supplementary lighting energy consumption, this study explored and optimized the structural design of cultivation frames in VF using parametric modeling, a light simulation platform, and a genetic algorithm. The optimal structure was stereoscopic, including four groups of cultivation trough units in the lower layer, two groups in the middle layer, and one group in the upper layer, with a layer height of 685 mm and a spacing of 350 mm between the cultivation trough units. A field experiment demonstrated lettuce in the middle and lower layers yielded 82.9% to 92.6% in the upper layer. The proposed natural light stereoscopic cultivation frame (NLSCF) for VF was demonstrated to be feasible through simulations and on-site lettuce cultivation experiments without supplementary lighting. These findings confirmed that the NLSCF could effectively reduce the energy consumption of supplemental lighting with the ensure of lettuce's regular growth. Moreover, the designing processes of the cultivation frame may elucidate further research on the enhancement of the sustainability and efficiency of VF systems.

Keywords: vertical farming; A-shape cultivation frame; parametric model; genetic algorithm; solar radiation; sunshine duration; light simulation

Citation: Jia, D.; Zheng, W.; Wei, X.; Guo, W.; Zhao, Q.; Gao, G. Parametric Design and Genetic Algorithm Optimization of a Natural Light Stereoscopic Cultivation Frame. *Agriculture* **2024**, *14*, 84. https://doi.org/10.3390/agriculture14010084

Academic Editor: Hyeon Tae Kim

Received: 22 November 2023
Revised: 28 December 2023
Accepted: 30 December 2023
Published: 30 December 2023

1. Introduction

Vertical farming (VF) has garnered global attention as a modern cultivation system that optimizes land utilization rates and ensures high product quality [1–4]. VF, often implemented in plant factories or urban skyscraper farms, relies on controlled environmental conditions for plant growth [5]. Light is one of the most critical factors in VF, which is closely related to the growth process of plants. Additionally, researchers have conducted various studies regarding crucial factors of VF, such as artificial lighting sources [6–8], environmental control [9,10], energy efficiency [11,12], nutrient solutions [13], and planting modes [14]. However, the overreliance on supplemental lighting in VF, accounting for 40% to 80% of the total electricity consumption [15–17], poses a substantial barrier to widespread implementation and commercial usage in skyscraper farms and plant factories [18,19].

In response to the challenges posed by artificial supplemental light, researchers have investigated alternative structures of stereoscopic cultivation frames and their interlayer shading influences. For example, Wang et al. compared three widely used vertical cultivation frames (also known as H-shape frames) for the stereo-cultivation of strawberries, showing that both of the three-layer arrangements were worse than the two-layer arrangement regarding light conditions, growth, and yield per plant [20]. Liao et al. found that a stereoscopic vertical cultivation frame could resolve the continuous cropping obstacle in growing *Panax notoginseng*, but the light intensity of the lower layer was only 1/10 of the upper layer even at noon [21]. Recent evidence showed that the staggered arrangement of cultivation frames (also known as the A-shape or trapezoidal frames) offers superior light conditions for crops compared to traditional vertical frames [22–24]. In addition Wang et al. even added a rotating mechanism to an A-shape cultivation frame, aligning with the sun's direction to significantly increase natural light utilization and the overall strawberry yield. However, the complex structure and high maintenance costs limit its practical applicability [25].

Taken together, these studies highlight that optimizing cultivation frame structures presents a cost-effective strategy for increasing crop yields while mitigating energy consumption. Nevertheless, existing studies have often employed predesigned cultivation frame structures, leaving the complex influence of structural parameters, such as arrangement directions, layer height, and the number of cultivation trough units (CTUs), in light distribution inadequately explored. Notably, challenges related to shading in the lower and middle layers of cultivation frames remain unresolved.

Recent academic discussions have emphasized the advantages of parametric design, which intelligently links essential parameters to performance optimization during the early stages of design. By simultaneously integrating and coordinating design components with environmental factors, specific software, plug-ins, and intelligent algorithms of the parametric design could be helpful in modifying and improving a design effectively. For example, light simulation software, such as Rhino 7/Grasshopper, were used to optimize the external shading facilities in building design [26,27]. Light simulation has been used in agricultural research in recent years, such as the radiance, to design the light absorption of individual plants [28] and the light distribution at the canopy of mango trees [29]. These studies offered promising ways of exploring the patterns of light distribution in the cultivation frame. Additionally, genetic algorithms have found widespread application in optimizing greenhouse climate factors and estimating the multivariate nature of the greenhouse system [30].

In light of these considerations, we make the assumption that the innate structure of the cultivation frame in VF relates directly to the shading issues of its lower layers. By investigating the effect of the growing frame structure on the light conditions of the crop, it is possible to use a low cost to achieve higher yields. Light simulation and a genetic algorithm become a feasible way to solve the shortage of light in the lower layer of the stereoscopic cultivation frame and to reduce the high energy consumption of the VF.

This study is driven by the overarching objective of optimizing the structural design of VF cultivation frames to address the challenges of supplementary lighting energy consumption. Specifically, the research aim is to introduce a natural light stereoscopic cultivation frame (NLSCF) that not only reduces energy consumption but also enhances light distribution for optimal crop growth. Parametric models and a light simulation platform were built for the NLSCF using Grasshopper and Ladybird plug-ins. Then, we analyzed and optimized other structural parameters of the NLSCF by the multi-objective genetic algorithm. Finally, we applied a field cultivation experiment to explore if the NLSCF could meet the lighting needs of lettuce. Therefore, the results are anticipated to provide valuable insights into resolving shading issues in the lower layers of the stereoscopic cultivation frame, reducing its energy consumption of supplemental light and improving the sustainability of VF. This research and the NLSCF are poised to benefit agricultural scientists, agricultural technology developers, and practitioners in vertical farming, offering practical guidance for optimizing crop growth environments and enhancing agricultural productivity.

2. Materials and Methods

2.1. Research Processes

To provide the lower layer crops with sufficient light to meet their growth requirements, this study combined the cultivation frame structure design demands, the minimum daily light integral for crop growth, parametric modeling, light simulation, and a genetic algorithm to design and optimize the structure of the stereoscopic cultivation frame. We combined cultivation verification experiments so as to ensure the completeness of this research path.

The research outline is shown in Figure 1. In the first stage, we analyzed the design requirements of the cultivation frame, including the structure of the CTU and the structure of the cultivation frame. In the second stage of conducting light simulation of the cultivation frame, we established the parameter model and light simulation platform for each layer of cultivation frames in Grasshopper and Ladybug. In the third stage, we designed the structure of the cultivation frames. We used Ladybug 1.5 to simulate and analyze the lighting of the cultivation frame, including the simulation analysis of the direction of the cultivation frame arrangement, the influence of the structure of the cultivation frame on the solar radiation of the middle and lower layers, and the calculation of the number of CTUs in each layer. The fourth stage was to optimize the structural parameters of the cultivation frame using a genetic algorithm. The fifth stage was to verify the feasibility of the design, and a lettuce cultivation experiment was used for verification. Through the above series of research processes, the design and optimization of the structural parameters were finally completed to meet the design requirements of the NLSCF.

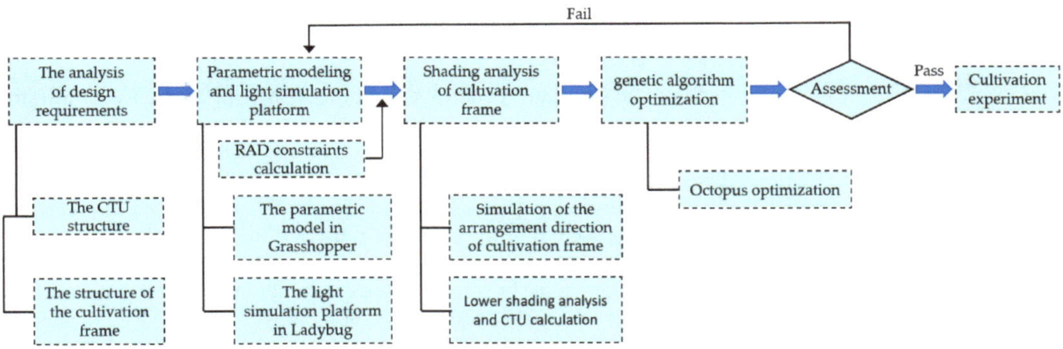

Figure 1. Design optimization flow chart.

2.2. The Software and Design Requirement Analysis

2.2.1. Software

The software Grasshopper used in this study is a parametric modeling plug-in in the 3D modeling software Rhino 7. It uses program algorithms to generate parametric models and render 3D models in Rh. Designers can model, simulate, and analyze in a single software, saving time in iterative modeling. We used the Ladybug 1.5 plug-in for light simulation analysis in Grasshopper, as it can simulate the sunshine duration and solar radiation to which an object is exposed, as well as work with parametric design software. Grasshopper also has a multi-objective genetic algorithm solver, which can optimize multiple target values [31].

2.2.2. The CTU Structure and the Cultivation Frame Parameter Design Requirements

As shown in Figure 2, the CTU in this study is composed of a planting board, a planting cup, slides, and a U-shaped cultivation trough. Slides are installed at both ends of the cultivation trough, facilitating the handling of the cultivation plates. The U-shaped cultivation troughs are used for easy interior cleaning. The width of the CTU is 350 mm, the spacing of the cultivation holes is 175 mm, and the height is 70 mm. Each cultivation board is capable of planting 6 lettuces. To facilitate manual operation, the structural parameters of the designed NLSCF need to meet the following conditions:

- The cultivation frame should have 3 layers, the layer height is over 300 mm, the number of CTUs in the lower layer is 4 groups, and the overall width is 1400 mm.
- The height of the cultivation frame should not exceed 2 m, and the between-group distance of the cultivation frame is 500 mm.
- The width of the middle and upper layers should not exceed the width of the bottom layer.

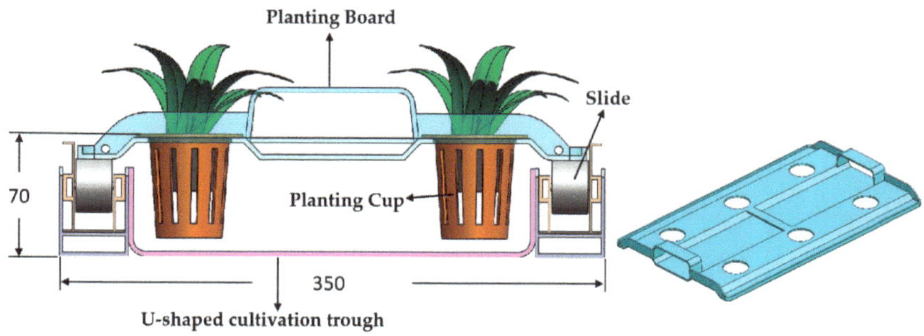

Figure 2. Structure diagram of the cultivation trough unit.

2.3. Establishment of the Parametric Model

To facilitate the simulation analysis, the bracket part of the cultivation frame was omitted in the parametric modeling processes. The parametric variables of the cultivation frame were the layer height (H), the number of CTUs in each layer (N), and the distance between the CTUs (d). Each CTU was simplified as a rectangle with a width of 350 mm, a height of 70 mm, and a length of 4000 mm. Figure 3 is the schematic diagram of the parametric modeling section of the cultivation frame. To comprehensively analyze the shading pattern of the lower layers, the four CTU groups were divided into inner and outer groups for modeling; the height of the lower layer of the CTUs was 400 mm from the ground. To restrict the total height of the cultivation frame to less than 2 m, the layer height (H) ranged from 300 mm to 700 mm, and the number of the CTUs was set, respectively, as N = 1, 2, 3, and 4. The spacing between the CTUs (d) ranged from 0 mm to 700 mm. After analyzing the modeling parameters of the cultivation frame, the complete model of each layer was constructed in Grasshopper, and the parameters were set as digital sliders. Taken together, Figure 4 summarizes the battery connection for the cultivation frame model in Grasshopper.

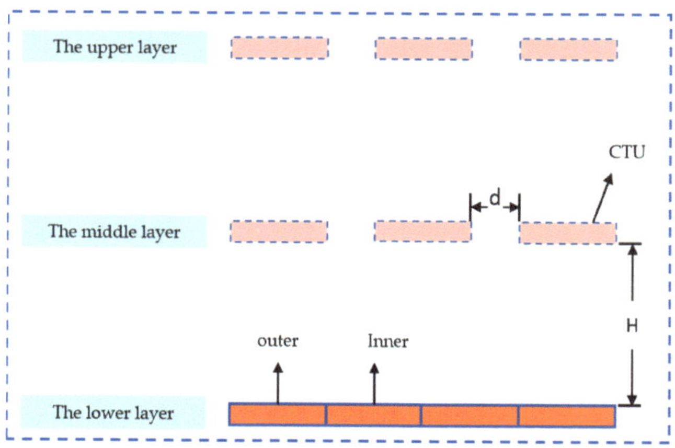

Figure 3. The schematic diagram of the parametric modeling section of the cultivation frame.

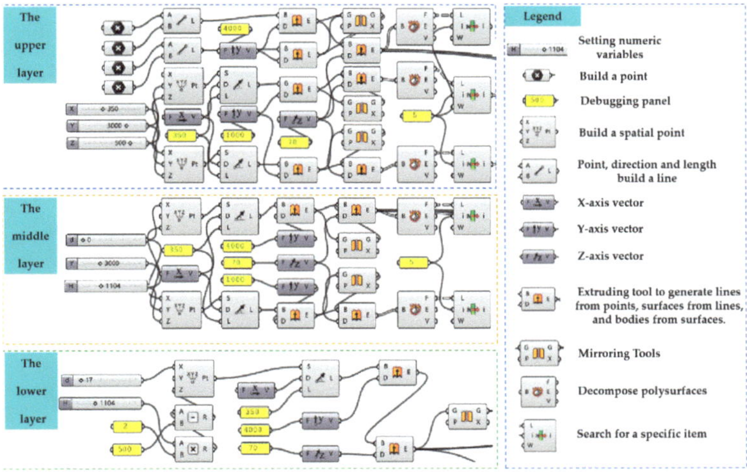

Figure 4. The battery connection diagram of the model in Grasshopper.

2.4. Establishment of the Light Simulation Platform

To simulate and analyze the sunshine duration and the solar radiation of the cultivation frame, we used the Ladybug 1.5 plug-in to build a light simulation platform for the cultivation frame. The location of the light simulation was Beijing, China. Firstly, we obtained the EPW meteorological data file of Beijing from https://www.ladybug.tools/epwmap/ (accessed on 27 July 2022), which contains the meteorological parameters of Beijing at any time of the year. Secondly, the standard EPW meteorological data file was imported into Ladybug 1.5 to construct the light simulation platform of sunshine duration and solar radiation (Figure 5). To make the results of the study more generalizable, the winter solstice (21 December), which has the worst light conditions in a year, was chosen for the cultivated frame light simulation. The solar trajectory of the winter solstice from 6 a.m. to 6 p.m. is shown in Figure 6. After the construction of the lighting simulation platform is completed, the simulation object (lower or middle layer of the cultivation frame) and the shadings (upper layer and its two sides of the cultivation frame) are, respectively, connected to the Geometry and Context ends of the sunshine duration module and the solar radiation module for simulation. Finally, we exported the simulation results for further analysis.

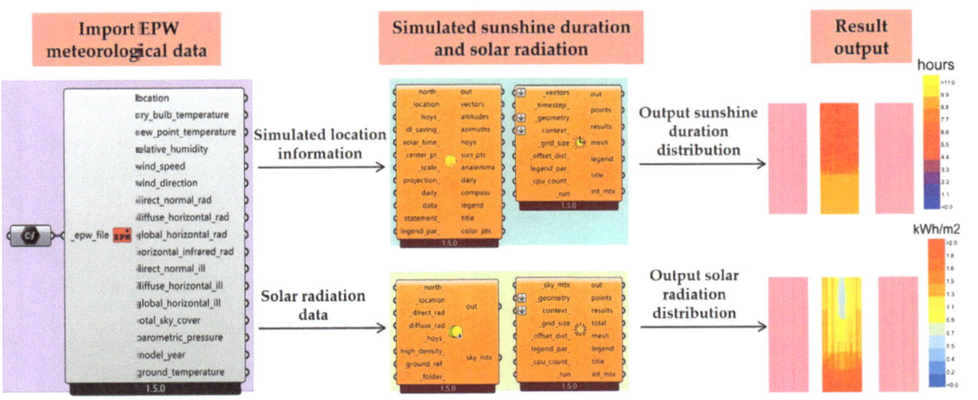

Figure 5. The description of the solar radiation analysis and sunshine duration analysis in Ladybug.

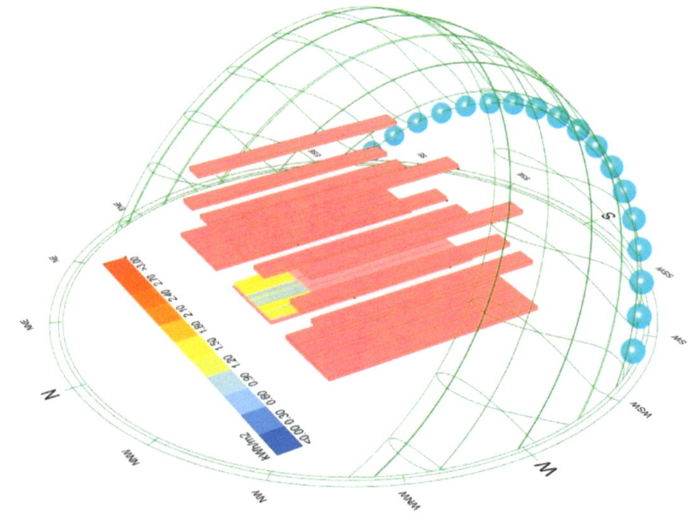

Figure 6. Sun path and parameterized model of the cultivation frame. The blue ball represents the sun, the green line represents the sun path, and the light pink rectangles represent the parametric model of the cultivation frame. The abbreviations at the bottom represent directions, such as NW representing northwest.

2.5. Calculation of the Solar Radiation Condition

To evaluate if the lower layer of the cultivation frame can obtain the light conditions necessary for crop growth, it is essential to determine the indicator of light conditions and the conversion relationship between this indicator and the simulated values of light. Previous studies have shown that the effect of light on plant growth is mainly reflected by the amount of light radiation received in plant photosynthesis. However, only a small range of the light spectrum—the photosynthetically active radiation between 400 and 700 nm—directly impacts plant development [32]. The accumulated quantity of photosynthetically active radiation reaching plants over 24 h is defined as the daily light integral (DLI), which has been shown to increase proportionally with the crop yield [33]. Therefore, we used the DLI as the light indicator of whether light is sufficient for normal crop growth in VF. The DLI can be converted to a simulated value of solar radiation (RAD) in Equation (1) [34]:

$$RAD = 5(DLI)/36 \quad (1)$$

DLI is the daily light integral ($mol \cdot m^{-2} \cdot day^{-1}$); RAD is the simulated value of solar radiation ($kWh \cdot m^{-2}$).

Lettuce was selected as the experimental crop, which DLI required for normal growth was 6–12 $mol \cdot m^{-2} \cdot day^{-1}$ [35]. Additionally, since the greenhouse skeleton and covering materials would shade the cultivation frame (e.g., the light transmittance of a general glass greenhouse is 60–90%) [36], we set the light transmittance of the solarization greenhouse in this study to 70%. Therefore, the light conditions required for lettuce growth were $DLI \geq 8.57$ $mol \cdot m^{-2} \cdot day^{-1}$. Combining the above conditions into Equation (1), the RAD required for lettuce growth should be over 1.19 $kWh \cdot m^{-2}$.

2.6. Genetic Algorithm

To homogenize and maximize the light conditions in the lower layer, this study optimally calculated the solar radiation in the lower layer of the cultivation frame for two objectives, the inner and outer CTU, using a genetic algorithm that is commonly used for multi-objective optimization problems. The instrument used in this study, Octopus, is a multi-objective optimization plug-in in Grasshopper. Octopus exhibits a fast solution speed, ease of operation for multi-objective optimization problems, and integrates Pareto optimality principles with a genetic algorithm. It features a visually intuitive 3D interface for observing and selecting optimal solutions, providing insights into the convergence of the genetic algorithm [37]. Based on the solution set obtained from the Pareto front distribution graph, we employed an iterative approach to optimize the results by adjusting the number of iterations until convergence and extracting the corresponding optimal solutions. Subsequently, we filtered out the optimized parameters and values that met the specified requirements.

As Octopus operators are designed to minimize objectives, we converted the desired maximization problem into a minimization problem through mathematical transformations. The optimization parameters in this study are the spacing of CTUs (d) and the layer height (H). The optimization objective aims to maximize the light intensity for both inner and outer cultivation trough units in the lower layer, with a preference for solutions minimizing the difference between the inner and outer light intensities. The optimization parameters were connected to the G-end, and the calculated solar radiation values of the optimization objectives were connected to the O-end (see Figure 7). The constraint variables of the genetic algorithm are cultivation trough unit spacing d and layer height H (see Table 1). The range of d is constrained to be 0–700 mm, and the range of H is constrained to be 300–700 mm. The key parameters incorporated into the genetic algorithm were Elitism, Mutation Probability, Mutation Rate, Crossover Rate, Population Size, Max Generations, Record interval, and Save interval. The values of each operational parameter were set as shown in Table 2.

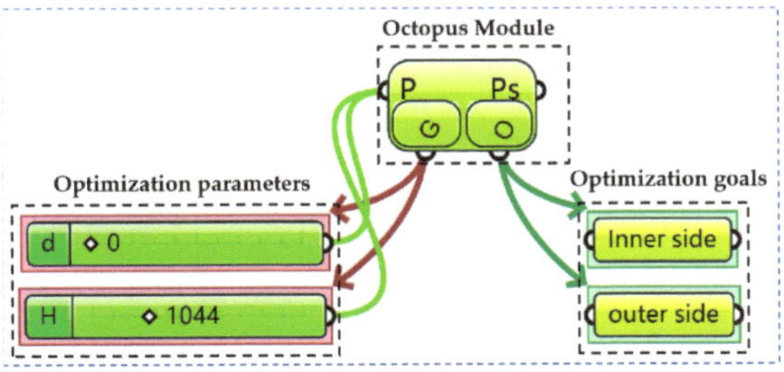

Figure 7. Multi-objective optimization components connection diagram in Grasshopper.

Table 1. Optimization variable and constraint condition-setting table.

Optimization Variable	Constraint Condition
d	[0 mm, 700 mm]
H	[300 mm, 700 mm]

Table 2. The operational parameter settings of Octopus.

Optimization Parameters	Value
Elitism	0.5
Mutation Probability	0.1
Mutation Rate	0.9
Crossover Rate	0.8
Population Size	100
Max Generation	0
Record interval	1
Save interval	0

Note: Elitism refers to the percentage of new solutions selected from the elite solutions of the previous generation. Mutation Probability represents the probability of mutation for each gene, affecting the convergence speed. Mutation Rate indicates the degree of gene mutation, where a higher value corresponds to more drastic mutations. Crossover Rate represents the probability of exchanging parameters between two generations. Population Size indicates the size of the population, influencing the computation time. Max Generation specifies the termination generation of genetic operation; 0 indicates an infinite process until convergence. Record Interval and Save Interval denote the production interval for storing historical records and the time interval for saving file data, respectively.

2.7. Cultivation Experiment

After the structural parameters of NLSCF were determined, a cultivation validation experiment was conducted to verify each layer's yield in the cultivation frame. The structure of the experimental NLSCF is shown in Figure 8a. The experiment location was in the glass greenhouse of the Beijing Academy of Agriculture and Forestry Sciences, Beijing, China (39.95° N, 116.28° E). The variety of lettuce selected for the experiment was Flandry from Rijk Zwaan Seed Company in the De Lier, Netherlands. The experimental procedure was as follows: Seedlings were nursed on 4 March 2023, and when the seedlings had 3 leaves and 1 heart on 25 March, the seedlings with the same growth conditions were selected for planting. All layers used identical irrigation conditions and nutrient solution during the planting period, and the temperature of the glasshouse was controlled between 18 and 25 °C. Plants were sampled and tested from the same location on the north side of the cultivation frame after 40 days of growth. The sampling layers were in order of upper, middle, inner, and outer lower layers. In accordance with Lei et al.'s [38] experiences in lettuce yield and quality, the indicators of design effectiveness include plant height (length from the base of the stem to the tallest part of the plant) and plant width (the widest part of the canopy) obtained by measuring live samples from the field, as well as the fresh weight of single plants (with roots removed) determined by taking samples in freshness and measuring them in a laboratory. Measuring equipment were electronic scales (range 0–500 g, accuracy 0.001 g) and straightedge (range 50 cm, accuracy 1 mm). The lettuce fresh weight measurements were retained with 2 valid digits, and plant width and height were retained with 1 valid digit. The average yield of five lettuces was calculated at each measurement and repeated three times (Figure 8b).

(**a**) (**b**)

Figure 8. Cultivation experiment. (**a**) Planting effect of NLSCF. (**b**) Lettuce fresh weight measurement. The total length of the cultivation frame was 20 m. The spacing between the middle CTUs was 350 mm, the height of all layers was 685 mm, and the height of the lower layer from the ground was 400 mm. The ground of the greenhouse installation site was hardened and leveled to ensure the stability of the cultivation frame installation and the stability of drainage from cultivation troughs. Pipeline installation was arranged at both ends of the cultivation frame and buried underground to prevent the pipes from interfering with the movement of the cultivation boards. The operational aisle width was set to 500 mm to ensure sufficient operational space while minimizing the use of aisle space resources.

3. Results and Discussion

3.1. Analysis of the Arrangement Direction of the Cultivation Frame

Traditionally, cultivation frames have been arranged in either the east–west or north–south direction based on the length of the cultivation frame, influencing the uniformity of crop growth [39]. To determine the optimal orientation for the cultivation frames, this study conducted a comparative simulation of sunshine duration distribution for cultivation frames placed in north–south and east–west directions under identical simulated conditions. Two types of shading structures were employed: other cultivation frames on both sides and mid-layer and upper-layer CTUs within the frame. The cultivation frame structural parameters were consistently set with a layer height of 600 mm, two sets of mid-layer CTUs spaced at 700 mm, and one CTU set in the upper layer. The results of the sunshine duration distribution for both orientations under these simulation conditions are depicted in Figure 9.

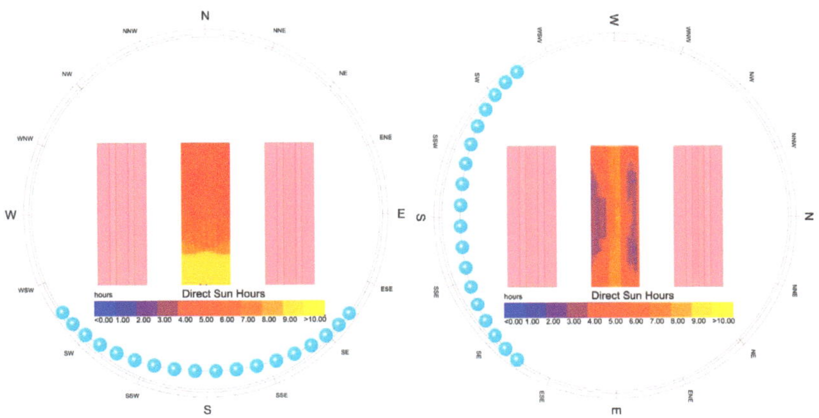

(a) (b)

Figure 9. Sunshine duration analysis chart. (**a**) The cultivation frame is placed in a north–south direction. (**b**) The cultivation frame is placed in the east–west direction. Blue circles represent the sun path, outer circles indicate direction, and inner rectangles represent two shading cultivation frames and one observation cultivation frame. The abbreviations for the periphery represent directions, such as NW representing northwest.

As seen in Figure 9, the north–south direction (Figure 9a) outperforms the east–west direction (Figure 9b) in terms of sunshine duration and uniformity in the lower CTUs. Specifically, under the north–south direction, the average daily sunshine duration in the lower layer was 6.16 h, whereas, under the east–west direction, it was only 4.81 h. Additionally, the north–south direction resulted in a gradual reduction in sunshine duration from south to north in the lower CTUs, ensuring uniformity in the east–west direction, with no discernible short sunlight areas. Conversely, the east–west direction exhibited distinct regions of short sunlight on both sides of the lower cultivation frame (depicted in blue), with the central area experiencing longer sunshine durations, leading to overall disparities in sunlight distribution.

Further analysis revealed the critical importance of the relationship between the direction of the sun's trajectory and the direction of the cultivation frame. When the sun's trajectory was perpendicular to the frame direction, the shading areas of the upper layers on the lower layer moved with the sun's trajectory, resulting in periodic exposure to sunlight for various areas of the lower layer, Periodic shading areas were formed in the east–west direction (as depicted in Figure 10 for the north–south direction). Additionally, due to the solar altitude angle, an unshaded area formed on the south side of the lower layer, gradually diminishing from south to north. Conversely, if the frame direction was parallel to the sun's trajectory (as depicted in Figure 11 for the east–west direction), fully shaded areas were created in the lower layer while the sunlight passing through the shading structure formed a brighter unshaded area in the lower layer, unaffected by changes in the sun's position.

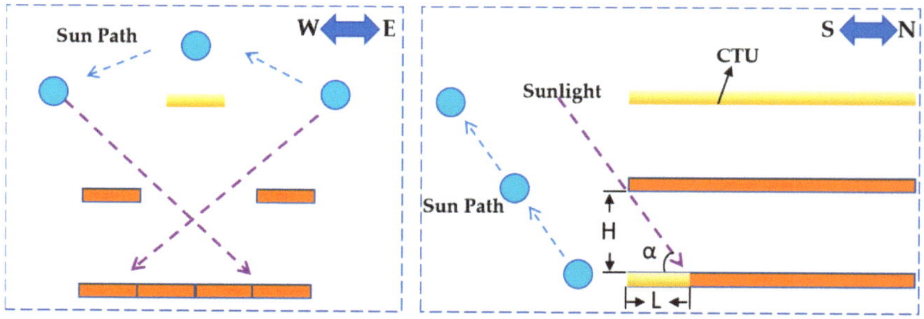

Figure 10. The north–south direction of the cultivation frame sunshine schematic diagram. Note: The yellow blocks represent unshaded areas of the cultivation troughs during the day, the orange-red blocks represent periodic shading areas, and the gray blocks represent fully shaded areas.

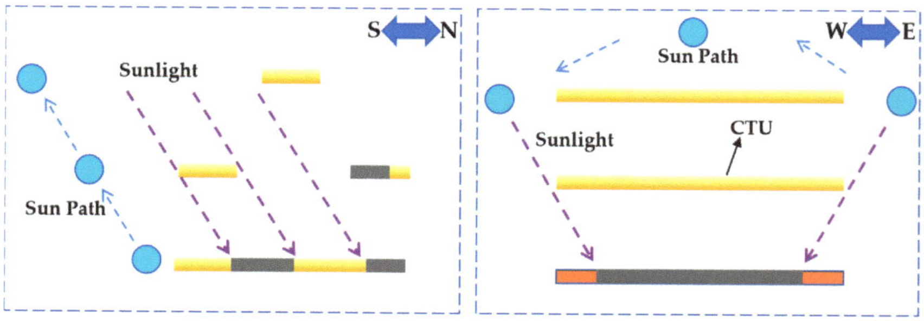

Figure 11. The east–west direction of the cultivation frame sunshine schematic diagram. Note: The yellow blocks represent unshaded areas of the cultivation troughs during the day, the orange-red blocks represent periodic shading areas, and the gray blocks represent fully shaded areas.

Based on this analysis, it is evident that the north–south direction, perpendicular to the sun's trajectory, ensures superior sunshine duration and uniformity in the lower cultivation frame. This aligns with the findings of Chaichana et al. [40] and Wang et al. [20]. Consequently, the designed direction of the NLSCF is consistently chosen as north–south, offering insights for the placement of the VF cultivation frame.

3.2. Analysis of the Influence of the Number of CTUs and the Layer Height on the Shading of the Lower Layer under a Single Shading Layer

3.2.1. Simulation Calculation of Lower Layer Solar Radiation

To investigate the impact of the number of CTUs and layer height on lower layer shading, this study employed CTU number (N = 1, 2, 3, and 4 groups) and layer height (H = 300 mm, 400 mm, 500 mm, 600 mm, and 700 mm) as independent variables. The study focused on simulating the northern 1 m region of the lower layer. The simulation aimed to calculate the maximum average solar radiation values obtainable in the northern 1m region of the lower layer. The simulation results are presented in Figure 12.

As depicted in Figure 12, it is evident that, with a constant layer height, the maximum average solar radiation in the lower layer exhibits a declining trend with an increase in the number of CTUs. Conversely, when the number of CTUs remains constant, the maximum average solar radiation in the lower layer shows an increasing trend with the elevation of the layer height. When N = 4, the simulated cultivation frame structure resembles an H-shape cultivation frame. When N < 4, the simulated structure resembles an A-shape cultivation frame. As observed in Figure 12, with a decrease in the number of CTUs, the maximum average solar radiation in the lower layer consistently demonstrates an increasing trend. This further substantiates that the structural pattern of the A-shape cultivation frame is more conducive to achieving superior light conditions compared to the novel H-shape cultivation frame structure.

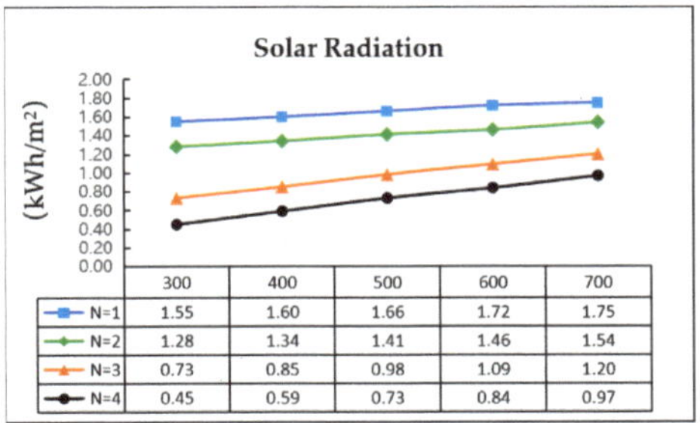

Figure 12. Simulation results of solar radiation.

3.2.2. Regression Model Construction and Determination of the Maximum Number of CTUs of the Shading Layers

To comprehensively analyze the influence patterns of the number of CTUs and layer height on the shading effect in the lower layer, SPSS 26.0 software was employed to calculate a linear regression model for the maximum average solar radiation (RAD) in the lower layer. Subsequent to the computations, the results are presented in Table 3, and Figure 13 illustrates the residual analysis. The regression equation is as follows in Formula (2).

$$RAD = 1.547 - 0.326 \cdot N + 0.001 \cdot H \quad (2)$$

Table 3. Linear regression analysis results of solar radiation.

Model	*B*	*t*	*p*	Collinearity Statistics	
				Tolerance	VIF
(Constant)	1.547	23.597	<0.001		
N	−0.326	−23.880	<0.001	1	1
H	0.001	8.419	<0.001	1	1

Dependent variable: RAD; notation: $R^2 = 0.971$, $p < 0.001$.

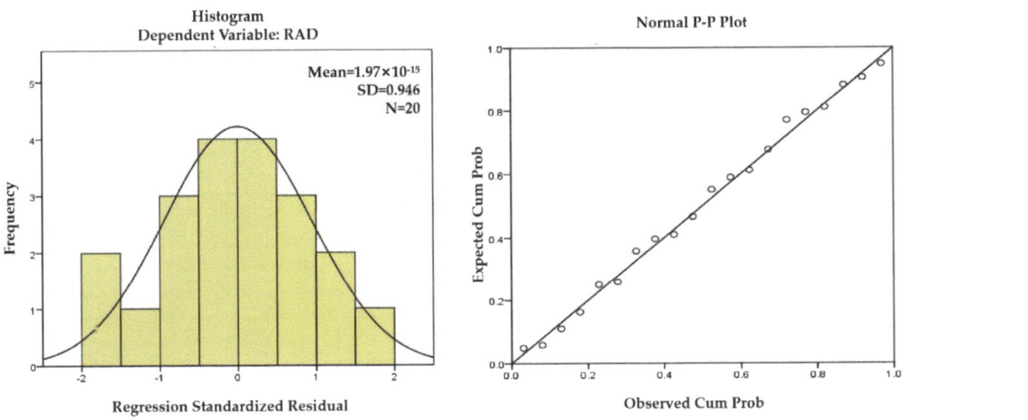

Figure 13. The solar radiation histogram of the model's residual distribution, and the P–P diagram of the model's normalized residuals.

RAD is the maximum average solar radiation ($kWh \cdot m^{-2}$), H is layer height (mm), and N is the number of CTUs. The results indicate a negative impact on the CTU numbers and a positive impact on the layer height. Specifically, when the independent variables are limited to practical ranges, for each additional CTU, the RAD decreases by 0.326 kWh·m^{-2}. Simultaneously, with every 100 mm increase in the layer height, the RAD increases by 0.1 kWh·m^{-2}.

Table 3 shows that the *p*-values of the number of CTUs (N) and layer height (H) are both less than 0.001, indicating significant influences of these two variables on the solar radiation. With an R^2 value of 0.971, the model demonstrates a satisfying fit. Residual analysis of the regression equation confirms its high accuracy. Previous researchers typically measured the solar radiation in the upper and lower layers of fixed-size cultivation frames directly using instruments. Subsequently, they studied the impact of sunlight on crop growth based on the yield and quality metrics [23,41]. However, these studies failed to capture the impact patterns of CTUs and layer height on the solar radiation in the lower layer. In contrast, the regression model obtained in this study provides an intuitive representation of the impact of the number of CTUs and layer height on the lower layer's solar radiation. Additionally, the model allows for the determination of the maximum number of CTUs in a single-layer cultivation frame. For instance, substituting RAD = 1.19 kWh·m^{-2} from Section 2.4 into Formula (2) yields the following results: when N = 2 and H = 295 mm, the total height of the cultivation frame is 1200 mm; when N = 3 and H = 621 mm, the total height of the cultivation frame is 1852 mm; and when N = 4 and H = 947 mm, the total of the cultivation frame is 2504 mm. If the cultivation frame is too high, it will lead to difficulty in the cultivation operation and higher input costs, which are not conducive to market promotion. In this study, the total height of the cultivation frame is not more than 2000 mm, and when the number of CTUs is two, the total height of the cultivation frame is the smallest and the solar radiation obtained is the best, so this study chooses the maximum number of CTUs of a single layer to be two groups.

3.3. Simulation Solution for the Number of CTUs in the Upper Layer

Based on the findings from Section 3.1, the distribution of CTUs is as follows: four sets in the lower layer, two sets in the middle layer, and two or one sets in the upper layer. As shown in Figure 14, the configuration with one set in the upper layer is designated as Mode 1, while the configuration with two sets in the upper layer is designated as Mode 2. The layout structures of the cultivation frame under both Mode 1 and Mode 2 are denoted by a letter, with d = 0 mm in the middle layer as a, d = 700 mm as b, and d = 700 mm in the upper layer as c. This study employs the average solar radiation of 1.19 kWh·m^{-2} in the northern 1 m region of the lower layer of the cultivation frame as a constraint. Through simulation, the required layer height values for both modes are determined, thereby establishing the number of CTUs in the upper layer of the cultivation frame. The simulation results are presented in Figure 15 and Table 4.

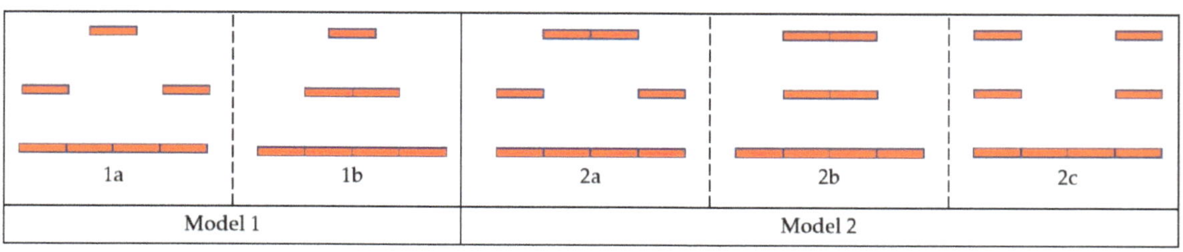

Figure 14. Cross-sectional view of the cultivation frame structures in two modes. Each orange rectangle represents a CTU.

Figure 15 illustrates that, when the solar radiation in the lower layer of the cultivation frame is set the same, Mode 1b outperforms the rest of the structures in uniformity of the solar radiation distribution. In regard to the solar radiation in the middle layer, Mode 1 (1a, 1b) shows higher solar radiation than that of Mode 2 (2a, 2b, 2c). The uniformity difference in the middle layer between Mode 1a and Mode 1b is not significant, while, in

Mode 2, Mode 2a is significantly better than that of Modes 2b and 2c. According to Table 4, Mode 1b also shows the lowest minimum layer height (670 mm) and the lowest total height (1950 mm), showing the best convenience and efficiency for manual operation. Based on the above analysis, we followed the design of Mode 1b by setting one group of CTUs in the center position of the upper layer.

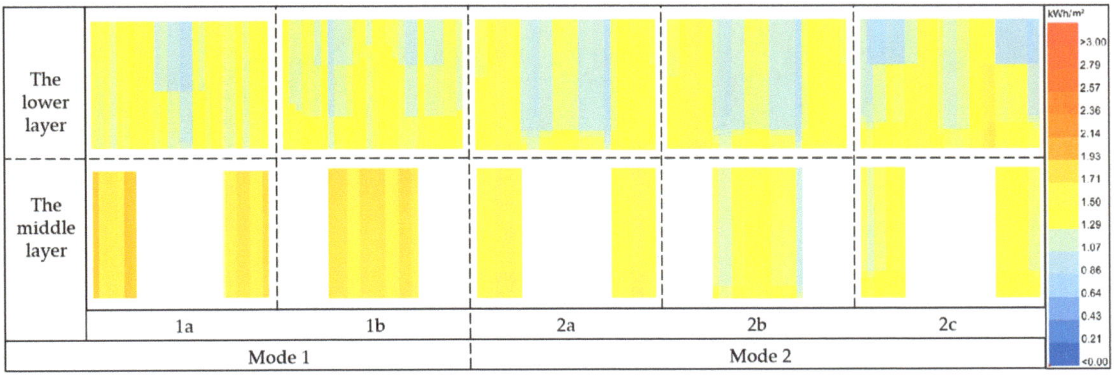

Figure 15. Solar radiation simulation results of Mode 1 and Mode 2.

Table 4. The results of the minimum layer height simulation calculations.

Model	The Average Solar Radiation of the Lower Layer ($kWh{\cdot}m^{-2}$)	The Average Solar Radiation of the Middle Layer ($kWh{\cdot}m^{-2}$)	Minimum Layer Height (mm)	The Total Height of the Cultivation Frame (mm)
1a	1.19	1.69	1020	2650
1b	1.19	1.64	670	1950
2a	1.19	1.49	1130	2870
2b	1.19	1.21	1080	2770
2c	1.19	1.24	1110	2830

3.4. Optimization of the Cultivation Frame Structure Using Octopus

From the solar radiation distribution diagram of the lower layer for Mode 1 in Figure 15, it is evident that the CTUs on the inner side have lower solar radiation compared to those on the outer side. To achieve a more balanced and maximized solar radiation on the lower layer of the cultivation frame, this study optimized the layer height (H) and the spacing between CTUs (d) within the Octopus plugin, and the results are presented in Figure 16.

After 11 iterations, Octopus achieved convergence in optimizing the average solar radiation of the CTUs on the inner and outer sides of the lower layer. Combining the optimization parameter values and optimization target values corresponding to each solution on the Pareto frontier in Figure 16, we selected the conditions that meet the design requirements of the cultivation frame, and the optimal simulation results are as follows. The spacing between CTUs is 348 mm; the layer height is 685 mm; and the maximum average solar radiation on the outer side of the lower layer is 1.26 $kWh{\cdot}m^{-2}$ (i.e., DLI $\approx$ 9.07 $mol{\cdot}m^{-2}{\cdot}day^{-1}$), on the inner side is 1.13 $kWh{\cdot}m^{-2}$ (i.e., DLI $\approx$ 8.14 $mol{\cdot}m^{-2}{\cdot}day^{-1}$), and in the middle layer is 1.66 $kWh{\cdot}m^{-2}$ (i.e., DLI $\approx$ 11.95 $mol{\cdot}m^{-2}{\cdot}day^{-1}$). The solar radiation on the inner and outer sides of the lower layer and the middle layer are 60.6%, 54.3%, and 79.8% of those on the upper layer, respectively. For convenience in the subsequent design, d is set to 350 mm, and L is set to 685 mm. Therefore, the optimized design of the NLSCF is shown in Figure 17.

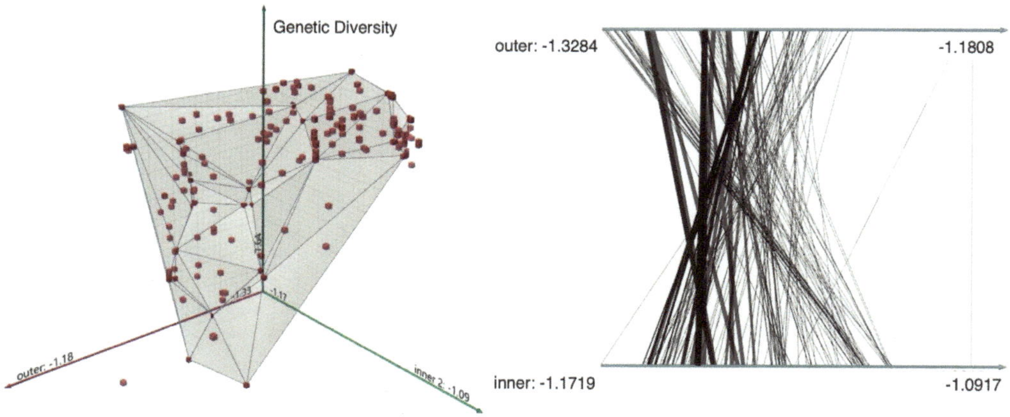

Figure 16. Pareto frontier images. Notes: The left image represents a spatial schematic of the optimal solution set obtained through the optimization processes, where the deep red cubes represent the Pareto frontier solution set. The right image illustrates the parameter distance diagram from the optimization processes, with each line representing an optimized solution. The intersection of the line with the axis indicates the value of that parameter. The smaller and more concentrated slopes of the line indicate a higher degree of optimization. Densely populated areas of lines represent the region of the optimal solution set, while dispersed areas signify eliminated solutions during the iteration processes.

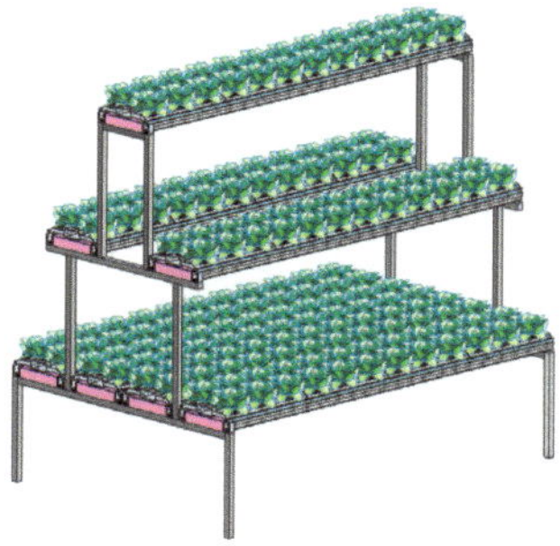

Figure 17. Structure diagram of the NLSCF.

By contrast, the simulated light conditions of the NLSCF are superior to those used by previous researchers. For example, Fang et al. [42] found that the DLIs of the middle and lower layers in a stereoscopic cultivation frame were only 1.30 mol·m^{-2}·day^{-1} and 1.26 mol·m^{-2}·day^{-1}, respectively, far lower than the results of this study: 11.95 and 8.61 mol·m^{-2}·day^{-1}. Chen et al. [24] studied the light–temperature effect of A-shape cultivation frames for strawberries, measuring that the photosynthetically active radiation in the middle and lower layers was only from 56.9% and from 39.3% of those in the upper layer on average, 22.9% and 18.2% lower than the present study. The NLSCF significantly increased the solar radiation of the lower and middle layers, making it possible to investigate the effects of natural lighting on lettuce yield and quality.

3.5. Lettuce Cultivation Experiment

In order to validate the feasibility of the designed NLSCF, we conducted an authentic cultivation experiment of lettuce without supplementary lighting. The average fresh weight, plant height, and plant width of lettuce in each layer are presented in Table 5.

Table 5. Effects of different cultivation layers on the fresh weight, plant height, and plant width of lettuce.

Variable	Fresh Weight (g)	Plant Width (cm)	Plant Height (cm)
The upper layer	140.68 ± 6.84 ^{a}	29.33 ± 2.38 ^{a}	20.47 ± 1.95 ^{a}
The middle layer	130.33 ± 11.63 ^{ab}	28.97 ± 1.95 ^{a}	18.46 ± 1.22 ^{ab}
The inner side of the lower layer	116.56 ± 9.40 ^{b}	29.27 ± 2.08 ^{a}	16.46 ± 1.94 ^{b}
The outer side of the lower layer	125.27 ± 7.51 ^{ab}	25.66 ± 2.37 ^{a}	17.20 ± 1.40 ^{b}

Values are given as the means ± standard error (n = 3). Different letter superscripts after the standard error indicate significant differences between treatments (Duncan's test, $p < 0.05$).

In the pure sunlight greenhouse experiment, the three lettuce yield indicators for each layer of the cultivation frame exhibited relatively consistent outcomes (see Table 5). The fresh weight, plant width, and height of lettuce decreased in the following order: upper layer > middle layer > outer side of the lower layer > inner side of the lower layer. This order is in line with expectations, as it is consistent with previous solar radiation simulation results. The differences in yield indicators among the layers of the cultivation frame were not pronounced. For instance, compared with lettuce grown in the upper layer, the average fresh weight per plant in the inner side of the lower layer, the outer side of the lower layer, and middle layer were 82.9%, 89.0%, and 92.6% of the lettuces in the upper layer, respectively. The average plant heights in these layers were 80.4%, 84.0%, and 90.2% of the upper layer. It is noteworthy that our cultivation experiment relied solely on natural sunlight and ambient scattered light within the greenhouse, without the use of artificial supplementary lighting. Therefore, the remaining yield differences could be explained by the remaining differences in light intensity. The better the light conditions, the more photosynthetic products there were, and the greater the fresh weight and plant height of lettuce. The insignificant difference in the plant width of lettuce may be due to the improvement of the light uniformity. These results emphasize the NLSCF's success in compensating for substantial differences in solar radiation among layers solely through harnessing natural light conditions. The results underscore the efficiency of the cultivation frame in creating a conducive environment for plant growth, effectively reducing the disparities in crop yields across different layers.

This experiment validates that the NLSCF improves interlayer light conditions, making the distribution of natural light more uniform, promoting biomass production, and consequently achieving yields in the middle and lower layers ranging from 82.9% to 92.6% of the upper layer. The optimized structure of the NLSCF could meet the normal growth requirements of lettuce. Compared with previous studies that optimized LED light recipes using the design of experiments (DoEs) methodology [43] and the Taguchi method [44], this study harnessed innovative computer simulation and modeling techniques, including parametric modeling, light simulation techniques, and the genetic algorithm method, to assess the impact of different design parameters on the lighting distribution and to utilize cost-efficient natural light in a better way.

Additionally, although the average simulated solar radiation (1.13 kWh·m^{-2}) on the inner side of the lower layer did not reach the required 1.19 kWh·m^{-2} for lettuce growth, the experimental results still supported the significant and effective light optimization effect achieved in the inner side of the lower layer. This may be attributed to two reasons. Firstly, this study intentionally selected the winter solstice day with the least light as the simulation target, indicating relatively favorable natural light conditions during actual cultivation, making it somewhat easier to meet the growth requirements of these lettuces. Secondly, the average plant width on the inner side of the lower layer was relatively large. This phenomenon might be due to the phototropic growth of plants, causing lettuce leaves in the inner side of the lower layer cultivation trough to grow outward, resulting in an increased plant width and effectively addressing the shading issue on the inner side of the lower layer.

This study involves several potential limitations. Due to the research purpose of designing a natural light cultivation frame for VF, the variability in plant responses to various light qualities and spectrums was not fully considered. Moreover, while providing practical insights, the cultivation experiment may have inherent variability due to real-world conditions. Factors such as other crop types, temperature variations, nutrient distribution, and other environmental conditions in the glass greenhouse might introduce variations in lettuce growth. By acknowledging these limitations, the study ensures transparency regarding potential sources of error or bias, contributing to a more comprehensive interpretation of the results and recommendations. Future research could delve deeper into these limitations, refining the methodologies for more robust outcomes. Future research should delve deeper into these limitations, exploring the use of cutting-edge research methods, such as advanced lighting technology, the Internet of Things (IoT) and sensing technology, and artificial intelligence, for more robust and fully considered outcomes.

4. Conclusions

1. Novel optimization methods for the structure design of VF cultivation frames were explored. By harnessing parametric modeling and light simulation techniques, our research introduced innovative approaches to designing and optimizing VF frames. The pivotal findings underscore the remarkable capability of these methods to swiftly and precisely simulate light characteristics across diverse frame structures. Notably, parametric modeling emerges as a key facilitator, streamlining design modifications with unprecedented convenience. These innovative methods provide technical support for the construction of VF cultivation systems, effectively reducing the design costs and design cycle of VF. This paper's primary contributions lie in expanding the technical toolkit for VF design and catalyzing practical advancements that propel the field toward enhanced sustainability and cost-effectiveness.
2. We designed a NLSCF to reduce supplementary lighting energy consumption in VF. This study fully considered the structural design requirements of the cultivation frame and the lighting needs of the lower layers. Through a combination of parametric modeling, light simulation, and genetic algorithm optimization, the structure of the cultivation frame was designed and optimized. Therefore, the NLSCF could meet the lighting design requirements for the middle and lower layers, even under no supplementary lighting conditions. The optimized structure consisted of four sets of CTUs for the lower layer, two sets for the middle layer, and one set for the upper layer, with a layer height of 685 mm and a spacing of 350 mm between CTUs.
3. We conducted cultivation experiments to validate the NLSCF. The results of lettuce cultivation under natural light verification experiments showed that the yields of the middle and lower layers could reach from 82.9% to 92.6% of the upper layer. Based on the simulated design, the practical effect of not requiring supplementary lighting was effectively verified.
4. Although the above research results provide a solution to reduce supplementary lighting energy consumption in VF, the planting density of the cultivation frame is lower than that of a plant factory. Further studies may apply the structural design methods and genetic algorithm to increase the height and number of layers of the cultivation frame and combine lifting and transporting equipment to achieve the goal of increasing planting density. The NLSCF system may also benefit from better light intensity range management, including avoiding photo saturation and photoinhibition and optimizing light distribution among layers.

Author Contributions: Conceptualization, D.J. and G.G.; methodology, D.J.; software, D.J.; validation, W.G. and Q.Z.; formal analysis, D.J.; investigation, D.J. and X.W.; resources, W.Z.; data curation, D.J.; writing—original draft preparation, D.J.; writing—review and editing, G.G.; visualization, X.W.; supervision, G.G.; project administration, W.Z.; funding acquisition, W.Z. All authors have read and agreed to the published version of the manuscript.

Funding: This research was funded by the Construction Project of Beijing Engineering Laboratory of Agricultural Internet of Things Technology (PT2023-29) and Beijing Innovation Team of the Modern Agricultural Research System (BAIC08-2023).

Institutional Review Board Statement: Not applicable.

Data Availability Statement: Data are contained within the article.

Conflicts of Interest: The authors declare no conflicts of interest.

References

1. SharathKumar, M.; Heuvelink, E.; Marcelis, L.F.M. Vertical Farming: Moving from Genetic to Environmental Modification. *Trends Plant Sci.* **2020**, *25*, 724–727. [CrossRef] [PubMed]
2. Van Gerrewey, T.; Boon, N.; Geelen, D. Vertical Farming: The Only Way Is Up? *Agronomy* **2021**, *12*, 2. [CrossRef]
3. Sambo, P.; Nicoletto, C.; Giro, A.; Pii, Y.; Valentinuzzi, F.; Mimmo, T.; Lugli, P.; Orzes, G.; Mazzetto, F.; Astolfi, S.; et al. Hydroponic Solutions for Soilless Production Systems: Issues and Opportunities in a Smart Agriculture Perspective. *Front. Plant Sci.* **2019**, *10*, 923. [CrossRef] [PubMed]
4. Benke, K.; Tomkins, B. Future Food-Production Systems: Vertical Farming and Controlled-Environment Agriculture. *Sustain. Sci. Pract. Policy* **2017**, *13*, 13–26. [CrossRef]
5. Beacham, A.M.; Vickers, L.H.; Monaghan, J.M. Vertical Farming: A Summary of Approaches to Growing Skywards. *J. Hortic. Sci. Biotechnol.* **2019**, *94*, 277–283. [CrossRef]
6. Montes Rivera, M.; Escalante-Garcia, N.; Dena-Aguilar, J.A.; Olvera-Gonzalez, E.; Vacas-Jacques, P. Feature Selection to Predict LED Light Energy Consumption with Specific Light Recipes in Closed Plant Production Systems. *Appl. Sci.* **2022**, *12*, 5901. [CrossRef]
7. Pennisi, G.; Pistillo, A.; Orsini, F.; Cellini, A.; Spinelli, F.; Nicola, S.; Fernandez, J.A.; Crepaldi, A.; Gianquinto, G.; Marcelis, L.F.M. Optimal Light Intensity for Sustainable Water and Energy Use in Indoor Cultivation of Lettuce and Basil under Red and Blue LEDs. *Sci. Hortic.* **2020**, *272*, 109508. [CrossRef]
8. Smirnov, A.A.; Semenova, N.A.; Dorokhov, A.S.; Proshkin, Y.A.; Godyaeva, M.M.; Vodeneev, V.; Sukhov, V.; Panchenko, V.; Chilingaryan, N.O. Influence of Pulsed, Scanning and Constant (16- and 24-h) Modes of LED Irradiation on the Physiological, Biochemical and Morphometric Parameters of Lettuce Plants (*Lactuca sativa* L.) While Cultivated in Vertical Farms. *Agriculture* **2022**, *12*, 1988. [CrossRef]
9. Zhang, Y.; Kacira, M. Analysis of Climate Uniformity in Indoor Plant Factory System with Computational Fluid Dynamics (CFD). *Biosyst. Eng.* **2022**, *220*, 73–86. [CrossRef]
10. Yeo, U.-H.; Lee, S.-Y.; Park, S.-J.; Kim, J.-G.; Cho, J.-H.; Decano-Valentin, C.; Kim, R.-W.; Lee, I.-B. Rooftop Greenhouse: (2) Analysis of Thermal Energy Loads of a Building-Integrated Rooftop Greenhouse (BiRTG) for Urban Agriculture. *Agriculture* **2022**, *12*, 787. [CrossRef]
11. Graamans, L.; Baeza, E.; van den Dobbelsteen, A.; Tsafaras, I.; Stanghellini, C. Plant Factories versus Greenhouses: Comparison of Resource Use Efficiency. *Agric. Syst.* **2018**, *160*, 31–43. [CrossRef]
12. Avgoustaki, D.D.; Xydis, G. Energy Cost Reduction by Shifting Electricity Demand in Indoor Vertical Farms with Artificial Lighting. *Biosyst. Eng.* **2021**, *211*, 219–229. [CrossRef]
13. Hosseini, H.; Mozafari, V.; Roosta, H.R.; Shirani, H.; van de Vlasakker, P.C.H.; Farhangi, M. Nutrient Use in Vertical Farming: Optimal Electrical Conductivity of Nutrient Solution for Growth of Lettuce and Basil in Hydroponic Cultivation. *Horticulturae* **2021**, *7*, 283. [CrossRef]
14. Büyüközkan, G.; Göçer, F.; Uztürk, D. A Novel Pythagorean Fuzzy Set Integrated Choquet Integral Approach for Vertical Farming Technology Assessment. *Comput. Ind. Eng.* **2021**, *158*, 107384. [CrossRef]
15. Rengasamy, N.; Othman, R.Y.; Che, H.S.; Harikrishna, J.A. Artificial Lighting Photoperiod Manipulation Approach to Improve Productivity and Energy Use Efficacies of Plant Factory Cultivated Stevia Rebaudiana. *Agronomy* **2022**, *12*, 1787. [CrossRef]
16. Watawana, B.; Isaksson, M. Design and Simulations of a Self-Assembling Autonomous Vertical Farm for Urban Farming. *Agriculture* **2023**, *13*, 112. [CrossRef]
17. Zhang, Y.; Kacira, M. Comparison of Energy Use Efficiency of Greenhouse and Indoor Plant Factory System. *Eur. J. Hortic. Sci.* **2020**, *85*, 310–320. [CrossRef]
18. Weidner, T.; Yang, A.; Hamm, M.W. Energy Optimisation of Plant Factories and Greenhouses for Different Climatic Conditions. *Energy Convers. Manag.* **2021**, *243*, 114336. [CrossRef]
19. Jürkenbeck, K.; Heumann, A.; Spiller, A. Sustainability Matters: Consumer Acceptance of Different Vertical Farming Systems. *Sustainability* **2019**, *11*, 4052. [CrossRef]
20. Wang, C.; Song, W.; Zhao, S.; Qu, S. Effect of different combinations of H-type cultivation frame on light and strawberry growth and yield. *Trans. Chin. Soc. Agric. Eng.* **2017**, *33*, 234–239. [CrossRef]
21. Liao, P.; Liu, P.; Wang, Y.; Huang, C.; Lan, L.; Yang, Y.; Cui, X. Stereoscopic Cultivation of Panax Notoginseng: A New Approach to Overcome the Continuous Cropping Obstacle. *Ind. Crops Prod.* **2018**, *126*, 38–47. [CrossRef]
22. Kempler, C. 'Out-of-season' greenhouse production of raspberry and strawberry. *Acta Hortic.* **2004**, *633*, 459–465. [CrossRef]

23. Lin, X.; Luo, Y.; Wang, H. Effect of light and temperature on strawberry in stereoscopic culture system. *J. China Agric. Univ.* **2014**, *19*, 67–73. [CrossRef]
24. Chen, Z.; Liu, P.; Zhang, B.; Wang, J.; Song, S.; Wang, H. Light and temperature and their effects on photosynthesis characteristics of stereoscopic cultivation in strawberry. *J. China Agric. Univ.* **2011**, *16*, 42–48.
25. Wang, C.; Sun, W.; Zhao, S.; Qu, M. Sun-tracking multi-layer stereo-cultivation system improving light and temperature environment and increasing yield for strawberry. *Trans. Chin. Soc. Agric. Eng.* **2017**, *33*, 187–194. [CrossRef]
26. Goharian, A.; Daneshjoo, K.; Yeganeh, M. Standardization of Methodology for Optimizing the Well Aperture as Device (Reflector) for Light-Wells; A Novel Approach Using Honeybee & Ladybug Plugins. *Energy Rep.* **2022**, *8*, 3096–3114. [CrossRef]
27. Eltaweel, A.; Su, Y. Controlling Venetian Blinds Based on Parametric Design; via Implementing Grasshopper's Plugins: A Case Study of an Office Building in Cairo. *Energy Build.* **2017**, *139*, 31–43. [CrossRef]
28. Saito, K.; Ishigami, Y.; Goto, E. Evaluation of the Light Environment of a Plant Factory with Artificial Light by Using an Optical Simulation. *Agronomy* **2020**, *10*, 1663. [CrossRef]
29. Jung, D.H.; Lee, J.W.; Kang, W.H.; Hwang, I.H.; Son, J.E. Estimation of Whole Plant Photosynthetic Rate of Irwin Mango under Artificial and Natural Lights Using a Three-Dimensional Plant Model and Ray-Tracing. *Int. J. Mol. Sci.* **2018**, *19*, 152. [CrossRef]
30. Badji, A.; Benseddik, A.; Bensaha, H.; Boukhelifa, A.; Hasrane, I. Design, Technology, and Management of Greenhouse: A Review. *J. Clean. Prod.* **2022**, *373*, 133753. [CrossRef]
31. Wang, S.; Zhang, Q.; Liu, P.; Liang, R.; Fu, Z. A Parameterized Design Method for Building a Shading System Based on Climate Adaptability. *Atmosphere* **2022**, *13*, 1244. [CrossRef]
32. Viršilė, A.; Brazaitytė, A.; Vaštakaitė-Kairienė, V.; Miliauskienė, J.; Jankauskienė, J.; Novičkovas, A.; Samuolienė, G. Lighting Intensity and Photoperiod Serves Tailoring Nitrate Assimilation Indices in Red and Green Baby Leaf Lettuce. *J. Sci. Food Agric.* **2019**, *99*, 6608–6619. [CrossRef] [PubMed]
33. Yan, Z.; He, D.; Niu, G.; Zhai, H. Evaluation of Growth and Quality of Hydroponic Lettuce at Harvest as Affected by the Light Intensity, Photoperiod and Light Quality at Seedling Stage. *Sci. Hortic.* **2019**, *248*, 138–144. [CrossRef]
34. Eaton, M.; Harbick, K.; Shelford, T.; Mattson, N. Modeling Natural Light Availability in Skyscraper Farms. *Agronomy* **2021**, *11*, 1684. [CrossRef]
35. Torres, A.P.; Lopez, R. Measuring Daily Light Integral in a Greenhouse. Purdue Extension. Ho-238-W. 2012. 1–10. Available online: https://mdc.itap.purdue.edu/item.asp?itemID=19338 (accessed on 29 July 2023).
36. Maraveas, C. Environmental Sustainability of Greenhouse Covering Materials. *Sustainability* **2019**, *11*, 6129. [CrossRef]
37. Xue, Y.; Liu, W. A Study on Parametric Design Method for Optimization of Daylight in Commercial Building's Atrium in Cold Regions. *Sustainability* **2022**, *14*, 7667. [CrossRef]
38. Lei, C.; Engeseth, N.J. Comparison of Growth Characteristics, Functional Qualities, and Texture of Hydroponically Grown and Soil-Grown Lettuce. *LWT* **2021**, *150*, 111931. [CrossRef]
39. Kwack, Y.; Kim, K.K.; Hwang, H.; Chun, C. Growth and Quality of Sprouts of Six Vegetables Cultivated under Different Light Intensity and Quality. *Hortic. Environ. Biotechnol.* **2015**, *56*, 437–443. [CrossRef]
40. Chaichana, C.; Man, A.; Wicharuck, S.; Mona, Y.; Rinchumphu, D. Modelling of Annual Sunlight Availability on Vertical Shelves: A Case Study in Thailand. *Energy Rep.* **2022**, *8*, 1136–1143. [CrossRef]
41. Lee, Y.Y.; Yik, J.; Koay, E.H.H.; Lau, H.H. Application of Light Redirecting Panels to Improve Sunlight Utilization in Multi-Tiered Vertical Farming Systems. *Clean Technol. Environ. Policy* **2023**. [CrossRef]
42. Fang, H.; Cheng, R.; Wu, G.; Tong, Y.; Liu, Q. Effects of multi-layer three-dimensional cultivation combined with artificial lighting on lettuce yield and quality in Greenhouse. *J. China Agric. Univ.* **2021**, *26*, 147–154. [CrossRef]
43. Barbieri, F.; Barbi, S.; Bertacchini, A.; Montorsi, M. Combined Effects of Different LED Light Recipes and Slow-Release Fertilizers on Baby Leaf Lettuce Growth for Vertical Farming: Modeling through DoE. *Appl. Sci.* **2023**, *13*, 8687. [CrossRef]
44. Farhangi, H.; Mozafari, V.; Roosta, H.R.; Shirani, H.; Farhangi, M. Optimizing Growth Conditions in Vertical Farming: Enhancing Lettuce and Basil Cultivation through the Application of the Taguchi Method. *Sci. Rep.* **2023**, *13*, 6717. [CrossRef] [PubMed]

agriculture

MDPI

Article

Detection of the Corn Kernel Breakage Rate Based on an Improved Mask Region-Based Convolutional Neural Network

Hongmei Zhang, Zhijie Li, Zishang Yang, Chenhui Zhu, Yinhai Ding, Pengchang Li and Xun He *

College of Mechanical and Electrical Engineering, Henan Agricultural University, Zhengzhou 450002, China; zhanghongmei0905@henau.edu.cn (H.Z.); lizhijie@stu.henau.edu.cn (Z.L.); zsyang@henau.edu.cn (Z.Y.); zhuchenhui@henau.edu.cn (C.Z.); dingyinhai@stu.henau.edu.cn (Y.D.); lipengchang@stu.henau.edu.cn (P.L.)

* Correspondence: hexun@henau.edu.cn

Abstract: Real-time knowledge of kernel breakage during corn harvesting plays a significant role in the adjustment of operational parameters of corn kernel harvesters. (1) Transfer learning by initializing the DenseNet121 network with pre-trained weights for training and validating a dataset of corn kernels was adopted. Additionally, the feature extraction capability of DenseNet121 was improved by incorporating the attention mechanism of a Convolutional Block Attention Module (CBAM) and a Feature Pyramid Network (FPN) structure. (2) The quality of intact and broken corn kernels and their pixels were found to be coupled, and a linear regression model was established using the least squares method. The results of the test showed that: (1) The $MAP^{b}50$ and $MAP^{m}50$ of the improved Mask Region-based Convolutional Neural Network (RCNN) model were 97.62% and 98.70%, in comparison to the original Mask Region-based Convolutional Neural Network (RCNN) model, which were improved by 0.34% and 0.37%, respectively; the backbone FLOPs and Params were 3.09 GMac and 9.31 M, and the feature extraction time was 206 ms; compared to the original backbone, these were reduced by 3.87 GMac and 17.32 M, respectively. The training of the obtained prediction weights for the detection of a picture of the corn kernel took 76 ms, so compared to the Mask RCNN model, it was reduced by 375 ms; based on the concept of transfer learning, the improved Mask RCNN model converged twice as quickly with the loss function using pre-training weights than the loss function without pre-training weights during training. (2) The coefficients of determination R^2 of the two models, when the regression models of the pixels and the quality of intact and broken corn kernels were analyzed, were 0.958 and 0.992, respectively. These findings indicate a strong correlation between the pixel characteristics and the quality of corn kernels. The improved Mask RCNN model was used to segment mask pixels to calculate the corn kernel breakage rate. The verified error between the machine vision and the real breakage rate ranged from −0.72% to 0.65%, and the detection time of the corn kernel breakage rate was only 76 ms, which could meet the requirements for real-time detection. According to the test results, the improved Mask RCNN method had the advantages of a fast detection speed and high accuracy, and can be used as a data basis for adjusting the operation parameters of corn kernel harvesters.

Keywords: Mask RCNN; machine vision; corn kernels; breakage rate

Citation: Zhang, H.; Li, Z.; Yang, Z.; Zhu, C.; Ding, Y.; Li, P.; He, X. Detection of the Corn Kernel Breakage Rate Based on an Improved Mask Region-Based Convolutional Neural Network. *Agriculture* **2023**, *13*, 2257. https://doi.org/10.3390/agriculture13122257

Academic Editor: Mustafa Ucgul

Received: 31 October 2023
Revised: 6 December 2023
Accepted: 8 December 2023
Published: 10 December 2023

1. Introduction

In China, corn is a significant feed, economic product, and food crop [1]. China has intensively marketed corn harvesters as mechanization has advanced. However, corn breakage can be easily caused by improperly adjusting the operational parameters of corn harvesters [2]. According to the national standard "GB/T 21962-2020" [3], corn harvester breakage rates must not exceed 5%. Therefore, to decrease the breakage rate when the corn breakage rate is high, the operational parameters of corn kernel harvesters must be changed. Traditional methods of measuring the corn kernel breakage rate include picking, weighing, and calculating by hand, which is labor- and time-intensive and does not allow for real-time detection of and feedback on the breakage rate. There is an urgent demand for innovative techniques and approaches. Using machine vision technology to recognize and segment intact and broken corn kernels to calculate the breakage rate not only improves efficiency but also has advantages such as precision and non-destructive testing. With recent advancements in machine vision technology, techniques like segmentation, object identification, and image classification have achieved prominence in the agricultural sector. The utilization of machine vision technology for crop detection has been the subject of extensive research conducted by both domestic and international scholars.

Regarding crop identification, Qiu [4] et al. used a convolutional neural network (CNN) to detect rice seeds of four different types and compared it to closest-neighbor KNN and support vector machine (SVM) methods with various training sample sizes. In the validation set, the CNN had a recognition accuracy of 87.0%; Han [5] et al. identified peanut samples containing aflatoxin using hyperspectral imaging and CNN techniques, with a recognition accuracy of 90.0%; Przybyo [6] and colleagues employed a CNN to learn both the local and global properties of acorns, which were then used to distinguish between healthy and broken acorns, with an accuracy of 85.0%; Zhang [7] et al. improved the AlexNet model to recognize five levels of peanut pods, with an average accuracy of 95.43%; Mao [8] et al. identified moderate, severe, and yellow dwarf illness using an improved faster RCNN, with an average accuracy of 91.06%.

In grain integrity recognition, Ni [9] et al. used a hierarchical system research method to establish a prototype detection visual system for corn kernels, which was used to distinguish between intact and broken corn kernels, with an accuracy of 87%. The detection time for a single corn kernel was between 1.5 s and 1.8 s; Steenhoek [10] et al. explored the detection of broken and moldy corn kernels. They utilized the color difference between broken and moldy corn kernels and intact kernels, and used the RGB pixel value of the image as a feature parameter. The recognition accuracy for broken and moldy corn kernels using neural networks reached 92%, but different lighting and camera depth effects could lead to image segmentation errors; Zhao [11] et al. used convolutional neural networks to recognize the integrity of peanut seeds, with a classification accuracy of 98.18% and an average detection time of 18 ms for a single peanut image; Chen [12] et al. adopted the multi-scale retinex with a color restoration algorithm to enhance the images and distinguish between intact and broken rice grains by color. The final segmentation result was then obtained using morphological processing. When encountering situations where impurities were similar to the color of the grains, classification and recognition errors arose readily; Song [13] et al. used the watershed approach and an improved object segmentation algorithm based on the OpenCV image processing package to extract the contours of soybean seeds in the image. Respectively, 95.2% and 91.25% of intact and broken soybean seeds were recognized. An image took an average time of 1.16 s to process, which is far too long.

In terms of calculation of the breakage rate, Mahirah et al. [14,15] proposed a dual light source illumination monitoring system for grain breakage and impurity content in the grain bin of a grain combine harvester, but did not calculate the grain breakage rate and impurity content; Yang [16] et al. created an online sampling tool for the corn kernel breakage rate, which was only used to sample grains in the grain bin of the corn kernel harvester and did not detect the status of the corn kernel breakage rate; Chen [17] et al. used an improved watershed algorithm to segment soybean seed images. The accuracy rates for intact and broken seeds were 87.26% and 86.45%, respectively. Based on the constructed quantitative model for the soybean seed breakage rate, the relative error between the calculated breakage rate and the actual breakage rate was 0.79%; Jin [18] et al. used an improved U-Net network to detect and classify intact and broken soybean seeds. The values of the comprehensive evaluation index were 95.50% and 91.88%, respectively. The bench test results showed that the absolute error of the mean for detecting the soybean seed breakage rate was 0.13% compared to the mean absolute error for manual detection; Liu [19] et al. segmented soybean seed images based on the DeepLabV3+segmentation network. The comprehensive evaluation index F1 values for intact and broken seeds were 89.49%

and 93.93%. The relative error between the breakage rate calculated by the constructed quantitative model and the actual breakage rate was 0.36%.

Machine vision technology is still immature in terms of its application to the detection of grain breakage rate. Most researchers have adopted the traditional machine learning method, but this form of recognition did not have a generalization ability, so it is necessary to develop a real-time method for the detection of the grain breakage rate using a deep learning algorithm. In this study, an improved Mask RCNN [20] model was proposed, which can reduce the computational burden (and the number of parameters), and improve the detection accuracy and detection speed. The improved Mask RCNN can realize fast and accurate identification and segmentation of corn kernels, and achieve the purpose of the real-time detection of the breakage rate.

2. Materials and Methods

2.1. Description of the Mask RCNN

The Mask RCNN model is a two-stage algorithm employed for a range of tasks, including target classification, detection, and semantic segmentation. Based on the Faster RCNN [21] model, it adds a pixel for classifying the region of interest, predicts the branches of the target mask, and realizes the goals of classification, localization, and segmentation. The backbone feature extraction network, the Region Proposal Network, the Region of Interest Align (ROI Align) layer, the fully connected network, and the fully convolutional neural networks comprise the majority of the network structure of the Mask RCNN, as shown in Figure 1. The image is first entered into the ResNet50 feature extraction network to obtain feature maps of various stages, and then it enters the Feature Pyramid Network (FPN) structure to fuse features of various scales to obtain a common feature map that has strong semantic and spatial information simultaneously. Second, the common feature map is used to create anchor boxes of various sizes and proportions under the impact of the Region Proposal Network. Multiple candidate regions of various sizes are created by computing the Intersection Over Union (IOU) between each anchor box and the annotated real box in the image. On the one hand, boundary box regression is conducted after classifying the candidate regions into foreground and background. On the other hand, the ROI Align layer receives the feature pictures corresponding to the candidate boxes, which are then maximum or average pooled using bilinear interpolation to make them uniform in size. The target categorization, bounding box regression, and exact pixel-level segmentation of the target are accomplished using fully connected neural networks and fully convolutional networks.

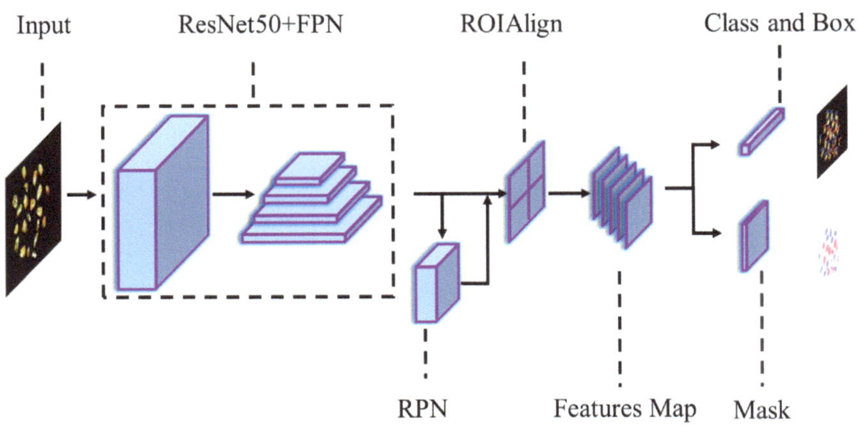

Figure 1. Mask RCNN model structure.

2.2. Improving the Mask RCNN

Despite the Mask RCNN model returning a high detection accuracy, its sluggish detection speed makes it challenging to achieve real-time detection of the corn kernel breakage rate, and it is strongly dependent on the computer hardware's computational capacity. This study aims to improve the Mask RCNN by reducing the parameters and improving the detection speed for the recognition and segmentation tasks of intact and broken corn kernels. First, replacing ResNet50_FPN with DenseNet121 [22] served as the backbone feature extraction network, accelerating the speed of feature extraction, and the Convolutional Block Attention Module (CBAM) [23] attention mechanism and FPN structure were added to improve the feature extraction on the basis of DenseNet121 to increase the detection accuracy, resulting in a DenseNet121_CBAM_FPN backbone feature extraction network (Figure 2). To reduce the training time and accelerate the convergence of the loss function, pre-training weights were loaded using the transfer learning concept

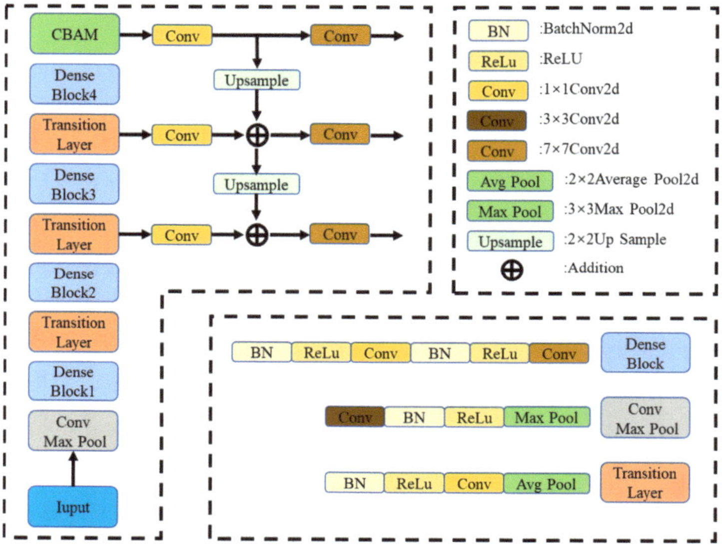

Figure 2. DenseNet121_CBAM_FPN backbone feature extraction network.

2.2.1. Replacing the Backbone Feature Extraction Network DenseNet

The complexity of the backbone feature extraction network increases in response to ongoing hardware upgrades. The ResNet network, which includes a residual structure as shown in Figure 3a, serves as the primary feature extraction network of the original Mask RCNN. The major benefit is that it can preserve the original features, prevent network deterioration, and overcome the gradient-vanishing issue often arising in deep neural network training by leveraging residual connections. However, a deep ResNet network necessitates intense computational power for training and reasoning, imposing onerous requirements in terms of the choice of hardware.

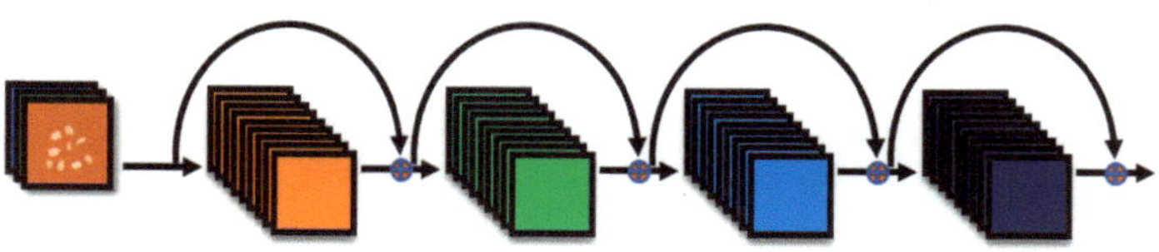

(**a**)

Figure 3. *Cont.*

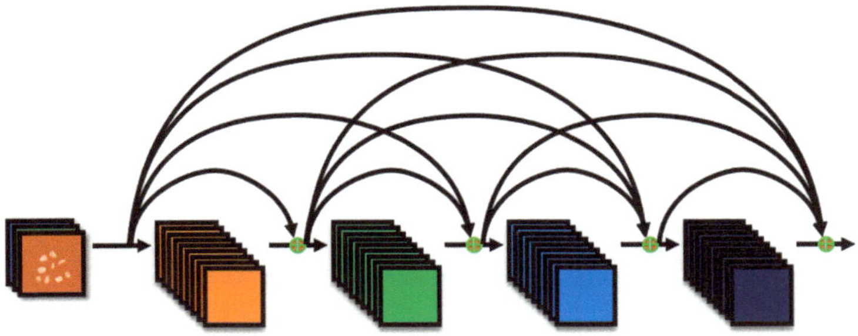

(b)

Figure 3. ResNet and DenseNet network structure. (**a**) ResNet, (**b**) DenseNet.

DenseNet, which begins with feature maps and uses a denser connection approach, is mostly based on the concept of the residual structure of ResNet. To ensure the maximal information flow between each layer (Figure 3b), forward propagation concatenates the feature maps of each layer with those of the other levels in the channel dimension.

DenseNet's network topology outperforms ResNet's in three key areas: (1) improving the feature propagation between networks; (2) implementing feature reuse while leveraging both low-level and high-level features; and (3) significantly lowering the number of parameters. Therefore, the study used DenseNet121 to replace ResNet50_ FPN to serve as the backbone feature extraction network of the Mask RNN for extracting specific features of the corn kernels. As shown in Figure 4, DenseBlock and transition structures comprise the majority of the topology of the DenseNet121 network.

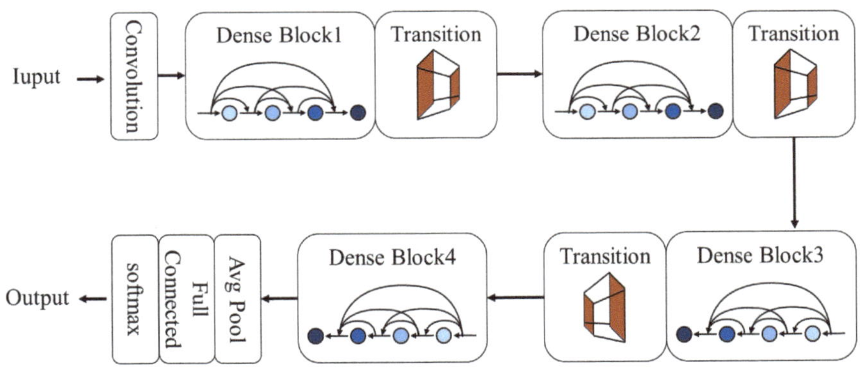

Figure 4. DenseNet network model structure.

The DenseBlock structure has a BN + ReLU + 1 × 1Conv + BN + ReLU + 3 × 3Conv composition, where the input of each layer comes from the feature maps of all the layers in front of it, and the output of each layer is directly connected to the input of all layers behind it to achieve feature reuse. To connect two neighboring DenseBlock structures, integrate the derived features, decrease the width and height of the feature map, and achieve the effect of downsampling and compressing the model, a transition layer structure with the formula BN + ReLU + 1 × 1Conv + 2 × 2 AvgPooling is employed.

2.2.2. CBAM Attention Mechanism

The attention mechanism is a network structure integrated into a model, designed to emphasize pertinent information, filter out less relevant details, and aid the model in choosing effective and appropriately sized features. This allows the model to extract features efficiently, allowing downstream tasks to focus on signals that are more closely related to the task at hand. As illustrated in Figure 5, the attention mechanism of the CBAM is a compact convolutional attention module that consists of the Channel Attention Module (CAM) and the Spatial Attention Module (SAM). The role of the CAM here is to process feature maps from different channels and focus on the meaningful feature map information After that, the channel compression weight matrix is output, and then multiplied by the original input feature characteristic matrix. When the feature map adjusted by the CAM enters the spatial attention sub-module, the SAM will process the feature region of meaningful information in the feature map, generate the spatial compression weight matrix, and perform the same multiplication operation. And finally, the refined feature map is obtained. It is challenging for the feature extraction network to extract useful information from slightly broken corn kernels since their characteristics resemble those of intact corn kernels. The CBAM attention mechanism helps the network pay more attention to the damage in the corn kernels; even if their characteristics are similar to intact corn kernels, it can achieve differentiation. To filter out the unimportant information in the feature layer and extract the more effective features of corn kernels, this paper added the CBAM attention mechanism layer after the fourth DenseBlock layer of DenseNet121 to obtain the DenseNet121_CBAM backbone feature extraction network.

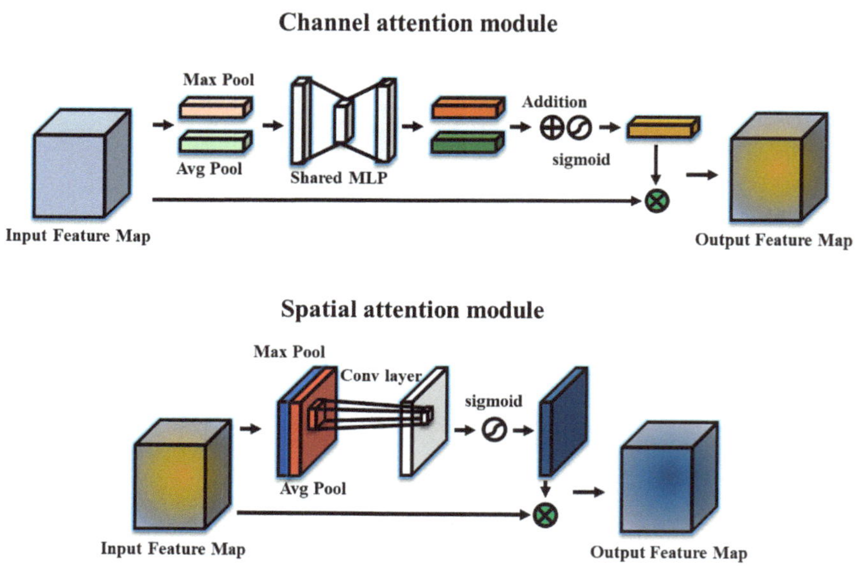

Figure 5. CBAM attention mechanism.

2.2.3. Multi-Scale Fusion FPN

The corn kernel feature maps extracted by the backbone feature network were relatively simple, and the lower-layer feature map had a higher resolution, containing more corn kernel locations and detailed information. However, due to the reduction in convolution, its semantics were lower and the noise greater. The high-level feature maps had strong semantic information, but the resolution was extremely low, and the perception of details was poor. The efficient fusion of the two-layer feature map can obtain more comprehensive corn kernel characteristics. To obtain more informative corn kernel feature maps, this paper adds an FPN structure after DenseNet121_CBAM to construct a DenseNet121_CBAM_FPN network (Figure 6). The corn kernel feature maps obtained from the second transition layer, the third transition layer, and the CBAM layer of the DenseNet_CBAM network were convolved and fused to obtain the M1, M2, and M3 feature maps. Following the convolution of the M1, M2, and M3 feature maps, P1, P2, and P3 feature maps with more abundant corn kernel information could be obtained.

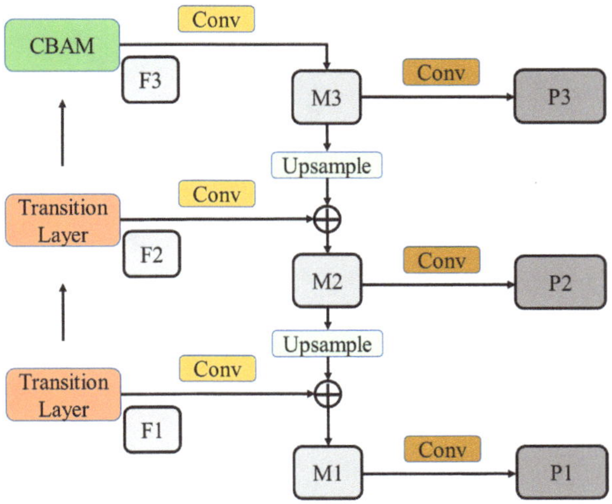

Figure 6. Feature pyramid network structure.

2.2.4. Transfer Learning

The lack of intact and broken corn kernel data made the training of an optimum recognition and segmentation model difficult. The issue of having fewer datasets can be resolved using the transfer learning method. Transfer learning is the process of using previously learned model weights and characteristics to detect new tasks. Large-scale datasets were used to train the weights to improve their ability to express features. Additionally, it could efficiently cut the training time and accelerate the loss function convergence. As a result, DenseNet121 weights that were trained on the expansive ImageNet dataset were used. The pre-training weights were loaded to train and validate the corn kernel dataset using transfer learning.

2.3. Construction of a Corn Kernel Dataset and Establishment of the Pixel–Mass Relationship

2.3.1. Data Collection

A Grain King TB70 grain combine harvester modified with a 4YB-4A corn header was used to gather the corn kernels for this study in Laowangji Town, Zhecheng County, Shangqiu City, Henan Province (34°07′3.11″ N, 115°30′23.84″ E, at an altitude of approximately 44 m above mean sea level), China. The variety is Xianyu 335. A black cloth was covered with the requisite grains after the corn harvester had unloaded the grain. A Shunhuali 500W industrial camera (8 mm focal length lens, Shenzhen Shunhuali Electronics Co., LTD., Shenzhen, China.) was used to randomly record 600 photographs of the grains at various exposure levels and with various contaminants (Figure 7). Table 1 displays the specific image distribution.

(a) (b) (c)

Figure 7. *Cont.*

(d) (e) (f)

Figure 7. Images of corn grains under different conditions. (**a**) Low-exposure image of small amount of impurities. (**b**) Low-exposure excess-impurity image. (**c**) Normal-exposure image of small amounts of impurities. (**d**) Normal exposure of excess impurity image. (**e**) High-exposure images of small amounts of impurities. (**f**) High-exposure images of excess impurities.

Table 1. Corn kernel images under different conditions.

Impurity Condition	Exposure Condition	Number of Images	Total
Minor impurity	Low exposure	43	600
	Normal exposure	80	
	High exposure	55	
Excess impurity	Low exposure	37	
	Normal exposure	60	
	High exposure	25	

Following the acquisition of the grain images, 200 intact and 160 broken corn kernels (Figure 8) were chosen and brought to the laboratory.

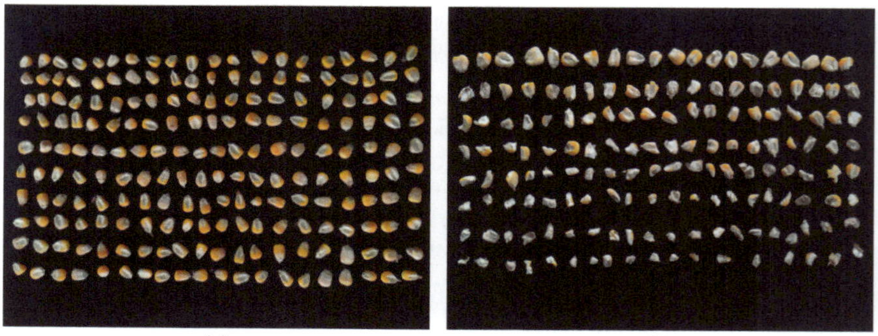

(a) (b)

Figure 8. Intact and broken corn kernels. (**a**) Intact corn kernels. (**b**) Broken corn kernels.

2.3.2. Data Preprocessing

For deep learning algorithms, the quality of the dataset directly affects the training and prediction performance of the network model. To improve the generalization ability and robustness of the network model and obtain more features of the corn kernels, 600 original images were appropriately expanded using traditional data augmentation methods such as adding noise, changing the brightness, horizontal flipping, vertical flipping, image shifts, etc. (Figure 9) to obtain 1200 enhanced images. The 1200 processed images were manually annotated using the Lableme tool software for the corn kernels. As this study aimed to achieve recognition and segmentation of the corn kernels, it was necessary to use the contour of corn kernels as a boundary for polygon annotation. Annotation targets can be divided into two categories, including intact and broken corn kernels. The broken corn kernels included cracks, root fractures, top damage, side damage, and so on. After image annotation, a JSON file containing the annotation information was obtained. Then, a specific script written in Python was used to convert the JSON file into a COCO dataset format that could be used to train the Mask RCNN model. The created dataset was then split (in a 7:2:1 ratio) into training, validation, and testing sets.

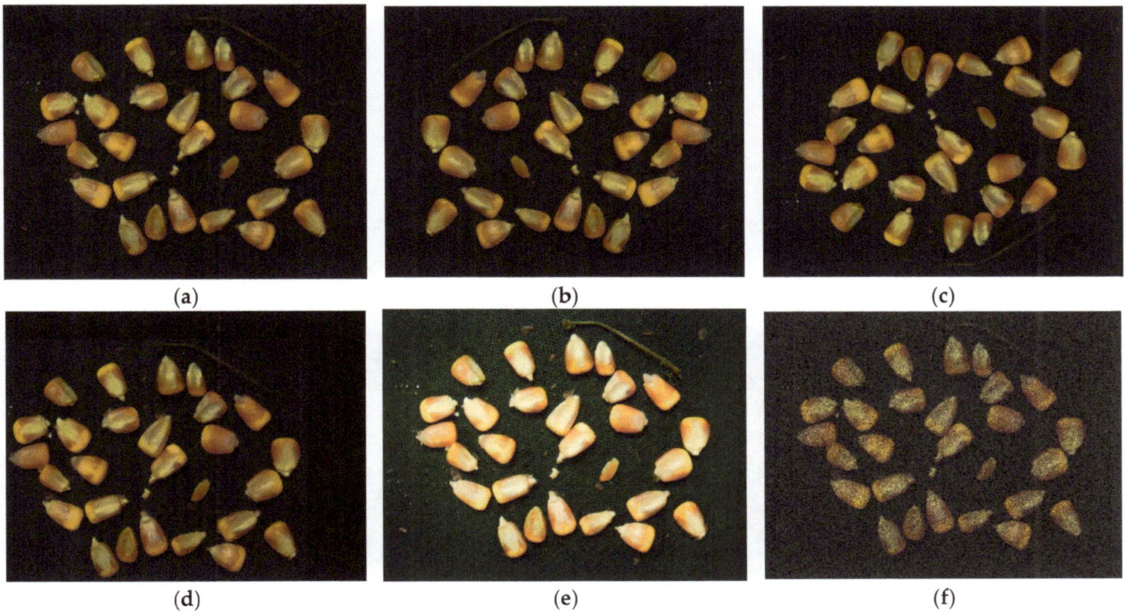

Figure 9. Traditional dataset enhancement methods. (**a**) Original image. (**b**) Horizontal flip. (**c**) Vertical flip. (**d**) Image shift. (**e**) Increased brightness. (**f**) Salt-and-pepper noise.

2.3.3. Establishment of the Pixel–Mass Relationship

Since the pixels and the quality of corn kernels fall into two different categories, it was necessary to develop a coupling relationship between the two to explain quality using pixels. Some of the corn kernels that were brought back were picked out and scattered evenly on a black cloth in the laboratory under natural daylight. Using a bracket, we secured the industrial camera above the velvet material and photographed the corn kernels with the lens some 150 mm vertically above the fabric (Figure 10). The corn kernels on the velvet material had to be removed after each image was captured, and a few more of the corn kernels that were dispersed there were chosen. We took 50 photographs of intact and broken corn kernels, respectively, and weighed each of them, recording the information associated with each photo using an electronic precision scale (to ± 0.001 g).

The captured corn kernel images were imported into the Adobe Photoshop software, the pixels of the intact kernels and broken kernels were obtained based on the histogram information, and then the coupling relationship between the pixels and quality was established and a one-way linear regression fit was plotted (Figure 11).

The results from the least squares regression analysis, examining the relationship between pixel and mass for the corn kernels, revealed coefficients of determination values (R^2) of 0.958 for the intact kernels and 0.992 for the broken kernels. The coefficient of determination measures the degree of correlation between the two variables, and the closer it is to 1, the stronger the correlation between the two variables, confirming that corn kernel pixels are highly correlated with quality. The univariate linear regression model was able to reflect the quantitative relationship between the kernel pixels and quality, which was expressed as:

$$M_b = 4.3273 \times 10^{-5} P_b - 0.1626 \quad (1)$$

$$M_f = 5.773 \times 10^{-5} P_f - 0.1644 \quad (2)$$

where M_b is the quality of the broken corn kernels, in g; P_b denotes the broken corn kernel pixel count, in pixels; M_f represents the quality of the intact corn kernels, in g; and P_f is the intact corn kernel pixel count, in pixels.

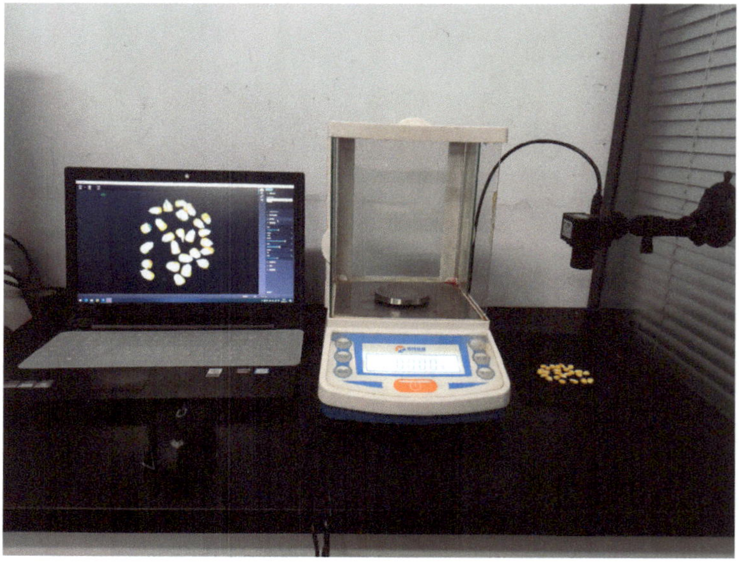

Figure 10. Corn kernel data acquisition experimental platform.

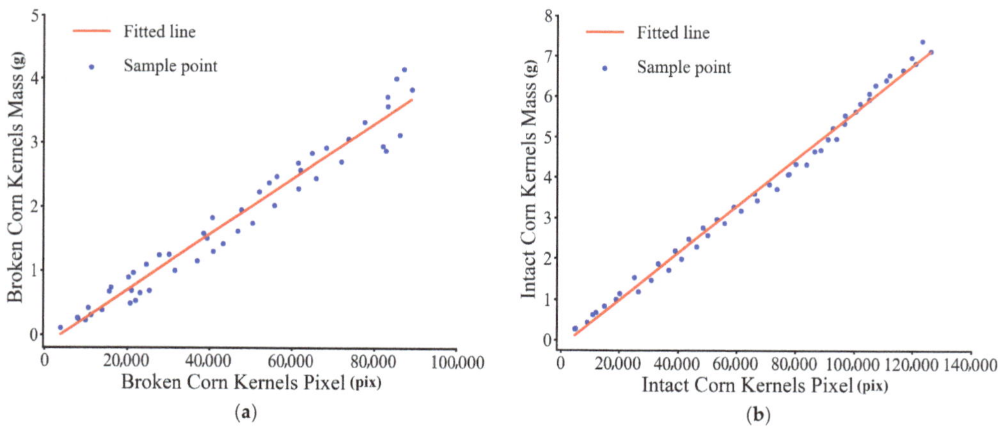

Figure 11. Linear regression analysis of pixel and mass. (**a**) The relationship between pixel and mass of broken corn kernels. (**b**) The relationship between pixel and mass of intact corn kernels.

2.4. Calculation Method of the Corn Kernel Breakage Rate

The corn kernel breakage rate may be calculated using Python language programming based on the established association between the corn kernel pixel counts and quality after segmenting the pixel mask of the intact and broken corn kernels using the improved Mask RCNN model, as given by:

$$Z_s = \frac{m_s}{m_i} \times 100\% \quad (3)$$

where Z_s is the grain breakage rate, %; m_s stands for the mass of broken corn kernels, g; m_i is the total mass of sample corn kernels, g.

3. Results and Discussion

3.1. Experimental Environment and Model Training Parameter Settings

The experiment used a 64-bit Windows 11 system, equipped with a 12th Gen Intel (R) Core (TM) i7-12700KF CPU, with a main frequency of 3.61 GHz, 32 GB of memory, an NVIDIA 3080Ti graphics processing unit (GPU), and 12 GB of graphics memory. The deep learning framework was Python 1.10.0+Cuda11.3, the programming language was Python 3.8.0, and the programming environment was PyCharm Community Edition. Based on the results of multiple tests, we set the number of iterations (epoch) to 50, the attenuation factor to 0.1, the initial learning rate of the model parameters to 0.008, the batch size of training data to 8, and the training set to be randomly mixed before each iteration. The optimizer selected SGD and the momentum factor was set to 0.9; mixed precision training was adopted.

3.2. Evaluation Indicators

The percentage of actual positive samples in the predicted sample to all positive samples is known as precision, whereas the percentage of actual positive samples in the predicted sample to all predicted samples is known as recall, as given by:

$$P = \frac{TP}{TP + FP} \tag{4}$$

$$R = \frac{TP}{TP + FN} \tag{5}$$

where, True Positive (*TP*) is the correct recognition of positive samples as positive samples; True Negative (*TN*) is the correct recognition of negative samples as negative samples; False Positive (*FP*) is a negative sample that is incorrectly recognized as a positive sample; a d False Negative (*FN*) is when a positive sample is mistakenly identified as a negative sample.

Mean Average Precision (*MAP*) is an important indicator used to measure the performance of network models, as given by:

$$AP = \int_0^1 p(r)dr \tag{6}$$

$$MAP = \frac{\sum_{i=1}^{k} AP_i}{k} \tag{7}$$

where, Average Precision (*AP*) is the average accuracy of identifying a certain class, and k is the total number of classifications.

Five indicators were used in this experiment to analyze and compare the experimental results with the original network: $\text{MAP}^{\text{b}}50$ (IoU = 0.50), $\text{MAP}^{\text{m}}50$ (IoU = 0.50), the computational complexity (FLOPs), the parameter complexity (Params) of the backbone feature extraction network, and the feature extraction time (Time).

3.3. Model Training

Five sets of experiments were designed to train various backbone feature extraction networks of the Mask RNN model on the corn seed dataset while using the same experimental environment and training parameters, to test the efficacy of using DenseNet121_CBAM_FPN as the backbone feature extraction network of the Mask RNN model. The apparent effects of various backbone feature extraction networks on detection performance are shown in Table 2.

Table 2. Experimental results of different feature extraction networks on corn kernel dataset.

Backbone	MAP^b50 (%)	MAP^m50 (%)	FLOPs (GMac)	Params (M)	Time (ms)
ResNet50_FPN	97.28	98.33	6.96	26.63	228
DenseNet121	97.16	98.18	2.88	6.95	170
DenseNet121_CBAM	97.32	98.29	2.88	7.08	176
DenseNet121_FPN	97.28	98.35	3.09	9.18	202
DenseNet121_CBAM_FPN	97.62	98.70	3.09	9.31	206

The results in Table 2 showed the following:

(1) The Mask RCNN model, which uses DenseNet121 as the backbone feature extraction network, was 97.16% and 98.18% on MAP^b50 and MAP^m50, respectively, its computation amount and parameter number were 2.88 GMac and 6.95 M, and the feature extraction time was 170 ms;

(2) The Mask RCNN model incorporating the CBAM attention mechanism into the DenseNet121 network structure achieved 97.32% and 98.29% on MAP^b50 and MAP^m50, respectively, with a computational and parameter load of 2.88 GMAc and 7.08 M, and a feature extraction time of 176 ms;

(3) The Mask RCNN model using an FPN structure in the DenseNet121 network structure had a performance of 97.28% and 98.35% on MAP^b50 and MAP^m50, respectively, with a computational and parameter load of 3.09 GMAc and 9.18 M, and a feature extraction time of 202 ms;

(4) The Mask RCNN model incorporating the CBAM attention mechanism and FPN structure into the DenseNet121 network structure achieved 97.62% and 98.70%, respectively, on MAP^b50 and MAP^m50, with a computational and parameter load of 3.09 GMAc and 9.31 M, and a feature extraction time of 206 ms. Compared to the ResNet50_FPN backbone feature extraction network, it improved by 0.34% and 0.37% on MAP^b50 and MAP^m50, respectively, and reduced the computational complexity by 3.87 GMac, the parameter size by 17.32 M, and the feature extraction time by 22 ms. The results showed that DenseNet121_CBAM_FPN, as the backbone feature extraction network, improved the accuracy of corn kernel recognition and segmentation compared to the original Mask RCNN model, and reduced the computational burden, parameter quantity, and feature extraction time, proving the effectiveness of the improved Mask RCNN model.

As shown in Figure 12, during the training process of the corn kernel dataset, the improved Mask RCNN model outperformed the original network model, even though MAP^b50 and MAP^m50 increased more slowly and tended to stabilize between 20 and 50 iterations.

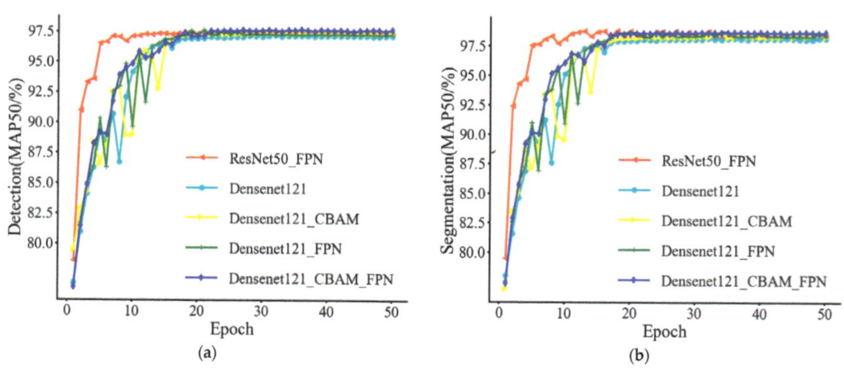

Figure 12. Mask RCNN recognition and segmentation curves of different backbone feature extraction networks. (**a**) Mask RCNN recognition curves of different backbone feature extraction networks. (**b**) Mask RCNN segmentation curves of different backbone feature extraction networks.

Using the idea of transfer learning to load the pre-trained weights of DenseNet121 for training, it can be seen from Figure 13a,b that the loss function of the model without pre-trained weights converged slowly, reaching complete convergence after about 40 iterations. The loss function of the model with pre-trained weights converged faster, reaching complete convergence after about 20 iterations. In comparison, the convergence of the loss function using pre-trained weights was twice as fast as that without pre-trained weights.

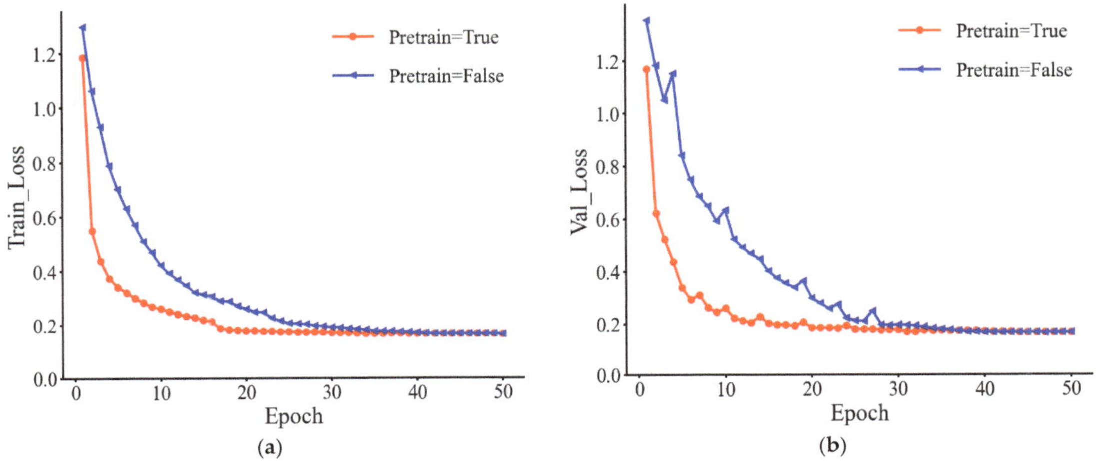

Figure 13. Training and validation loss curves with and without pre-training weights. (**a**) Training loss curves with and without pre-training weights. (**b**) Validation loss curves with and without pre-training weights.

3.4. Visualization Analysis of Different Segmented Networks

The corn kernel dataset was trained using the original Mask RCNN, improved Mask RCNN, and YOLACT (You Only Look At CoefficienTs) [24] network models. YOLACT is a classic real-time segmentation model, which has the characteristics of a good segmentation effect and rapid detection; the results are displayed in Table 3. The improved Mask RCNN model had an increase of 0.34% and 0.37% in $MAP^{b}50$ and $MAP^{m}50$ compared to the original Mask RCNN model, and a reduction of 375 ms in the testing time for one image, as shown by the findings in the table. The $MAP^{b}50$ and $MAP^{m}50$ exhibited increases of 0.83% and 0.92%, respectively, and a reduction in the testing time of 17 ms for one image when compared to the YOLACT network model.

Table 3. Training and verification results of three different network models.

Network Model	$MAP^{b}50$ (%)	$MAP^{m}50$ (%)	Time (ms)
Original Mask RCNN	97.28	98.33	451
Improved Mask RCNN	97.62	98.70	76
Yolact	96.21	97.34	93

Figure 14 displays the visualization outcomes for the three network models. The segmentation of intact and broken corn kernels by the original Mask RCNN model and the improved Mask RCNN model both yielded masks that were better suited to corn kernels. The detection performance of the YOLACT network model was subpar, with recognition mistakes and evident rough edges visible on the segmentation mask that deviated somewhat from the corn kernels. The improved Mask RCNN model outperformed the YOLACT network model in terms of the visual effect. Compared with the traditional machine learning algorithm of Yang [25], the improved Mask RCNN increased the recognition accuracy of corn kernels and optimized the segmentation of corn kernels.

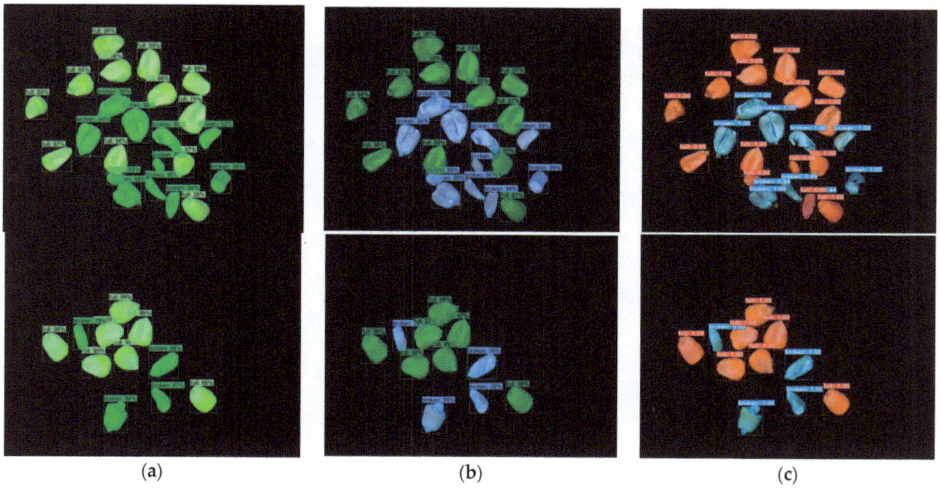

Figure 14. Test results of different network models. (**a**) Original Mask RCNN. (**b**) Improved Mask RCNN. (**c**) Yolact.

Images with various exposure intensities and other contaminants were chosen for segmentation and recognition to better demonstrate the robustness of the enhanced Mask RCNN model. As shown in Figure 15, in photos in challenging settings, intact and broken corn kernels could still be recognized and separated, and the pixel segmentation result was quite accurate, demonstrating the improved durability of the Mask RCNN model.

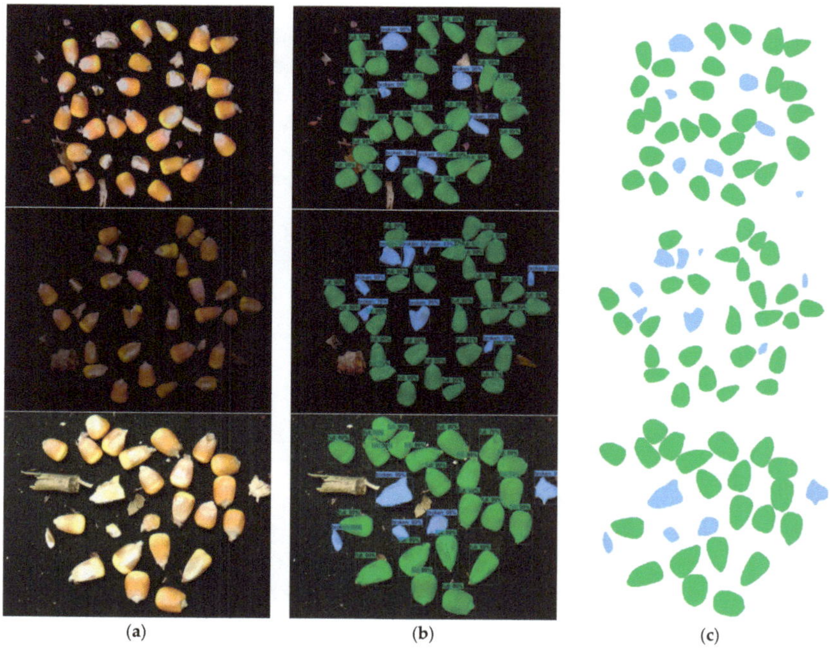

Figure 15. Detection results of images under different conditions. (**a**) Original image under complex conditions. (**b**) Detection image. (**c**) Split mask image.

3.5. Real-Time Detection of the Corn Kernel Breakage Rate

We chose five times at random and selected any hybrid corn kernels, both healthy or damaged, then photographed them using an industrial camera positioned 150 mm above them, and sent the images to an enhanced Mask RCNN model for breakage rate detection. Then, corn kernels were manually selected to determine the breakage rate, and the time was logged. Table 4 lists a comparison of the outcomes. The statistics in Table 4 show that the detection was substantially faster than that realized using manual calculation, and the error range between the machine vision detection and manual calculation was between 0.72% and 0.65%. Therefore, human computation of the corn kernel breakage rates can be replaced by machine vision detecting technology.

Table 4. Comparison results of different detection methods of breakage rate.

Serial Number	Machine Vision (%)	Detection Time (ms)	Actual Calculation (%)	Calculation Time (ms)
1	12.36	64	11.89	6725
2	8.57	88	9.29	8452
3	15.29	69	14.91	7684
4	20.16	79	19.51	7332
5	6.52	78	7.05	9246

4. Conclusions

Building upon the Mask RCNN concept, this study introduced the DenseNet121 backbone feature extraction network as the fundamental framework. The DenseNet121_CBAM_FPN backbone feature extraction network was established by adding the CBAM attention mechanism and FPN structure. The improved Mask RCNN model outperformed the baseline model on $MAP^{b}50$ and $MAP^{m}50$ by 0.34% and 0.37%, respectively. The computational and parameter quantities dropped by 3.87 GMAc and 17.32 M, respectively, when compared to the ResNet50_FPN backbone feature extraction network. The time taken to extract features lowered by 22 ms, while the time taken to detect images decreased by 395 ms. The improved Mask RCNN model had a higher accuracy of recognition and faster speed of detection, which proved the effectiveness of the improved network.

The relationship between the quality of corn kernels and pixel characteristics was established. Machine vision had a detection error of between −0.72% and 0.65% for the corn kernel breakage rate. The typical detection time for a given image of corn kernels was 76 ms, which can be used to replace the manual measurement of the breakage rate of corn kernels with machine vision detection technology.

In the next research work, a corn kernel sampling device fixed to a corn kernel harvester will be designed to detect the breakage rate of corn kernels in real time in conjunction with the improved Mask RCNN model.

This study did not address the overlap among the corn kernels, focusing solely on the classification and division of broken corn kernels on the surface. Investigating this situation in greater detail is suggested as a fruitful direction for future research.

Author Contributions: Conceptualization, H.Z. and X.H.; methodology, H.Z. and Z.L.; investigation, Z.Y. and C.Z.; Visualization, Y.D. and P.L.; writing—original draft preparation, Z.L. All authors have read and agreed to the published version of the manuscript.

Funding: This research was funded by Henan Province Modern Agricultural Industrial Technology System Maize Whole-process Mechanization Special Project (HARS-22-02-G4); National Key Research and Development Program (2018YFD0300704); and Henan Province Science and Technology Research (222102110457).

Institutional Review Board Statement: Not applicable.

Data Availability Statement: The data presented in this study are available on request from the authors.

Acknowledgments: The authors would like to thank their college and the laboratory, as well as gratefully appreciate the reviewers who provided helpful suggestions for this manuscript.

Conflicts of Interest: The authors declare no conflict of interest.

References

1. Cheng, S.; Han, H.; Qi, J.; Ma, Q.; Liu, J.; An, D.; Yang, Y. Design and Experiment of Real-Time Grain Yield Monitoring System for Corn Kernel Harvester. *Agriculture* **2023**, *13*, 294. [CrossRef]
2. Li, X.; Zhang, W.; Xu, S.; Du, Z.; Ma, Y.; Ma, F.; Liu, J. Low-Damage Corn Threshing Technology and Corn Threshing Devices: A Review of Recent Developments. *Agriculture* **2023**, *13*, 1006. [CrossRef]
3. *GB/T 21962-2020*; Corn Combine Harvester. China Standards Press: Beijing, China, 2020.
4. Qiu, Z.J.; Chen, J.; Zhao, Y.Y.; Zhu, S.S.; He, Y.; Zhang, C. Variety identification of single rice seed using hyperspectral imaging combined with convolutional neural network. *Appl. Sci.* **2018**, *8*, 212. [CrossRef]
5. Han, Z.Z.; Gao, J.Y. Pixel-level aflatoxin detecting based on deep learning and hyperspectral imaging. *Comput. Electron. Agric.* **2019**, *164*, 104888. [CrossRef]
6. Przybyło, J.; Jabłoński, M. Using deep convolutional neural network for oak acorn viability recognition based on color images of their sections. *Comput. Electron. Agric.* **2019**, *156*, 490–499. [CrossRef]
7. Zhang, R.Q.; Li, Z.W.; Hao, J.J.; Sun, L.; Li, H.; Han, P. Image recognition of peanut pod grades based on transfer learning with convolutional neural network. *Trans. Chin. Soc. Agric. Eng.* **2020**, *36*, 171–180.
8. Mao, R.; Zhang, Y.C.; Wang, Z.X.; Gao, S.C.; Zhu, T.; Wang, M.L.; Hu, X.P. Recognizing stripe rust and yellow dwarf of wheat using improved Faster-RCNN. *Trans. Chin. Soc. Agric. Eng.* **2022**, *38*, 176–185.
9. Ni, B.; Paulsen, M.R.; Reid, J.F. Corn kernel crown shape identification using image processing. *Trans. ASAB* **1997**, *40*, 833–838. [CrossRef]
10. Steenhoek, L.W.; Misra, M.K.; Batchelor, W.D.; Davidson, J.L. Probabilistic Neural Networks for Segmentation of Features in Corn Kernel Images. *Proc. Nutr. Soc.* **2001**, *41*, 225–234. [CrossRef]
11. Zhao, Z.H.; Song, H.; Zhu, J.B.; Lu, L.; Sun, L. Identification algorithm and application of peanut kernel integrity based on convolution neural network. *Trans. Chin. Soc. Agric. Eng.* **2018**, *34*, 195–201.
12. Chen, J.; Gu, Y.; Lian, Y.; Han, M.N. Online recognition method of impurities and broken paddy grains based on machine vision. *Trans. Chin. Soc. Agric. Eng.* **2018**, *34*, 187–194.
13. Song, C.X.; Yu, C.Y.; Xing, Y.C.; Li, S.M.; He, H.; Yu, H.; Feng, X.Z. Algorith for acquiring multi-phenotype parameters of soybean seed based on OpenCV. *Trans. Chin. Soc. Agric. Eng.* **2022**, *38*, 156–163.
14. Mahirah, J.; Kazuya, Y.; Munenori, M.; Naoshi, K.; Yuichi, O.; Tetsuhito, S.; Harshana, H.; Usman, A. Double Lighting Machine Vision System to Monitor Harvested Paddy Grain Quality during Head-Feeding Combine Harvester Operation. *Machines* **2015**, *3*, 352–363.
15. Mahirah, J.; Kazuya, Y.; Munenori, M.; Naoshi, K.; Yuichi, O.; Tetsuhito, S.; Harshana, H.; Usman, A. Monitoring harvested paddy during combine harvesting using a machine vision-Double lighting system. *Eng. Agric. Environ. Food.* **2017**, *10*, 140–149. [CrossRef]
16. Yang, L.; Wang, Z.; Bai, X.P.; Gao, L.; Hu, W.B. Design of on-line sampling device for maize kernel broken rate. *J. Agric. Mech. Res.* **2019**, *41*, 121–124+129.
17. Chen, M.; Ni, Y.L.; Jin, C.Q.; Xun, J.S.; Zhang, G.Y. Online Monitoring Method of Mechanized Soybean Harvest Quality Based on Machine Vision. *Trans. Chin. Soc. Agric. Eng.* **2021**, *52*, 91–98.
18. Jin, C.Q.; Liu, S.K.; Chen, M.; Yang, T.X.; Xu, J.S. Online quality detection of machine-harvested soybean based on improved U-Net network. *Trans. Chin. Soc. Agric. Eng.* **2022**, *38*, 70–80.
19. Liu, S.K.; Jin, C.Q.; Chen, M.; Yang, T.X.; Xu, J.S. Online detection method of detecting crushing rate of soybean harvest based on DeepLabV3+ network. *J. Chin. Agric. Mech.* **2023**, *44*, 170–175.
20. He, K.M.; Gkioxari, G.; Dollár, P.; Girshick, R. *Mask R-CNN*; ICCV: Venice, Italy, 2017; pp. 2980–2988.
21. Ren, S.Q.; He, K.M.; Girshick, R.; Sun, J. Faster R-CNN: Towards Real-Time Object Detection with Region Proposal Networks. *IEEE Trans. Pattern Anal. Mach. Intell.* **2017**, *39*, 1137–1149. [CrossRef] [PubMed]
22. Huang, G.; Liu, Z.; Laurens, V.D.M.; Kilian, Q.W. *Densely Connected Convolutional Networks*; CVPR: Honolulu, HI, USA, 2017; pp. 2261–2269.
23. Woo, S.; Park, J.; Lee, J.Y.; Kweon, I.S. *CBAM: Convolutional Block Attention Module*; CVPR: Long Beach, CA, USA, 2018; pp. 3–19.
24. Bolya, D.; Zhou, C.; Xiao, F.Y.; Lee, Y.J. *YOLACT: Real-Time Instance Segmentation*; ICCV: Seoul, Republic of Korea, 2019; pp. 9156–9165.
25. Yang, L. Research on Online Inspection Equipment and Method for Broken Rate of Maize Kernels. Master's Thesis, Shenyang Institute of Automation, Chinese Academy of Sciences, Shengyang, China, 2018.

MDPI AG
Grosspeteranlage 5
4052 Basel
Switzerland
Tel.: +41 61 683 77 34

Agriculture Editorial Office
E-mail: agriculture@mdpi.com
www.mdpi.com/journal/agriculture

www.ingramcontent.com/pod-product-compliance
Lightning Source LLC
Chambersburg PA
CBHW041420210926
51850CB00013B/3321
* 9 7 8 3 7 2 5 8 3 9 2 5 4 *